Interplant Resource Integration

Interplant Resource Integration

Optimization and Allocation

Chuei-Tin Chang, Cheng-Liang Chen,
and Jui-Yuan Lee

CRC Press
Taylor & Francis Group
Boca Raton London New York

CRC Press is an imprint of the
Taylor & Francis Group, an **informa** business

First edition published 2021
by CRC Press
6000 Broken Sound Parkway NW, Suite 300, Boca Raton, FL 33487-2742

and by CRC Press
2 Park Square, Milton Park, Abingdon, Oxon, OX14 4RN

Library of Congress Cataloging-in-Publication Data

Names: Chang, Chuei-Tin, author. | Chen, Cheng Lung, 1931- author. | Lee, Jui-Yuan, author.
Title: Interplant resource integration : optimization and allocation / authored by Chuei-Tin Chang, Cheng-Liang Chen, Jui-Yuan Lee.
Description: First edition. | Boca Raton, FL : CRC Press, 2021. | Includes bibliographical references and index. | Summary: "This book presents an introduction to the planning and implementation methods for interplant resource integration. The analytic tools provided in the book can be used for the tasks of formulating mathematical programming model(s) to maximize the achievable overall saving and also devising the "fair" distribution scheme(s) to allocate individual financial benefits among the participating plants. The superstructures of various resource-exchange networks, their model formulations, and the corresponding allocation algorithms are given in sufficient detail to facilitate practical applications"-- Provided by publisher.
Identifiers: LCCN 2021004226 (print) | LCCN 2021004227 (ebook) | ISBN 9780815346432 (hbk) | ISBN 9781351170406 (ebk)
Subjects: LCSH: Industrial ecology. | Plant performance. | Power resources. | Industrial districts. | Public goods.
Classification: LCC TS161 .C442 2021 (print) | LCC TS161 (ebook) | DDC 363.73/1--dc23
LC record available at https://lccn.loc.gov/2021004226
LC ebook record available at https://lccn.loc.gov/2021004227

ISBN: 978-0-8153-4643-2 (hbk)
ISBN: 978-1-0320-3329-7 (pbk)
ISBN: 978-1-351-17040-6 (ebk)

Typeset in Times
by Deanta Global Publishing Services, Chennai, India

Contents

Preface

This book presents an introduction to the planning and implementation strategies for interplant resource integration in an industrial park. The analytic tools provided in the book can be used for the daunting tasks of formulating mathematical programming model(s) to maximize the achievable overall saving and also devising the "fair" distribution scheme(s) to allocate individual financial benefits among the participating plants. The superstructures of various resource-exchange networks, their model formulations, and the corresponding allocation algorithms are given in sufficient detail to facilitate practical applications. It is therefore appropriate for use as an industrial reference and also as a textbook for a graduate course on advanced process design.

Traditionally, heat, water, and various other resources are integrated within a single process to save the operating cost and also for pollution abatement. Since it is generally believed that greater cost saving and waste reduction can be achieved by expanding the scope of integration, the emphases of recent studies have been gradually shifted to the development of resource-exchange schemes across plant boundaries. In these pioneering works, the common objective was simply to minimize the total resource usage. Such an all-encompassing approach tried essentially to maximize the overall benefit of a group of participants, but neglected the economic incentives of individual plants. Consequently, the resulting interplant integration arrangements may not always be acceptable to all parties involved. To facilitate implementable exchanges among plants in a realistic environment, each and every business entity must be allowed to pursue its own benefit as much as possible. In such a multi-agent system, if all behavioral patterns of interactions between agents are the results of players acting according to the available strategies in a game, then the mathematical tools of game theory are applicable for analyzing the decisions of agents. On the basis of this argument, the authors attempt in the present book to address the aforementioned optimization and allocation issues with a variety of rigorous mathematical programs.

The contents in each chapter are briefly summarized below.

Chapter 1 provides a general introduction to the model-based optimization and allocation strategies for interplant resource integration. The so-called "resource" in this book refers to either utilities or process materials. The heat-exchange network (HEN) and water network (WN) are examples of the former category, while biomass and petroleum supply chains are those of the latter.

Chapter 2 presents a mathematical programming approach to generating multi-plant HEN designs for continuous processes. Without addressing the benefit allocation issues, the traditional sequential and simultaneous approaches for generating the HEN in a single plant can both be applied, with minor modifications, to produce the multi-plant HEN designs. These modifications are detailed in the model formulations given in this chapter. The conceptual superstructures, the corresponding model formulation, two existing solution strategies, i.e., the sequential and simultaneous synthesis procedures, and several simple illustrative examples are all included therein to ensure thorough understanding. Although it can be shown in the examples presented in this chapter that a significant total annual cost (TAC) saving can usually be obtained via total-site heat integration, the subsequent allocation issues are still unsettled.

Chapter 3 presents a mathematical technique for synthesizing an indirect batch HEN that achieves the minimum utility consumption. In the indirect heat integration scheme, a heat transfer medium (HTM) is recirculated between storage tanks to absorb heat from hot streams and reject the heat to cold streams. This allows heat to be transferred from process streams in earlier time periods to those in later time intervals, thus relaxing the time constraint on heat exchange. Based on a superstructure, the mathematical model for indirect batch HEN synthesis is formulated as a mixed integer nonlinear program (MINLP).

Chapter 4 outlines the benefit distribution strategies for total-site heat integration (TSHI) in an industrial park. An existing non-cooperative-game-theory based sequential optimization strategy has been improved on the basis of the individual negotiation power of every participating plant to stipulate the "fair" price for each energy trade, to determine the "reasonable" proportions of capital cost to be shouldered by the involved parties of every interplant heat exchanger, and to produce an acceptable distribution of TAC savings. Also, to address various safety and operational concerns about direct heat transfers across plant boundaries, the interplant heat flows have been facilitated in the proposed integration schemes with either the available utilities or an extra intermediate fluid. Because of these additional practical features in interplant heat integration, realization of the resulting financial and environmental benefits should become more likely.

In Chapter 5, a rigorous synthesis procedure is introduced to produce a realistic cost-sharing plan for interplant heat integration in the spirit of a cooperative game. In addition, a comprehensive design procedure is also proposed to revamp the multi-plant HENs for lowering the overall utility consumption level and to divide and distribute the resulting cost saving fairly among all members of the interplant heat integration scheme.

Chapter 6 presents a mathematical optimization model for interplant water network (IPWN) synthesis, with process units operating in mixed continuous and batch modes. The proposed model can be implemented in two stages, and is dedicated to the realistic case where there are more continuous units than batch units.

A successful total-site water integration (TSWI) project should be facilitated by not only an optimal design of the multi-plant water network (MPWN) but also a "fair" revenue-sharing scheme. Chapter 7 proposes an approach to address both issues in two stages. The minimum TAC of every potential coalition in an industrial park is first determined using the conventional methods. The benefit allocation problem is dealt with in the second stage on the basis of three criteria of the cooperative game theory (i.e., the core, the Shapley value, and the risk-based Shapley value).

A large supply chain in the chemical industry can usually be broken down into several processes owned by different companies. Since the product of an upstream plant in this supply chain may be used as the feedstock of one or more downstream plant, the "resource" in this case is a process material. Two typical examples (i.e., petroleum and biomass) are discussed in Chapters 8–10.

Chapter 8 considers the petroleum supply network, operated by one or more oil companies, in which crude oil is used to produce various petrochemical products. A single mixed-integer linear program (MILP) is formulated in this chapter to coordinate various planning and scheduling decisions for optimizing the supply chain performance. For the sake of illustration convenience, the subsequent benefit allocation issues in the petroleum supply network are addressed separately in the next chapter. Chapter 9 integrates three cooperative game-based methods to fairly allocate benefits among participating members of a petroleum supply network while maximizing the overall profit, and also to assess the impacts of (1) supply-chain structures, (2) import prices of intermediate products, and (3) their import and export price differences.

Chapter 10 presents linear mathematical models for regional bioenergy planning and biomass supply chain network synthesis. The objective is to determine the optimal allocation of biomass with minimum carbon emissions from biomass transport. A simple, superstructure-based model that considers biomass availability and demand, as well as all possible connections between the biomass sources and sinks, is introduced first. This basic model is then modified to account for topological constraints, biomass allocation rate limits, and the location of biomass conversion plants. Operational flexibility under multiple biomass supply scenarios arising from uncertain factors can also be taken into account. This approach facilitates decision-making and contributes to the cleaner production of bioenergy.

Authors

Chuei-Tin Chang earned his BS degree in Chemical Engineering from National Taiwan University, and PhD degree in Chemical Engineering from Columbia University, New York City, USA, in 1976 and 1982, respectively. He worked as a process engineer in FMC Corporation (Princeton, New Jersey, USA) from 1982 to 1985, and also as an assistant professor at the Department of Chemical Engineering at the University of Nebraska (Lincoln, Nebraska, USA) from 1985 to 1989. He later joined the faculty of the Chemical Engineering Department of National Cheng Kung University (Tainan, Taiwan) in 1989, became a full professor in 1993, and later received the Distinguished Professor Award (2008–2011). He was a visiting research scholar at Northwestern University (1992), Lehigh University (1998), and Georgia Institute of Technology (2006). He is a senior member of AIChE and TwIChE. His current research interests are mainly concerned with process systems engineering (PSE), which includes process integration, process safety assessment and fault diagnosis, etc. He is author of 143 Scopus-indexed publications, with a total of 2207 citations and an h-index of 20.

Cheng-Liang Chen is a professor in the Department of Chemical Engineering, National Taiwan University (NTU), Taiwan. Dr Chen earned his PhD in Chemical Engineering from NTU in 1987, and then joined NTU directly. Professor Chen served concurrently as director of the Petrochemical Industry Research Center of NTU, board member of Taiwan Petroleum Corporation and Sino-American Petrochemical Corporation, and deputy executive director of the National Energy Research Program. His research focuses on chemical process simulation and design, process integration, and optimization, with particular research interests in sustainable energy systems. He is the author of 130 Scopus-indexed publications.

Jui-Yuan Lee is an associate professor in the Department of Chemical Engineering and Biotechnology, National Taipei University of Technology (Taipei Tech), Taiwan. Dr Lee earned his PhD in Chemical Engineering from National Taiwan University (NTU) in January 2011, and conducted postdoctoral research at NTU (March 2011–September 2013), University of the Witwatersrand, Johannesburg, South Africa (October 2013–May 2014), and National Tsing Hua University, Taiwan (July–December 2014), before joining Taipei Tech in February 2015. His research centers on chemical process design, integration, and optimization, with particular research interests in sustainable energy systems. He is the author of 95 Scopus-indexed publications, with a total of 1009 citations and an h-index of 19. Dr Lee works closely with collaborators in Malaysia, the Philippines, Mainland China, and South Africa.

1 Introduction

1.1 BACKGROUND

The 2030 Agenda for Sustainable Development has 17 goals and 169 targets (United Nations, 2020). These goals comprehensively address various economic, social, and environmental issues. In particular, goal 6 aims to ensure availability and sustainable management of water, while goal 13 is to take urgent action to combat climate change. For the latter, energy efficiency plays an important role in CO_2 emissions reductions especially in the industry, buildings, and transport sectors. In chemical processes, energy efficiency can be improved from heat recovery by exchanging heat between hot and cold process streams. This entails the synthesis of a heat exchanger network (HEN) and is also known as heat integration. On the other hand, integrated water resource management in the process industries for water conservation involves the synthesis of water networks.

1.2 DEVELOPMENT OF PROCESS INTEGRATION

Within the chemical engineering community, a special research area known as process integration has emerged since the 1970s for the development of systematic design tools for various resource conservation problems. Process integration may be defined as "a holistic approach to design and operation that emphasizes the unity of the process" (El-Halwagi and Foo, 2014). The first trend of process integration was developed for the synthesis of HENs, in response to the world oil crisis in the mid-1970s (Linnhoff and Flower, 1978a, b). The techniques were then extended to address various energy-intensive processes in the 1980s (Linnhoff et al., 1982; Smith, 2016). Since the late 1990s, in response to the need for pollution prevention, process integration techniques were extended into mass exchanger network (MEN) synthesis and other mass integration domains (El-Halwagi, 2017). An important branch of mass integration for integrated water resources management was established in the mid-1990s, termed water network synthesis or water minimization (Wang and Smith, 1994). Different water recovery schemes (e.g., direct reuse/recycling and regeneration reuse/recycling) may be explored.

Systematic procedures, including those for insight-based pinch analysis, and mathematical programming techniques have been developed. The main philosophy of pinch analysis is to establish performance targets (e.g., the minimum utility consumption and the minimum freshwater flowrate) ahead of design. Pinch analysis also identifies the bottleneck in heat/water recovery, or the pinch point. Such insights are useful for guiding the subsequent network design. Mathematical programming, on the other hand, allows cost considerations, design constraints, and multiple quality indices to be effectively incorporated, and is preferred for complex process integration problems. After four decades of development, these techniques are considered mature and can be found in several books (El-Halwagi, 2017; Foo, 2013; Klemeš, 2013).

1.3 SINGLE- AND MULTI-PLANT PROCESS INTEGRATION

Process integration has been considered not only in single process plants but also among multiple process plants of industrial complexes. In addition to in-plant integration for the individual plants, interplant integration provides further scope for resource conservation with more potential benefits. Typical examples are found in eco-industrial parks.

1.3.1 Multi-Plant Heat Integration

Approaches to HEN synthesis in single plants are in general applicable to multi-plant HEN design, with minor modifications. Multi-plant HENs can be synthesized using a sequential or simultaneous approach. The former may consist of three consecutive optimization steps to determine (1) the minimum total utility cost of the entire site, (2) the minimum number of heat exchange matches, and (3) the minimum total capital cost. Alternatively, a simultaneous approach may be used to explore the trade-off between utility and capital costs, with the objective of minimizing the total annual cost. Unlike a single process plant, there are multiple hot and cold utilities available in a total site. Therefore, the possible matches between the utility and process streams also need to be considered.

Heat integration in process plants has direct and indirect heat exchange schemes. In the direct scheme, heat is transferred directly from hot streams to cold streams; in the indirect scheme, heat is transferred through one or more heat transfer media. While direct heat integration allows the maximum heat recovery potential, the heat transfer medium serves as a buffer and allows better operational flexibility. Furthermore, for heat integration in batch process plants, the indirect scheme is less schedule-sensitive with the use of a heat transfer medium for heat storage. This allows the time constraint on heat exchange between time-dependent process streams to be bypassed.

1.3.2 Multi-Plant Water Integration

Optimization methods for single-plant water network design have been extensively developed. Concurrently, multi-plant or total site water network synthesis for improved water conservation has become a popular research topic in recent years. Apart from in-plant water recovery, opportunities for interplant water integration may be explored when considering an industrial complex with multiple process plants. Similar to the case of multi-plant heat integration, methods for single-plant water integration can be modified for multi-plant applications. A common objective for multi-plant water network synthesis is to minimize the total annual cost. Most published works on interplant water network synthesis were developed for continuous processes; however, there are also cases where process plants contain batch units. For example, in petrochemical complexes most processes operate continuously, while some operate batch-wise.

1.3.3 Benefit Allocation

Interplant heat integration arrangements designed using conventional methods may not be implementable in practice because the profit margin may not be acceptable to each participating party. Conventionally, the design objective of interplant heat integration is simply to minimize the total cost. A viable interplant heat integration scheme should, however, allow each plant to maximize its own benefit while achieving the maximum possible overall cost saving. To address the benefit allocation problem, non-cooperative game theory may be used to generate fair interplant integration schemes (Cheng et al., 2014). If the participating plants in a total site heat integration project are operated by different companies, the corresponding benefit allocation problem can be assumed to be analogous to a non-cooperative game. This assumption may not always be valid, in which case total site heat integration may preferably be treated as a cooperative game (Jin et al., 2018).

Similarly, developing a fair and stable benefit allocation scheme that facilitates the implementation of designed arrangements is the key to a successful total site water integration project. The benefit allocation problem associated with interplant water integration can also be solved by applying game theory.

1.4 SUPPLY CHAIN MANAGEMENT

Apart from process integration, supply chain management problems are often encountered in the process industries, particularly in petrochemical complexes. The general goal of supply chain management is to establish an efficient coordination framework to allow the suppliers, manufacturers, distributors, and retailers in the supply network to work together closely for raw materials to be acquired and converted into products for delivery. A petroleum supply chain can be divided into four segments: (1) exploration, (2) transportation, (3) refining, and (4) distribution. Crude oil is sourced, transported to refineries, and converted to various products. The refineries are often interconnected to achieve a high degree of flexibility for fluctuating market demand. The products are then delivered to customers via pipelines, trucks, or rail cars.

The refining segment of a petroleum supply chain requires production planning and scheduling, which are hierarchically linked. In the planning stage, the goal is to make higher-level decisions, such as the purchase amount of crude oil and the inventory and production levels of various products, according to the market demand forecast over a relatively long time horizon (e.g., months). The lower-level scheduling tasks, on the other hand, are performed in shorter time periods (e.g., weeks) to determine the timings of unit operations to meet the production levels set at the planning stage. Although conventionally considered sequentially for petrochemical processes, the planning and scheduling issues should be addressed simultaneously to ensure a feasible solution. In addition, for real-world applications, all the important petrochemical products produced in conversion refineries should be considered.

As with multi-plant process integration, there is a benefit allocation problem in petroleum supply chain management. The general objective is to allocate benefits fairly among the participating members while maximizing the overall profit. Again, game theory can be applied to facilitate the development of long-term and stable cooperative relations among the supply chain members.

1.5 SUMMARY

This book deals with multi-plant resource conservation and supply chain management. The corresponding benefit allocation problems are also addressed. In the next chapter, multi-plant HEN design and optimization for continuous processes from a total-site perspective are considered first. Before addressing benefit allocation issues, conventional sequential and simultaneous approaches for single-plant HENs can be applied to multi-plant HEN design with minor modifications. In Chapter 3, the design of indirect batch HENs with heat storage is addressed. The proposed indirect heat integration scheme and the formulation for single batch plants is applicable to multi-plant cases. Benefit allocation issues for interplant heat integration are then addressed in Chapters 4 and 5, based on non-cooperative and cooperative games, respectively. In Chapter 6, a two-stage approach for the synthesis of interplant water networks involving continuous and batch units has been presented. The models can be solved either sequentially for maximum water recovery, or simultaneously for minimum total cost. In Chapter 7, a systematic methodology is developed to produce the distribution scheme of TAC saving achieved by a total site water integration project. In Chapter 8, an integrated mixed-integer linear programming model is presented to coordinate various planning and scheduling decisions for optimizing the performance of a comprehensive petroleum supply network in typical conversion refineries. The corresponding benefit allocation problem is then addressed in Chapter 9. Finally, superstructure-based mathematical programming models for the synthesis of regional bioenergy supply chain networks are developed in Chapter 10.

REFERENCES

Cheng, S.-L., Chang, C.-T., Jiang, D. 2014. A game-theory based optimization strategy to configure inter-plant heat integration schemes. *Chemical Engineering Science* 118: 60–73.

El-Halwagi, M.M. 2017. *Sustainable Design Through Process Integration*. Waltham, MA: Elsevier.

El-Halwagi, M.M., Foo, D.C.Y. 2014. Process synthesis and integration. In: Seidel, A., Bickford, M., editors. *Kirk-Othmer Encyclopedia of Chemical Technology*. John Wiley & Sons, Inc, pp. 1–24. https://doi.org/10.1002/0471238961.1618150308011212.a01.pub2.

Foo, D.C.Y. 2013. *Process Integration for Resource Conservation*. Boca Raton, FL: CRC Press.

Jin, Y., Chang, C.-T., Li, S., Jiang, D. 2018. On the use of risk-based Shapley values for cost sharing in interplant heat integration programs. *Applied Energy* 211: 904–920.

Klemeš, J., editor. 2013. *Handbook of Process Integration (PI): Minimisation of Energy and Water Use, Waste and Emissions.* Cambridge: Woodhead Publishing.

Linnhoff, B., Flower, J.R. 1978a. Synthesis of heat exchanger networks: I. Systematic generation of energy optimal networks. *AIChE Journal* 24(4): 633–642.

Linnhoff, B., Flower, J.R. 1978b. Synthesis of heat exchanger networks: II. Evolutionary generation of networks with various criteria of optimality. *AIChE Journal* 24(4): 642–654.

Linnhoff, B., Townsend, D.W., Boland, D., Hewitt, G.F., Thomas, B.E.A., Guy, A.R., Marshall, R.H. 1982. *A User Guide on Process Integration for the Efficient Use of Energy.* Rugby: IChemE.

Smith, R. 2016. *Chemical Process Design and Integration.* 2nd ed. West Sussex, England: John Wiley & Sons, Inc.

United Nations 2020. The 17 goals. https://sdgs.un.org/goals.

Wang, Y.P., Smith, R. 1994. Wastewater minimisation. *Chemical Engineering Science* 49(7): 981–1006.

2 Multi-Plant HEN Designs for Continuous Processes *Optimization from a Total-Site Perspective*

Without addressing the benefit allocation issues, the traditional approaches for synthesizing the HEN in a single plant are in general applicable for producing the multi-plant HEN designs. In fact, only minor modifications are required in the conventional models. These needed changes are presented in the following.

2.1 SEQUENTIAL SYNTHESIS

A series of three optimization problems are solved consecutively in the design procedure for generating HEN designs and these steps are summarized below:

1. The minimum acceptable total utility cost of the entire site is first determined with a linear program, which can be formulated by modifying the conventional transshipment model (Papoulias and Grossmann, 1983).
2. By fixing the interplant heat-flow patterns determined in the previous step, the minimum total number of matches and the corresponding heat duties can be determined with an extended version of the conventional MILP model (Papoulias and Grossmann, 1983).
3. By following the approach suggested by Floudas et al. (1986), the superstructure of total-site HEN designs can be built to facilitate the matches identified in Step 2. A nonlinear programming model (Floudas et al., 1986; Floudas and Grossmann, 1987) can be constructed accordingly and then solved to generate the optimal multi-plant HEN configuration that minimizes the total capital cost.

2.1.1 Minimum Total Utility Cost

Traditionally, a transshipment model has been used to compute the minimum total utility cost of a single-plant HEN. This model can be easily modified for multi-plant applications by incorporating interplant heat flows in energy balance constraints.

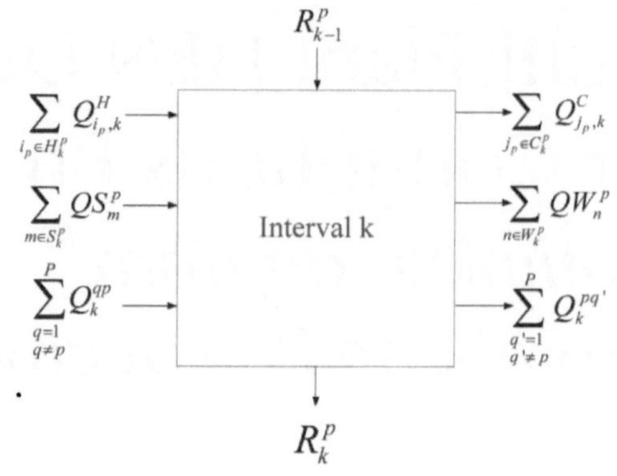

FIGURE 2.1 Input-output structure of a temperature interval in plant *p*. (Adapted with permission from *Chem. Eng. Sci.* 2014, 118: 60–73. Copyright 2014 Elsevier.)

Specifically, let us consider temperature interval *k* of plant *p* in Figure 2.1. The corresponding LP model can be formulated accordingly as follows.

$$\min \sum_{p=1}^{P} Z'_p \tag{2.1}$$

s.t.

- Energy balance around temperature interval *k* in plant *p*

$$R_k^p - R_{k-1}^p - \sum_{m \in S_k^p} QS_m^p - \sum_{\substack{q=1 \\ q \neq p}}^{P} Q_k^{qp} + \sum_{n \in W_k^p} QW_n^p + \sum_{\substack{q'=1 \\ q' \neq p}}^{P} Q_k^{pq'} = \Delta H_k^p \tag{2.2}$$

$$\Delta H_k^p = \sum_{i_p \in H_k^p} Q_{i_p,k}^H - \sum_{j_p \in C_k^p} Q_{j_p,k}^C = \left(\sum_{i_p \in H_k^p} F_{i_p}^H - \sum_{j_p \in C_k^p} F_{j_p}^C \right) \Delta T_k \tag{2.3}$$

where $k = 1, 2, \cdots, K$ and $p = 1, 2, \cdots, P$.

- Residual heat flows

$$R_k^p \geq 0 \tag{2.4}$$

where $k = 1, 2, \cdots, K - 1$, $R_0^p = R_K^p = 0$, and $p = 1, 2, \cdots, P$.

- Total utility costs

$$Z'_p = \sum_{m \in S^p} c_m^H QS_m^p + \sum_{n \in W^p} c_n^C QW_n^p \tag{2.5}$$

$$\bar{Z}_p - Z'_p \geq 0 \tag{2.6}$$

where $p = 1, 2, \cdots, P$.

- Interplant heat flows

$$Q_k^{q,p} \geq 0 \tag{2.7}$$

$$Q_k^{p,q'} \geq 0 \tag{2.8}$$

where $k = 1, 2, \cdots, K$; $p, q, q' = 1, 2, \cdots, P$ ($p \neq q$ and $p \neq q'$). All notations in this model are defined below.

Sets:

C^p Set of cold streams in plant p
C_k^p Set of cold streams in plant p that can consume heat from interval k of plant p
H^p Set of hot streams in plant p
H_k^p Set of hot streams in plant p that can supply heat to interval k of plant p
S^p Set of hot utilities in plant p
S_k^p Set of hot utilities in plant p that can supply heat to interval k of plant p
W^p Set of cold utilities in plant p
W_k^p Set of cold utilities in plant p that can consume heat from interval k of plant p

Parameters:

c_m^H Unit cost of hot utility m
c_n^C Unit cost of cold utility n
$F_{i_p}^H$ Heat capacity flow rate of hot stream i_p in plant p
$F_{j_p}^C$ Heat capacity flow rate of cold stream j_p in plant p
$Q_{i_p,k}^H$ Heat supply rate of hot stream i_p in plant p to interval k of plant p
$Q_{j_p,k}^C$ Heat consumption rate of cold stream j_p in plant p from interval k of plant p
$\bar{Z}_p$ Minimum total utility cost of standalone plant p
ΔT_k Temperature difference of upper and lower bounds of interval k

Variables:

R_k^p Residual heat flow rate from temperature interval k in plant p
QS_m^p Consumption rate of hot utility m in plant p

QW_n^p Consumption rate of cold utility n in plant p
$Q_k^{q,p}$ Heat flow rate from plant q to interval k in plant p
$Q_k^{p,q'}$ Heat flow rate from interval k in plant p to plant q'
Z_p' Minimum total utility cost of plant p after interplant heat integration

Example 2.1

Let us consider the stream and utility data presented in Table 2.1 and Table 2.2, respectively. Note that, in Table 2.1, T^{in} and T^{out} denote the initial and target temperatures of the corresponding process stream, respectively, and F_{cp} is the heat capacity flow rate. By using a minimum temperature approach (ΔT_{min}) of 10 K, the corresponding heat-flow cascades and pinch points of three standalone plants in this example are presented in Figure 2.2. The hot utility consumption rates of plants P1, P2, and P3 can be found to be 800 kW, 100 kW, and 255 kW respectively, while the cold utility consumption rates are 210 kW, 160 kW, and 670 kW respectively. The corresponding energy costs of these plants are 66,100 USD/yr, 6600 USD/yr, and 30,300 USD/yr, respectively.

Figure 2.3 shows the integrated heat-flow cascade obtained by solving the aforementioned LP model. Notice that plant P1 in this scheme does not consume hot utilities and P2 does not need cold utilities. The high-pressure steam is used primarily in P2 (220 kW) and fuel in P3 (440 kW), while the cooling waters are called for only in P1 (485 kW) and P3 (60 kW). To facilitate better understanding of the economic implications of interplant heat integration, a cost analysis is presented in Table 2.3. Notice that the total utility cost savings in the last column are not evenly distributed and, in particular, this figure for plant P2 is nil. It should also be noted that, although the high-pressure steam of P2 is the cheapest hot utility in this example, the upper

TABLE 2.1
Stream Data Used in Example 2.1

Plant #	Stream #	T^{in} (°C)	T^{out} (°C)	F_{cp} (kW/°C)
P1	H1	150	40	7
	C1	60	140	9
	C2	110	190	8
P2	H1	200	70	5.5
	C1	30	110	3.5
	C2	140	190	7.5
P3	H1	370	150	3.0
	H2	200	40	5.5
	C1	110	360	4.5

TABLE 2.2
Utility Data Used in Example 2.1

Plant #	Utility Type	Temperature (°C)	Unit Cost (USD/kW·yr)	Capacity (kW)
P1	Cooling water (CW)	25	10	1000
	HP steam @ 240 psig	200	90	1000
	Fuel	500	80	1000
P2	Cooling water (CW)	25	22.5	1000
	HP steam @ 240 psig	200	30	1000
	Fuel	500	120	1000
P3	Cooling water (CW)	25	30	1000
	HP steam @ 240 psig	200	60	1000
	Fuel	500	40	1000

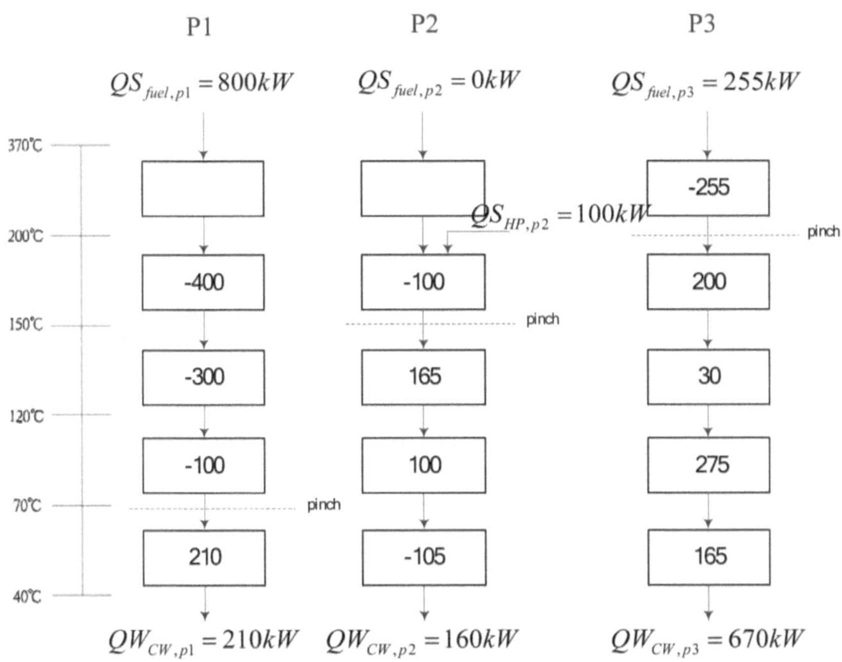

FIGURE 2.2 Standalone heat flow cascades obtained in Example 2.1. (Adapted with permission from *Chem. Eng. Sci.* 2014, 118: 60–73. Copyright 2014 Elsevier.)

bound in equation (2.6) forbids its consumption rate exceeding 220 kW. If this constraint is removed, the disparity in cost savings becomes even more serious.

Figure 2.4 shows the integrated heat-flow cascade obtained by removing all upper bounds on total utility costs. Notice that the hot utilities of P1 and cold utility of P3 are still not used due to their relatively high costs. The consumption levels of

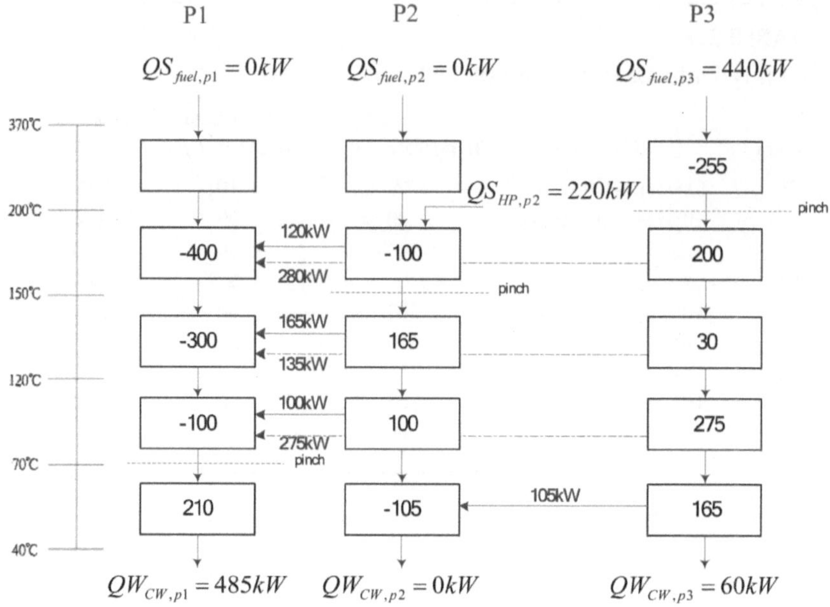

FIGURE 2.3 Integrated heat flow cascade obtained by imposing upper bound on total utility cost of every plant in Example 2.1. (Reproduced with permission from *Chem. Eng. Sci.* 2014, 118: 60–73. Copyright 2014 Elsevier.)

TABLE 2.3
Utility Cost Estimates for Interplant Heat Integration with Cost Upper Bounds in Example 2.1

Plant #	Utility Type	Annual Cost without Integration (USD/yr)	Annual Cost with Integration (USD/yr)	Individual Cost Saving (USD/yr)	Overall Cost Saving (USD/yr)
P1	Fuel	64,000	0	64,000	61,250
	CW	2100	4850	–2750	
P2	Fuel	0	0	0	0
	HP Steam	3000	6600	–3600	
	CW	3600	0	3600	
P3	Fuel	10,200	17,600	–7400	10,900
	CW	20,100	1,800	18,300	

high-pressure steam (P2: 405 kW), fuel (P3: 255 kW), and cooling waters (P1: 485 kW; P2: 60 kW) are selected mainly according to their costs also. Since there are no cost upper bounds in this case, the cheapest hot utility, i.e., the high-pressure steam of P2, is utilized as much as possible in Figure 2.4. As a result, although P1 and P3 reap large savings, P2 ends up with a higher total utility cost after interplant heat integration (see Table 2.4).

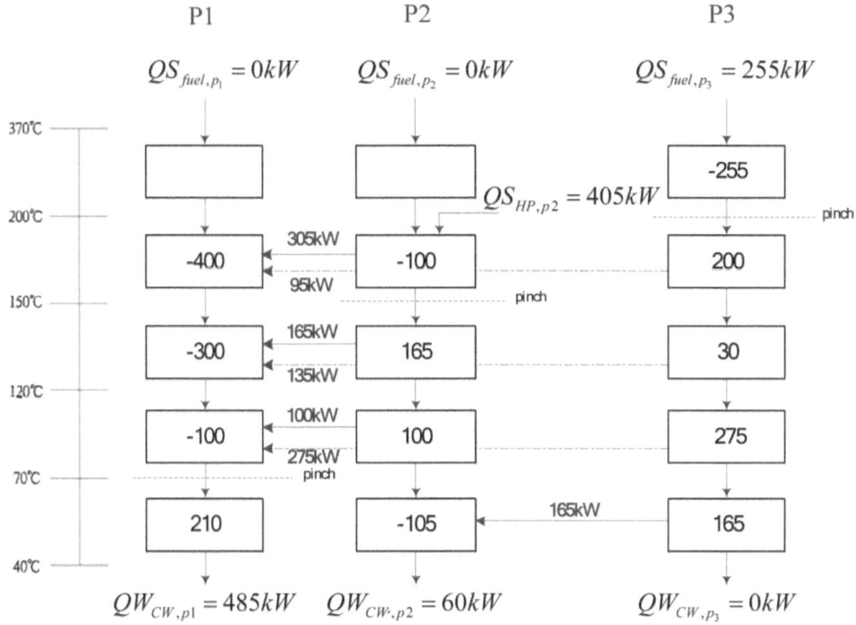

FIGURE 2.4 Integrated heat flow cascade obtained without upper bounds on total utility costs of all plants in Example 2.1.

TABLE 2.4
Utility Cost Estimates for Interplant Heat Integration without Cost Upper Bounds in Example 2.1

Plant #	Utility Type	Annual Cost without Integration (USD/yr)	Annual Cost with Integration (USD/yr)	Individual Cost Saving (USD/yr)	Overall Cost Saving (USD/yr)
P1	Fuel	64,000	0	64,000	61,250
	CW	2100	4850	–2,750	
P2	Fuel	0	0	0	–6900
	HP Steam	3000	12,150	–9,150	
	CW	3600	1350	2250	
P3	Fuel	10,200	10,200	0	20,100
	CW	20,100	0	20,100	

2.1.2 Minimum Total Number of Matches

To incorporate the interplant heat flows determined in the previous step, it is necessary to consider the heat-flow pattern inside the temperature interval, i.e., the "warehouse" in the transshipment model, given in Figure 2.1. Specifically, the internal heat flows of energy warehouse can be depicted in detail as shown in Figure 2.5.

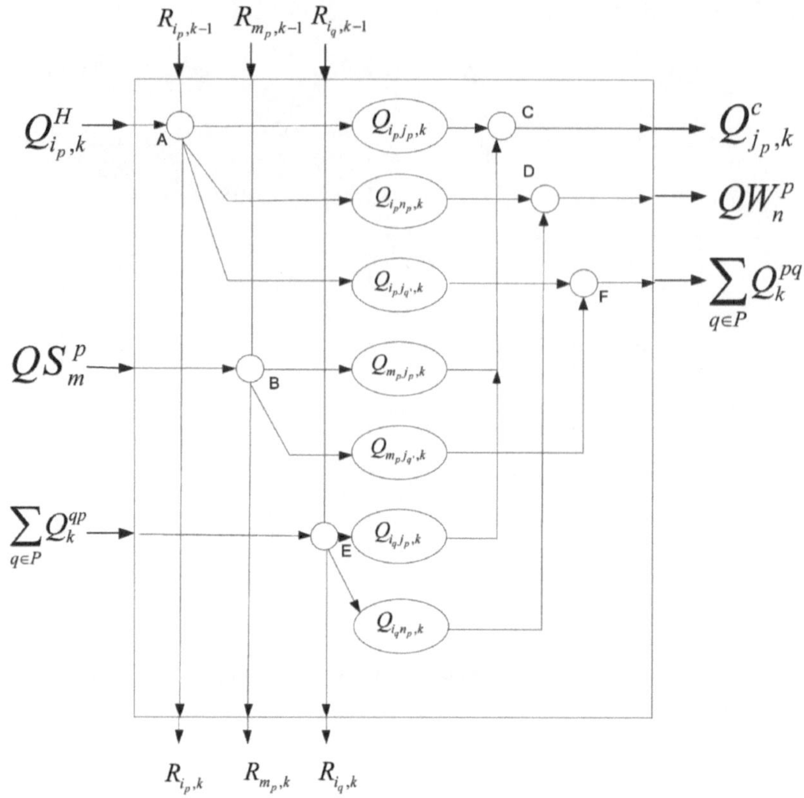

FIGURE 2.5 Heat flow pattern within temperature interval k in plant p. (Adapted with permission from *Chem. Eng. Sci.* 2014, 118: 60–73. Copyright 2014 Elsevier.)

To facilitate formulation of the corresponding MILP model, additional variables should be defined accordingly, i.e.

$R_{i_p,k}$	Residual heat of hot stream i_p in plant p from interval k
$R_{m_p,k}$	Residual heat of hot utility m_p in plant p from interval k
$R_{i_q,k}$	Residual heat of hot stream (or hot utility) i_q in plant q from interval k
$Q_{i_p,j_p,k}$	Heat exchange rate within interval k between hot stream i_p and cold stream j_p in plant p
$Q_{i_p,n_p,k}$	Heat exchange rate within interval k between hot stream i_p and cold utility n_p in plant p
$Q_{i_p,j_{q'},k}$	Heat exchange rate within interval k between hot stream i_p in plant p and cold stream (or cold utility) $j_{q'}$ in plant q'
$Q_{m_p,j_p,k}$	Heat exchange rate within interval k between hot utility m_p and cold stream j_p in plant p

$Q_{m_p,j_{q'},k}$ Heat exchange rate within interval k between hot utility m_p in plant p and cold stream $j_{q'}$ in plant q'

$Q_{i_q,j_p,k}$ Heat exchange rate within interval k between hot stream (or hot utility) i_q in plant q and cold stream j_p in plant p

$Q_{i_q,n_p,k}$ Heat exchange rate within interval k between hot stream i_q in plant q and cold utility n_p in plant p

Following is a list of energy balances around all nodes in interval k $(k = 1,2,\cdots,K)$ of plant p $(p = 1,2,\cdots,P)$:

- Node A:

$$R_{i_p,k} - R_{i_p,k-1} + \sum_{j_p \in C_k^p} Q_{i_p,j_p,k} + \sum_{n_p \in W_k^p} Q_{i_p,n_p,k} + \sum_{\substack{q'=1 \\ q' \neq p}}^{P} \sum_{j_{q'} \in C_k^{q'} \cup W_k^{q'}} Q_{i_p,j_{q'},k} = Q_{i_p,k}^{H} \quad (2.9)$$

  where $i_p \in H_k^p$; $Q_{i_p,k}^H$ is a model parameter representing the heat supply rate of hot stream i_p in plant p to interval k of plant p.

- Node B:

$$R_{m_p,k} - R_{m_p,k-1} + \sum_{j_p \in C_k^p} Q_{m_p,j_p,k} + \sum_{\substack{q'=1 \\ q' \neq p}}^{P} \sum_{j_{q'} \in C_k^{q'}} Q_{m_p,j_{q'},k} = QS_{m_p}^{p} \quad (2.10)$$

  where $m_p \in S_k^p$; $QS_{m_p}^p$ is the consumption rate of hot utility m_p in plant p, which can be determined by solving the transshipment model in the previous step.

- Node C:

$$\sum_{i_p \in H_k^p} Q_{i_p,j_p,k} + \sum_{m_p \in S_k^p} Q_{m_p,j_p,k} + \sum_{\substack{q=1 \\ q \neq p}}^{P} \sum_{i_q \in H_k^q \cup S_k^q} Q_{i_q,j_p,k} = Q_{j_p,k}^{C} \quad (2.11)$$

  where $j_p \in C_k^p$; $Q_{j_p,k}^C$ is a model parameter denoting the heat consumption rate of cold stream j_p in plant p from interval k of plant p.

- Node D:

$$\sum_{i_p \in H_k^p} Q_{i_p,n_p,k} + \sum_{\substack{q=1 \\ q \neq p}}^{P} \sum_{i_q \in H_k^q} Q_{i_q,n_p,k} = QW_{n_p}^{p} \quad (2.12)$$

  where $n_p \in W_k^p$; $QW_{n_p}^p$ is the consumption rate of cold utility n_p in plant p, which can be determined by solving the transshipment model in the previous step.

- Node E:

$$R_{i_q,k} - R_{i_q,k-1} + \sum_{j_p \in C_k^p} Q_{i_q,j_p,k} + \sum_{n_p \in W_k^p} Q_{i_q,n_p,k} = Q_{i_q,k}^{HE_1} \tag{2.13}$$

where $i_q \in H_k^q$.

$$R_{i_q,k} - R_{i_q,k-1} + \sum_{j_p \in C_k^p} Q_{i_q,j_p,k} = Q_{i_q,k}^{HE_2} \tag{2.14}$$

where $i_q \in S_k^q$.

Notice that the right sides of the above two equations respectively represent the heat inputs from hot stream and hot utility in plant q $(q = 1,\cdots,p-1,p+1,\cdots,P)$. Notice also that these heat inputs must satisfy the following constraints:

$$\sum_{i_q \in H_k^q} Q_{i_q,k}^{HE_1} + \sum_{i_q \in S_k^q} Q_{i_q,k}^{HE_2} = Q_k^{q,p} \tag{2.15}$$

As explained before, $Q_k^{q,p}$ is the heat flow rate from plant q to interval k in plant p and its value can be determined by solving the transshipment model in the previous step.

- Node F:

$$\sum_{i_p \in H_k^p} Q_{i_p,j_{q'},k} + \sum_{m_p \in S_k^p} Q_{m_p,j_{q'},k} = Q_{j_{q'},k}^{CE_1} \tag{2.16}$$

where $j_{q'} \in C_k^{q'}$.

$$\sum_{i_p \in H_k^p} Q_{i_p,j_{q'},k} = Q_{j_{q'},k}^{CE_2} \tag{2.17}$$

where $j_{q'} \in W_k^{q'}$.

Notice again that the right sides of the above two equations respectively represent the heat outputs to cold stream and cold utility in plant q' $(q' = 1,\cdots,p-1,p+1,\cdots,P)$, and these outputs must satisfy the following constraints:

$$\sum_{j_{q'} \in C_k^{q'}} Q_{j_{q'},k}^{CE_1} + \sum_{j_{q'} \in W_k^{q'}} Q_{j_{q'},k}^{CE_2} = Q_k^{p,q'} \tag{2.18}$$

As also explained before, $Q_k^{p,q'}$ is the heat flow rate from interval k in plant p to plant q' and its value can be determined by solving the LP model in the previous step. It should also be noted that the sum of the aforementioned residual heat flow rates from interval k in plant p must be the same as the total residual heat flow rate determined previously with the transshipment model, i.e.

$$R_k^p = \sum_{i_p \in H_k^p} R_{i_p,k} + \sum_{m_p \in S_k^p} R_{m_p,k} + \sum_{\substack{q=1 \\ q \neq p}}^{P} \sum_{i_q \in H_k^q \cup S_k^q} R_{i_q,k} \tag{2.19}$$

Finally, since the goal in present step is to minimize the total match number, the objective function of the MILP model can be formulated as

$$\begin{aligned} Z = & \sum_{p=1}^{P} \sum_{i_p \in H^p} \sum_{j_p \in C^p} z_{i_p,j_p} + \sum_{p=1}^{P} \sum_{i_p \in S^p} \sum_{j_p \in C^p} z_{i_p,j_p} + \sum_{p=1}^{P} \sum_{i_p \in H^p} \sum_{j_p \in W^p} z_{i_p,j_p} \\ & + \sum_{p=1}^{P} \sum_{\substack{q=1 \\ q \neq p}}^{P} \sum_{i_p \in H^p} \sum_{j_q \in C^q} z_{i_p,j_q} + \sum_{p=1}^{P} \sum_{\substack{q=1 \\ q \neq p}}^{P} \sum_{i_p \in S^p} \sum_{j_q \in C^q} z_{i_p,j_q} \\ & + \sum_{p=1}^{P} \sum_{\substack{q=1 \\ q \neq p}}^{P} \sum_{i_p \in H^p} \sum_{j_q \in W^q} z_{i_p,j_q} \end{aligned} \tag{2.20}$$

where the binary variables can be determined by imposing the following inequality constraints:

$$\sum_{k=1}^{K} Q_{i_p,j_q,k} - z_{i_p,j_q} U_{i_p,j_q} \leq 0 \tag{2.21}$$

where $p,q = 1,2,\cdots,P$; $z_{i_p,j_q} \in \{0,1\}$; U_{i_p,j_q} is a large enough constant; $i_p \in H^p \cup S^p$; $j_q \in C^q \cup W^q$.

Example 2.2

Based on the optimization results given in Figure 2.3 and Figure 2.4, two corresponding MILP models can be constructed according to equations (2.9)–(2.21). Table 2.5 shows the optimum solution of the former model, and it can be observed that a total of 16 matches are needed in this case. On the other hand, from the optimum solution of the later model in Table 2.6, one can see that the minimum number of matches is lowered to 14. Notice also that the interplant matches are marked in red. Again, although better matches can be identified without imposing upper bounds on total utility costs, the latter scheme may still be infeasible due to the unsettled benefit allocation issues.

2.1.3 Minimum Total Capital Cost

As mentioned before, the third step of the sequential approach to synthesize the multi-plant HEN is exactly the same as that in single-plant applications. By applying the method suggested by Floudas et al. (1986), a superstructure can be built to facilitate the matches identified in Step 2. An NLP model (Floudas et al., 1986; Floudas

TABLE 2.5
Optimal Matches with Utility Cost Upper Bounds

Cold Stream \ Hot Stream	P1_H1	P2_H1	P3_H1	P3_H2	P2_HP	P3_Fuel
P1_C1	285	100	60	275		
P1_C2		285	275	80		
P2_C1		175		105		
P2_C2		155			220	
P3_C1			325	360		440
P1_CW	485					
P3_CW				60		

TABLE 2.6
Optimal Matches without Utility Cost Upper Bounds

Cold Stream \ Hot Stream	P1_H1	P2_H1	P3_H1	P3_H2	P2_HP	P3_Fuel
P1_C1	285	160		275		
P1_C2		380		230	30	
P2_C1		175		105		
P2_C2					375	
P3_C1			660	210		255
P1_CW	485					
P2_CW				60		

and Grossmann, 1987) can then be constructed accordingly and then solved to generate the optimal HEN configuration that minimizes the total capital cost.

The basic features of a superstructure are listed below:

- Each exchanger in HEN corresponds to a match predicted by the MILP model and its heat duty should also be the same as that predicted in the previous step.
- All possible interconnections among the heat exchangers on a particular process stream are incorporated in a distinct superstructure. For any given HEN synthesis problem, there may be more than one superstructure, and each is associated with a unique hot or cold process stream.
- The flow rates and temperatures of stream interconnections in the above superstructures are treated as unknowns which must be determined by solving the NLP model.

Let us first use a simple example to illustrate the construction procedure of superstructure:

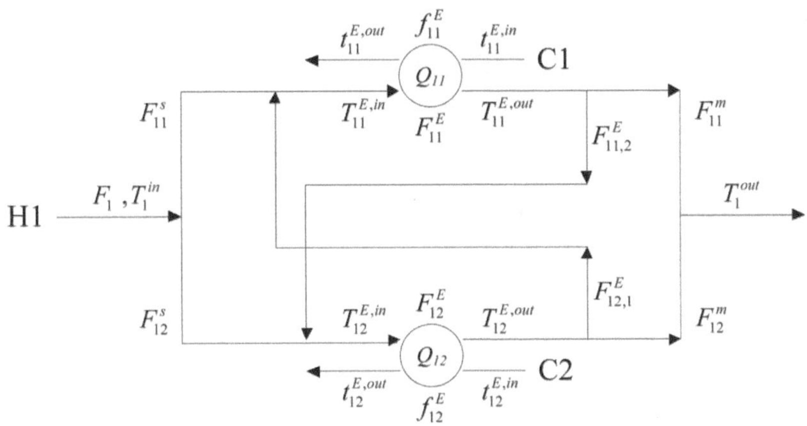

FIGURE 2.6 Superstructure of Example 2.3.

Example 2.3

To facilitate clearer illustration, let us consider one hot stream (H1) and two cold streams (C1 and C2). The optimum solution of MILP shows that the minimum number of matches is two, i.e., (H1,C1) and (H1,C2). Therefore, one can produce the superstructure in Figure 2.6 accordingly.

Next, an NLP model can be formulated according to the superstructures. To present a concise formulation without loss of generality, all notations in the superstructure and the corresponding model formulation are devoid of labels for distinguishing the plant origins. Specifically,

Sets:

C Set of all cold streams in a multi-plant HEN
$\hat{C}_i$ Set of all cold streams which are matched with hot stream $i \in H$
H Set of all hot streams in a multi-plant HEN
$\hat{H}_j$ Set of all hot streams which are matched with cold stream $j \in C$

Parameters:

$c_{i,j}$ Capital cost coefficient of heat exchanger (i, j)
F_i, f_j Heat capacity flow rates of hot stream i and cold stream j
$Q_{i,j}$ Heat duty of heat exchanger (i, j)
T_i^{in}, t_j^{in} Initial temperatures of hot stream i and cold stream j
T_i^{out}, t_j^{out} Target temperatures of hot stream i and cold stream j
$U_{i,j}$ Overall heat transfer coefficient of heat exchanger (i, j)
ΔT_{min} Minimum temperature difference in every heat exchanger
β Model parameter of heat exchanger capital cost

Variables:

$F^s_{i,j}, f^s_{j,i}$	Heat capacity flow rates of hot stream i and cold stream j from their respective inlet splitters to heat exchanger (i, j)
$F^m_{i,j}, f^m_{j,i}$	Heat capacity flow rates of hot stream i and cold stream j from heat exchanger (i, j) to their respective outlet mixers
$F^E_{i,j}, f^E_{j,i}$	Heat capacity flow rates of hot stream i and cold stream j in heat exchanger (i, j)
$F^{E,s}_{i,j,n}$	Heat capacity flow rate of hot stream i from the outlet splitter of heat exchanger (i, j) to heat exchanger (i, n), where $n \in C_i$ and $n \neq j$
$f^{E,s}_{j,i,n}$	Heat capacity flow rate of cold stream j from the outlet splitter of heat exchanger (i, j) to heat exchanger (m, j), where $m \in H_j$ and $m \neq i$
$T^{E,in}_{i,j}, t^{E,in}_{j,i}$	Hot stream i and cold stream j temperatures at inlets of heat exchanger (i, j)
$T^{E,out}_{i,j}, t^{E,out}_{j,i}$	Hot stream i and cold stream j temperatures at outlets of heat exchanger (i, j)
$\theta^1_{i,j}, \theta^2_{i,j}$	Hot and cold end temperatures of heat exchanger (i, j)

The generalized NLP model can be formulated as follows:

$$\min C = \sum_{i \in H} \sum_{j \in C} c_{i,j} \left[\frac{Q_{i,j}}{U_{i,j} \left[\theta^1_{i,j} \theta^2_{i,j} (\theta^1_{i,j} + \theta^2_{i,j}) / 2 \right]^{\frac{1}{3}}} \right]^{\beta} \tag{2.22}$$

Subject to:

- Mass balances at inlet splitters and outlet mixers of HEN

$$F_i = \sum_{j \in C_i} F^s_{i,j} = \sum_{j \in C_i} F^m_{i,j}, \quad \forall i \in H \tag{2.23}$$

$$f_j = \sum_{i \in H_j} f^s_{j,i} = \sum_{i \in H_j} f^m_{j,i}, \quad \forall j \in C \tag{2.24}$$

- Energy balances at outlet mixers of HEN

$$F_i T^{out}_i = \sum_{j \in C_i} F^m_{i,j} T^{E,out}_{i,j}, \quad \forall i \in H \tag{2.25}$$

$$f_j t^{out}_j = \sum_{i \in H_j} f^m_{j,i} t^{E,out}_{j,i}, \quad \forall j \in C \tag{2.26}$$

- Mass and energy balances at inlet mixers of heat exchanger (i, j)
  - For hot stream $i \in \mathrm{H}$

$$F_{i,j}^{E} = F_{i,j}^{s} + \sum_{\substack{n \in C_i \\ n \neq j}} F_{i,n,j}^{E,s}, \quad \forall j \in C_i \tag{2.27}$$

$$F_{i,j}^{E} T_{i,j}^{E,in} = F_{i,j}^{s} T_{i}^{in} + \sum_{\substack{n \in C_i \\ n \neq j}} F_{i,n,j}^{E,s} T_{i,n}^{E,out}, \quad \forall j \in C_i \tag{2.28}$$

  - For cold stream $j \in \mathrm{C}$

$$f_{j,i}^{E} = f_{j,i}^{s} + \sum_{\substack{m \in H_j \\ m \neq i}} f_{j,m,i}^{E,s}, \quad \forall i \in H_j \tag{2.29}$$

$$f_{j,i}^{E} t_{ji}^{E,in} = f_{j,i}^{s} t_{j}^{in} + \sum_{\substack{m \in H_j \\ m \neq i}} f_{j,m,i}^{E,s} t_{j,m}^{E,out}, \quad i \in H_j \tag{2.30}$$

- Mass balances at outlet splitters of heat exchanger (i, j)
  - For hot stream $i \in \mathrm{H}$

$$F_{i,j}^{E} = F_{i,j}^{m} + \sum_{\substack{n \in C_i \\ n \neq j}} F_{i,j,n}^{E,s}, \quad \forall j \in C_i \tag{2.31}$$

  - For cold stream $j \in \mathrm{C}$

$$f_{j,i}^{E} = f_{j,i}^{m} + \sum_{\substack{m \in H_j \\ m \neq i}} f_{m,j,i}^{E,s}, \quad \forall i \in H_j \tag{2.32}$$

- Energy balances around heat exchanger (i, j)
  - For hot stream $i \in \mathrm{H}$

$$Q_{i,j} = F_{i,j}^{E}\left(T_{i,j}^{E,in} - T_{i,j}^{E,out}\right), \quad \forall j \in C_i \tag{2.33}$$

  - For cold stream $j \in \mathrm{C}$

$$Q_{i,j} = f_{j,i}^{E}(t_{j,i}^{E,out} - t_{j,i}^{E,in}), \quad \forall i \in H_j \tag{2.34}$$

- Temperature differences at hot and cold ends of heat exchanger (i, j)

$$\theta_{i,j}^{1} = T_{i,j}^{E,in} - t_{i,j}^{E,out} \tag{2.35}$$

$$\theta^2_{i,j} = T^{E,out}_{i,j} - t^{E,in}_{i,j} \tag{2.36}$$

- Lower bound of temperature differences

$$\theta^1_{i,j}, \theta^2_{i,j} \geq \Delta T_{\min} \tag{2.37}$$

- Nonnegative heat capacity flow rates

$$F^s_{i,j}, F^m_{i,j}, F^{E,s}_{i,j,m}, F^E_{i,j}, f^s_{j,i}, f^m_{j,i}, f^{E,s}_{j,i,n}, f^E_{j,i} \geq 0 \tag{2.38}$$

To illustrate the model building step more clearly, let us consider the superstructure obtained in Example 2.3.

Example 2.4

An NLP model can be written explicitly according to the superstructure in Figure 2.6 as follows:

$$\min C = c_{1,1}\left[\frac{Q_{1,1}}{U_{1,1}\left[\theta^1_{1,1}\theta^2_{1,1}(\theta^1_{1,1}+\theta^2_{1,1})/2\right]^{\frac{1}{3}}}\right]^{\beta} + c_{1,2}\left[\frac{Q_{1,2}}{U_{1,2}\left[\theta^1_{1,2}\theta^2_{1,2}(\theta^1_{1,2}+\theta^2_{1,2})/2\right]^{\frac{1}{3}}}\right]^{\beta} \tag{2.39}$$

subject to:

$$F_1 = F^s_{1,1} + F^s_{1,2} = F^m_{1,1} + F^m_{1,2} \tag{2.40}$$

$$f_1 = f^s_{1,1} = f^m_{1,1} \tag{2.41}$$

$$f_2 = f^s_{2,1} = f^m_{2,1} \tag{2.42}$$

$$F_1 T_1^{out} = F^m_{1,1} T^{E,out}_{1,1} + F^m_{1,2} T^{E,out}_{1,2} \tag{2.43}$$

$$f_1 t_1^{out} = f^m_{1,1} t^{E,out}_{1,1} \tag{2.44}$$

$$f_2 t_2^{out} = f^m_{2,1} t^{E,out}_{2,1} \tag{2.45}$$

$$F^E_{1,1} = F^s_{1,1} + F^{E,s}_{1,2,1} \tag{2.46}$$

$$F^E_{1,2} = F^s_{1,2} + F^{E,s}_{1,1,2} \tag{2.47}$$

$$f^E_{1,1} = f^s_{1,1} \tag{2.48}$$

$$f^E_{2,1} = f^s_{2,1} \tag{2.49}$$

$$F^{E}_{1,1}T^{E,in}_{1,1} = F^{s}_{1,1}T^{in}_{1} + F^{E,s}_{1,2,1}T^{E,out}_{1,2} \tag{2.50}$$

$$F^{E}_{1,2}T^{E,in}_{1,2} = F^{s}_{1,2}T^{in}_{1} + F^{E,s}_{1,1,2}T^{E,out}_{1,1} \tag{2.51}$$

$$f^{E}_{1,1}t^{E,in}_{1,1} = f^{s}_{1,1}t^{in}_{1} \tag{2.52}$$

$$f^{E}_{2,1}t^{E,in}_{2,1} = f^{s}_{2,1}t^{in}_{2} \tag{2.53}$$

$$F^{E}_{1,1} = F^{m}_{1,1} + F^{E,s}_{1,1,2} \tag{2.54}$$

$$F^{E}_{1,2} = F^{m}_{1,2} + F^{E,s}_{1,2,1} \tag{2.55}$$

$$f^{E}_{1,1} = f^{m}_{1,1} \tag{2.56}$$

$$f^{E}_{2,1} = f^{m}_{2,1} \tag{2.57}$$

$$Q_{1,1} = F^{E}_{1,1}\left(T^{E,in}_{1,1} - T^{E,out}_{1,1}\right) = f^{E}_{1,1}\left(t^{E,out}_{1,1} - t^{E,in}_{1,1}\right) \tag{2.58}$$

$$Q_{1,2} = F^{E}_{1,2}\left(T^{E,in}_{1,2} - T^{E,out}_{1,2}\right) = f^{E}_{2,1}\left(t^{E,out}_{2,1} - t^{E,in}_{2,1}\right) \tag{2.59}$$

$$\theta^{1}_{1,1} = T^{E,in}_{1,1} - t^{E,out}_{1,1} \tag{2.60}$$

$$\theta^{1}_{1,2} = T^{E,in}_{1,2} - t^{E,out}_{2,1} \tag{2.61}$$

$$\theta^{2}_{1,1} = T^{E,out}_{1,1} - t^{E,in}_{1,1} \tag{2.62}$$

$$\theta^{2}_{1,2} = T^{E,out}_{1,2} - t^{E,in}_{2,1} \tag{2.63}$$

$$\theta^{1}_{1,1}, \theta^{1}_{1,2}, \theta^{2}_{1,1}, \theta^{2}_{1,2} \geq \Delta T_{\min} \tag{2.64}$$

$$F^{s}_{1,1}, F^{s}_{1,2}, F^{m}_{1,1}, F^{m}_{1,2}, F^{E,s}_{1,1,2}, F^{E,s}_{1,2,1}, F^{E}_{1,1}, F^{E}_{1,2}, f^{s}_{1,1}, f^{s}_{2,1}, f^{m}_{1,1}, f^{m}_{2,1}, f^{E}_{1,1}, f^{E}_{2,1} \geq 0 \tag{2.65}$$

Let us next follow the aforementioned approach to build the superstructures and the corresponding NLP models according to the optimization results obtained in Examples 2.1 and 2.2. From the optimum solutions, one can then determine if the interplant heat integration programs are indeed financially feasible. For computation convenience, let us use the following fictitious model parameters: $U_{i,j} = 1\ \text{W}/\text{m}^2\text{K}$, $c_{i,j} = 670\ \text{USD}/\text{m}^2$, and $\beta = 0.83$. Let us assume that the hot utilities experience no temperature drops in heaters and the temperature of cold utility is always raised by 5°C in each cooler. Let us further assume that the annualization factor (*Af*) is 0.1349, which is associated with an annual interest rate of 5.85% over a period of 10 years. Finally, by following the traditional sequential procedure to synthesize the optimal HEN of each individual plant, the minimum total capital costs of P1, P2, and P3 were found to be 43,670, 43,447, and 58,302 USD, respectively.

Example 2.5

The multi-plant HEN design in Figure 2.7 can be obtained with an NLP model built according to the data presented in Tables 2.1, 2.2, and 2.5. The heat-transfer areas of all units in this HEN are shown in Table 2.7, and the corresponding total capital cost is 171,692 USD. Notice that, without interplant heat integration, the sum of minimum total capital costs of the above three plants is 145,419 (= 43,670 + 43,477 + 58,302) USD. However, the annualized extra capital expenditure, i.e., (171,692 − 145,419) × 0.1349 = 3,544 USD/yr, is still much lower than the total annual saving in utility cost, i.e., 72,150 USD/yr (see Table 2.3). Thus, the interplant heat integration projects should be feasible as a whole. The only unsolved issue is obviously concerned with benefit allocation.

On the other hand, Tables 2.1, 2.2, and 2.6 can be utilized to produce the multi-plant HEN design in Figure 2.8 and Table 2.8. It is anticipated that this design is less practical since plant P2 is asked to shoulder not only extra utility cost but also extra capital expenditure.

2.2 SIMULTANEOUS SYNTHESIS

Although it is relatively easy to understand and implement the aforementioned sequential approach to synthesize multi-plant HENs, the obvious drawback of a stepwise optimization procedure is the inherent difficulty in evaluating the tradeoffs

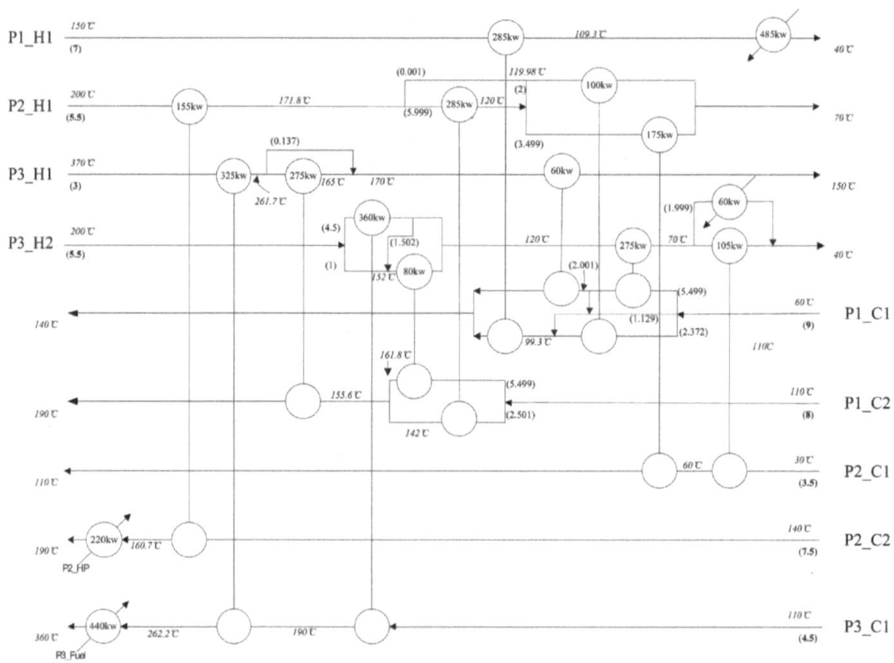

FIGURE 2.7 Multi-plant HEN design obtained by imposing upper bound on total utility cost of every plant in Example 2.1.

TABLE 2.7
Heat-Transfer Areas of All Units in Figure 2.6 (m^2)

Cold Stream \ Hot Stream	P1_H1	P2_H1	P3_H1	P3_H2	P2_HP	P3_Fuel
P1_C1	28.494	7.627	1.726	27.528		
P1_C2		28.529	8.927	8.008		
P2_C1		17.495		10.505		
P2_C2		4.373			5.116	
P3_C1			3.673	35.985		2.384
P1_CW	12.672					
P3_CW				2.356		

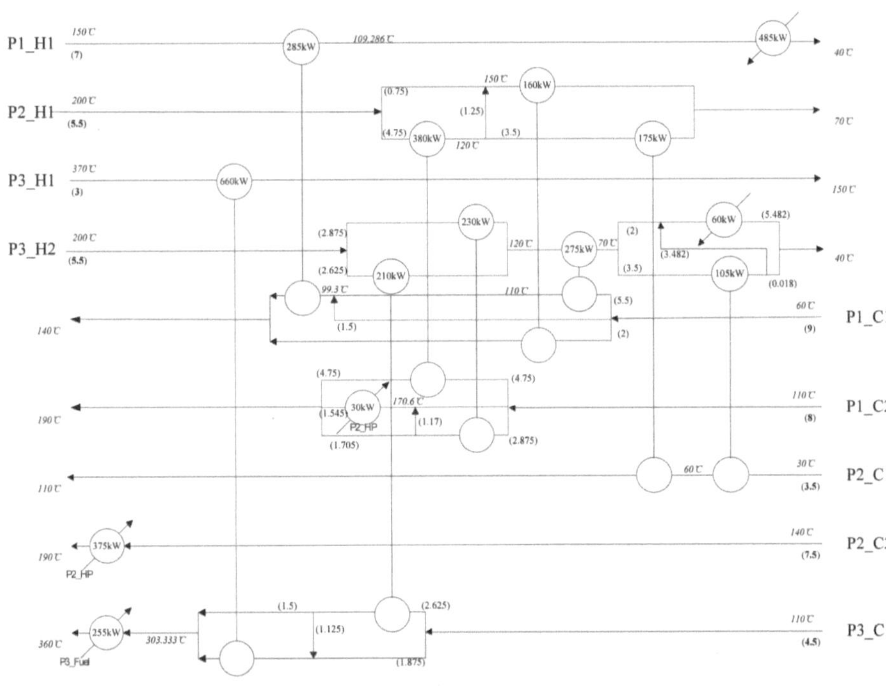

FIGURE 2.8 Multi-plant HEN design obtained without upper bounds on total utility costs of all plants in Example 2.1. (Adapted with permission from *Chem. Eng. Sci.* 2014, 118: 60–73. Copyright 2014 Elsevier.)

between utility and capital costs. To address this legitimate issue properly, let us consider the simultaneous HEN synthesis strategy originally developed for single-plant applications (Yee et al., 1990; Yee and Grossmann, 1990). For the purpose of generating interplant heat integration schemes, the conventional superstructure could be slightly modified, and a fictitious version is presented in Figure 2.9 for illustration

TABLE 2.8
Heat-Transfer Areas of All Units in Figure 2.7 (m^2)

Cold Stream \ Hot Stream	P1_H1	P2_H1	P3_H1	P3_H2	P2_HP	P3_Fuel
P1_C1	28.5	16		27.5		
P1_C2		38		23	1.67	
P2_C1		17.5		10.5		
P2_C2					13.592	
P3_C1			66	21		1.193
P1_CW	12.672					
P2_CW				3.37		

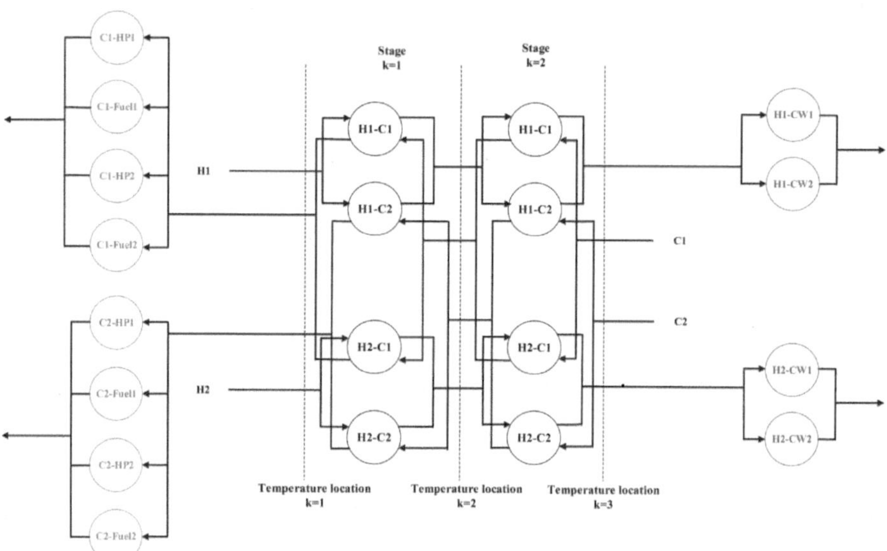

FIGURE 2.9 A constrained superstructure for simultaneous synthesis of multi-plant HENs.

convenience. Note that the number of stages in this structure is usually set to be the larger one of the hot and cold stream numbers. Note also that, to facilitate concise model formulation, all process and utility streams in this superstructure are without labels denoting the plant origins. Within each stage of the superstructure, all possible matches between hot and cold process streams may take place. In order to simplify the model formulation, let us temporarily assume that the outlets of heat exchangers in each stage must be isothermally mixed and, in addition, bypasses and split streams with more than two matches on each branch are all excluded. Unlike the single-plant HENs, there may be multiple utility streams onsite. Thus, the potential matches between the available hot utilities and each cold stream and also between the available cold utilities and each hot stream must be placed at the corresponding exits of the superstructure.

On the basis of this superstructure, an MINLP model can be formulated according to the notations defined below.

Sets:

C	Set of all cold process streams in a multi-plant HEN
CU	Set of all cold utilities in a multi-plant HEN
H	Set of all hot process streams in a multi-plant HEN
HU	Set of all hot utilities in a multi-plant HEN
ST	Set of all stages in a superstructure

Parameters:

TIN_i, TIN_j	Inlet temperatures of hot stream $i \in H$ and cold stream $j \in C$
$TOUT_i, TOUT_j$	Outlet temperatures of hot stream $i \in H$ and cold stream $j \in C$
F_i, F_j	Heat capacity flowrates of hot stream $i \in H$ and cold stream $j \in C$
TI_m, TI_n	Inlet temperatures of hot utility $m \in HU$ or cold utility $n \in CU$
$TT_{i,n}$	Outlet temperature of cold utility $n \in CU$, when it exchanged heat with hot stream $i \in H$
$TT_{m,j}$	Outlet temperature of hot utility $m \in HU$, when it exchanged heat with cold stream $j \in C$
$U_{i,j}$	Overall heat transfer coefficient between hot stream $i \in H$ and cold stream $j \in C$
$U_{i,n}$	Overall heat transfer coefficient between hot stream $i \in H$ and cold utility $n \in CU$
$U_{m,j}$	Overall heat transfer coefficient between cold stream $j \in C$ and hot utility $m \in HU$
CCU_n	Unit cost of cold utility $n \in CU$
CHU_m	Unit cost of hot utility $m \in HU$
$CF_{i,j}$	Fixed cost of exchanger (i, j), where $i \in H$ and $j \in C$
$CF_{i,n}$	Fixed cost of exchanger (i, n), where $i \in H$ and $n \in CU$
$CF_{m,j}$	Fixed cost of exchanger (m, j), where $m \in HU$ and $j \in C$
$CA_{i,j}$	Area cost coefficient of exchanger (i, j), where $i \in H$ and $j \in C$
$CA_{i,n}$	Area cost coefficient of exchanger (i, n), where $i \in H$ and $n \in CU$
$CA_{m,j}$	Area cost coefficient of exchanger (m, j), where $m \in HU$ and $j \in C$
β	Exponent for area cost
NOK	Total number of stages
NST	Upper bound of split streams in each stage
ΔT_{min}	Minimum temperature difference in a heat exchanger
$\Omega_{i,j}$	Maximum heat duty of exchanger (i, j), where $i \in H$ and $j \in C$
$\Omega_{i,n}$	Maximum heat duty of exchanger (i, n), where $i \in H$ and $n \in CU$
$\Omega_{m,j}$	Maximum heat duty of exchanger (m, j), where $m \in HU$ and $j \in C$

$\Gamma_{i,j}$ Upper bound of temperature difference in exchanger (*i*, *j*), where $i \in H$ and $j \in C$

$\Gamma_{i,n}$ Upper bound of temperature difference in exchanger (*i*, *n*), where $i \in H$ and $n \in CU$

$\Gamma_{m,j}$ Upper bound of temperature difference in exchanger (*m*, *j*), where $m \in HU$ and $j \in C$

Variables:

$t_{i,k}$ Temperature of hot stream $i \in H$ at hot end of stage $k \in ST$

$t_{j,k}$ Temperature of cold stream $j \in C$ at hot end of stage $k \in ST$

$q_{i,j,k}$ Heat exchange rate between hot stream $i \in H$ and cold process stream $j \in C$ in stage $k \in ST$

$q_{i,n}$ Heat exchange rate between hot stream $i \in H$ and cold utility $n \in CU$

$q_{m,j}$ Heat exchange rate between hot utility $m \in HU$ and cold stream $j \in C$

$z_{i,j,k}$ Binary variable to denote existence of match (*i*, *j*) in stage *k*, where $i \in H$ and $j \in C$

$z_{i,n}$ Binary variable to denote existence of match (*i*, *n*), where $i \in H$ and $n \in CU$

$z_{m,j}$ Binary variable to denote existence of match (*m*, *j*), where $m \in HU$ and $j \in C$

$dt_{i,j,k}$ Temperature approach for match (*i*, *j*) at hot end of stage *k*, where $i \in H$, $j \in C$ and $k \in ST$

$dtin_{i,n}$ Temperature approach at the hot end of match (*i*, *n*), where $i \in H$ and $n \in CU$

$dtout_{i,n}$ Temperature approach at the cold end of match (*i*, *n*), where $i \in H$ and $n \in CU$

$dtin_{m,j}$ Temperature approach at the cold end of match (*m*, *j*), where $m \in HU$ and $j \in C$

$dtout_{m,j}$ Temperature approach at the hot end of match (*m*, *j*), where $m \in HU$ and $j \in C$

The model constraints can be written according to the above definitions as follows:

- **Overall heat balance for each process stream.** This equality constraint is needed to ensure sufficient heating or cooling of each process stream. Specifically, the heat load of each stream must equal the sum of the heat it exchanges with other process streams in all stages plus that with the utility streams.

$$\left(TIN_i - TOUT_i\right)F_i = \sum_{k \in ST}\sum_{j \in C} q_{i,j,k} + \sum_{n \in CU} q_{i,n}; \tag{2.66}$$

$$\left(TOUT_j - TIN_j\right)F_j = \sum_{k \in ST}\sum_{i \in H} q_{i,j,k} + \sum_{m \in HU} q_{m,j}. \tag{2.67}$$

where $i \in H$ and $j \in C$.

- **Heat balance at each stage.** These energy balances are used to determined $NOK + 1$ boundary temperatures of the NOK stages. Note that, due to the isothermal mixing assumption, the heat capacity flow rates of the split streams do not appear explicitly in the following equality constraints.

$$\left(t_{i,k} - t_{i,k+1}\right) F_i = \sum_{j \in C} q_{i,j,k}; \tag{2.68}$$

$$\left(t_{j,k} - t_{j,k+1}\right) F_j = \sum_{i \in H} q_{i,j,k}. \tag{2.69}$$

where $i \in H$, $j \in C$, and $k \in ST$.

- **Cold and hot utility loads.** A cooling utility may be adopted to lower the temperature of hot stream i from the cold end of stage NOK to its target. A heating utility may be used to raise the temperature of cold stream j from the hot end of stage 1 to its target.

$$\left(t_{i,NOK+1} - TOUT_i\right) F_i = \sum_{n \in CU} q_{i,n}; \tag{2.70}$$

$$\left(TOUT_j - t_{j,1}\right) F_j = \sum_{m \in HU} q_{m,j}. \tag{2.71}$$

where $i \in H$ and $j \in C$.

- **Match selections.** Logic constraints and binary variables are used to determine if match (i, j) is present in stage k, i.e.

$$q_{i,j,k} - \Omega_{i,j} z_{i,j,k} \le 0. \tag{2.72}$$

where $i \in H$; $j \in C$; $k \in ST$; $z_{i,j,k} \in \{0,1\}$; $\Omega_{i,j}$ is a model parameter which must be computed in advance according to the following formula:

$$\Omega_{i,j} = \min\left\{F_i\left(TIN_i - TOUT_i\right), F_j\left(TOUT_j - TIN_j\right)\right\}. \tag{2.73}$$

- **Utility selections.** In this modified model, the sums on the right sides of equations (2.70) and (2.71) are used to respectively replace the cold and hot utility loads on hot stream i and cold stream j in the original formulation. Note that these sums also appear in the second terms on the right sides of equations (2.66) and (2.67). In addition, logic constraints and binary variables are used to determine if matches (i, n) and (m, j) are present to bring the process streams to their target temperatures, i.e.

$$q_{i,n} - \Omega_{i,n} z_{i,n} \le 0; \tag{2.74}$$

$$q_{m,j} - \Omega_{m,j} z_{m,j} \le 0. \tag{2.75}$$

where $i \in H$; $j \in C$; $m \in HU$; $n \in CU$. The model parameters can be calculated as follows:

$$\Omega_{i,n} = F_i\left(TIN_i - TOUT_i\right); \tag{2.76}$$

$$\Omega_{m,j} = F_j\left(TOUT_j - TIN_j\right). \tag{2.77}$$

To facilitate selection of *only one* utility among all possible options, the following logic constraints must also be imposed:

$$\sum_{n \in CU} z_{i,n} \le 1; \tag{2.78}$$

$$\sum_{m \in HU} z_{m,j} \le 1. \tag{2.79}$$

where $z_{i,n}, z_{m,j} \in \{0,1\}$.

- **Temperature precedence on each process stream.** Inequality constraints are needed to ensure a monotonic temperature decrease at each successive stage. Fixed supply temperatures of the process streams are assigned as the inlet temperatures of the superstructure, while a bound is also imposed on the outlet temperatures of the superstructure at the respective stream's target temperature.

$$TIN_i = t_{i,1}; \tag{2.80}$$

$$t_{i,k} \ge t_{i,k+1}; \tag{2.81}$$

$$t_{i,NOK+1} \ge TOUT_i; \tag{2.82}$$

$$TOUT_j \ge t_{j,1}; \tag{2.83}$$

$$t_{j,k} \ge t_{j,k+1}; \tag{2.84}$$

$$t_{j,NOK+1} = TIN_j. \tag{2.85}$$

where $i \in H$, $j \in C$, and $k \in ST$.

- **Temperature approach in each heat exchanger.** The required area of each heat exchanger is incorporated in the objective function for computing the corresponding capital cost. To facilitate calculation of this area, the corresponding temperature approaches at the hot and cold ends must be determined a priori. To ensure feasible driving force in every selected match, a binary variable is adopted to activate/deactivate the following constraints for approach temperatures:

$$dt_{i,j,k} \le t_{i,k} - t_{j,k} + \Gamma_{i,j}\left(1 - z_{i,j,k}\right); \tag{2.86}$$

$$dt_{i,j,k+1} \leq t_{i,k+1} - t_{j,k+1} + \Gamma_{i,j}\left(1 - z_{i,j,k}\right); \tag{2.87}$$

$$dt_{i,j,k} \geq \Delta T_{min}. \tag{2.88}$$

where $i \in H$, $j \in C$ and $k \in ST$. The model parameter $\Gamma_{i,j}$ in the above inequalities can be determined as follows:

$$\Gamma_{i,j} = \max\left\{0, TIN_j - TIN_i, TOUT_j - TIN_i, TIN_j - TOUT_i, TOUT_j - TOUT_i\right\} + \Delta T_{min}.$$

- **Temperature approach in each cooler.** The heat-transfer area of each cooler is needed for computing the corresponding capital cost. To facilitate calculation of this area, the temperature approaches at the hot and cold ends should also be determined a priori. To ensure feasible driving force, a binary variable is adopted to activate/deactivate the following constraints for imposing the approach temperatures:

$$dtin_{i,n} \leq t_{i,NOK+1} - TT_{i,n} + \Gamma_{i,n}\left(1 - z_{i,n}\right); \tag{2.89}$$

$$dtin_{i,n} \geq \Delta T_{min}; \tag{2.90}$$

$$dtout_{i,n} \leq TOUT_i - TI_n + \Gamma_{i,n}\left(1 - z_{i,n}\right); \tag{2.91}$$

$$dtout_{i,n} \geq \Delta T_{min}. \tag{2.92}$$

where $i \in H$, $j \in C$, and $n \in CU$. The model parameter $\Gamma_{i,n}$ can be determined as follows:

$$\Gamma_{i,n} = \max\left\{0, TI_n - TIN_i, TI_n - TOUT_i, TT_{i,n} - TIN_i, TT_{i,n} - TOUT_i\right\} + \Delta T_{min}.$$

- **Temperature approach in each heater.** The required heat-transfer area of each heater is used in the objective function for computing the corresponding capital cost. To facilitate calculation of this area, the temperature approaches at the hot and cold ends should be determined a priori. To ensure feasible driving force, a binary variable is adopted to activate/deactivate the following constraints for the approach temperatures.

$$dtin_{m,j} \leq TT_{m,j} - t_{j,1} + \Gamma_{m,j}\left(1 - z_{m,j}\right); \tag{2.93}$$

$$dtin_{m,j} \geq \Delta T_{min}; \tag{2.94}$$

$$dtout_{m,j} \leq TI_m - TOUT_j + \Gamma_{m,j}\left(1 - z_{m,j}\right); \tag{2.95}$$

$$dtout_{m,j} \geq \Delta T_{min}. \tag{2.96}$$

where $j \in C, m \in HU$, and $\Gamma_{m,j} = \max\{0, TIN_j - TI_m, TIN_j - TT_{m,j}, TOUT_j - TI_m, TOUT_j - TT_{m,j}\} + \Delta T_{min}$.

- **Maximum number of split streams in each stage.** There may be a need to impose an upper bound on the total number of split streams in each stage so as to simplify the network structure. Following are these optional constraints:

$$\sum_{j \in C} z_{i,j,k} \le NST \tag{2.97}$$

$$\sum_{i \in H} z_{i,j,k} \le NST \tag{2.98}$$

where$i \in H; j \in C; k \in ST$; NST is a constant positive integer.

- **Objective function.** The objective function of the proposed MINLP model is the total annual cost (*TAC*), which can be approximated with the sum of the total annual utility cost (*TAUC*) and the total annualized capital cost (*TACC*). Specifically,

$$TAUC = \sum_{i \in H}\sum_{n \in CU} CCU_n q_{i,n} + \sum_{j \in C}\sum_{m \in HU} CHU_m q_{m,j} \tag{2.99}$$

$$TACC = TACC_{fix} + TACC_{area} \tag{2.100}$$

The total fixed cost and variable cost in equation (2.100) can be repressed respectively as follows

$$TACC_{fix} = \sum_{i \in H}\sum_{j \in C}\sum_{k \in ST} CF_{i,j,k} z_{i,j,k} + \sum_{i \in H}\sum_{n \in CU} CF_{i,n} z_{i,n} + \sum_{j \in C}\sum_{m \in HU} CF_{m,j} z_{m,j} \tag{2.101}$$

$$\begin{aligned} TACC_{area} = & \sum_{i \in H}\sum_{j \in C}\sum_{K \in ST} CA_{i,j} \left(\frac{q_{i,j,k}}{U_{i,j}\left(dt_{i,j,k} dt_{i,j,k+1}\left(dt_{i,j,k} + dt_{i,j,k+1}\right)/2\right)^{1/3}} \right)^{\beta} \\ & + \sum_{i \in H}\sum_{n \in CU} CA_{i,n} \left(\frac{q_{i,n}}{U_{i,n}\left(dtin_{i,n} dtout_{i,n}\left(dtin_{i,n} + dtout_{i,n}\right)/2\right)^{1/3}} \right)^{\beta} \\ & + \sum_{j \in C}\sum_{m \in HU} CA_{m,j} \left(\frac{q_{m,j}}{U_{m,j}\left(dtin_{m,j} dtout_{m,j}\left(dtin_{m,j} + dtout_{m,j}\right)/2\right)^{1/3}} \right)^{\beta} \end{aligned} \tag{2.102}$$

Finally, it should be noted that the log-mean temperature difference (LMTD) of each heat-transfer unit in the above equation is calculated with the empirical formula developed by Chen (1987).

Example 2.6

A simple example is presented here to illustrate the aforementioned approach to synthesizing multi-plant HEN. The stream and utility data used in this example are given in Table 2.9 and Table 2.10 respectively.

A minimum temperature approach (ΔT_{min}) of 10°C was assumed in designing heat exchangers, and the annualized capital investment of every unit was calculated according to the following formula: $10000 + 670 \times area^{0.83}$ ($/yr). Three single-plant HENs were first generated on the basis of the aforementioned MINLP model (see

TABLE 2.9
Stream Data Used in Example 2.6

Plant	Stream	*TIN* (°C)	*TOUT* (°C)	*F*(kW/°C)
P1	H1	150	40	7.0
P1	C1	60	140	9.0
P1	C2	110	190	8.0
P2	H1	200	70	5.5
P2	C1	30	110	3.5
P2	C2	140	190	7.5
P3	H1	370	150	3.0
P3	H2	200	40	5.5
P3	C1	110	360	4.5

TABLE 2.10
Utility Data Used in Example 2.6

Plant	Utility	*TI* (°C)	*CQ* ($/kW·yr)
P1	Cooling water	25	100
P1	HP steam	200	800
P1	Hot oil	500	1200
P2	Cooling water	25	150
P2	HP steam	200	900
P2	Hot oil	500	1300
P3	Cooling water	25	80
P3	HP steam	200	850
P3	Hot oil	500	1100

Figures 2.10–2.12), and their minimum TACs were found to be 725,433.4, 168,593.8, and 404,900.8 USD/yr respectively. On the other hand, the TAC of optimal design in the three-plant scenario (Figure 2.13) was reduced to 887,932.4 USD/yr. Although a significant 31.6% TAC saving is achieved in this case via total-site heat integration, it should be noted that the subsequent allocation issues are still unsettled.

It has been well established that the simplifying assumption of isothermal mixing at the stage outlets is realistic only when the desired HEN network is without stream splits and bypasses. In practical applications when these network configurations are required, the aforementioned MINLP model should be modified by introducing flow variables. Specifically, the following additional equality constraints should be formulated according to Figure 2.14 and then introduced into the original model.

- **Heat balance of each exchanger.** Since the isothermal mixing assumption is removed, it becomes necessary to determine the outlet temperature of each heat exchanger in every stage. Consequently, the corresponding heat balances should be included in the mathematical programming model:

$$\left(t_{i,k} - ths_{i,j,k}\right) rh_{i,j,k} F_i = q_{i,j,k} \tag{2.103}$$

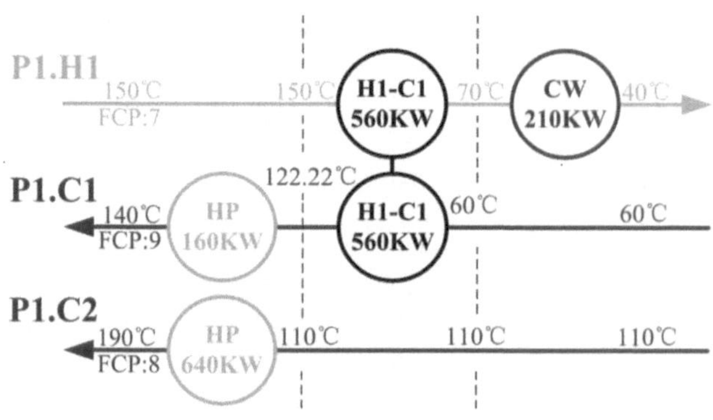

FIGURE 2.10 Optimal HEN design for plant P1. (Reproduced with permission from *Applied Energy* 2018, 211: 904–920. Copyright 2018 Elsevier.)

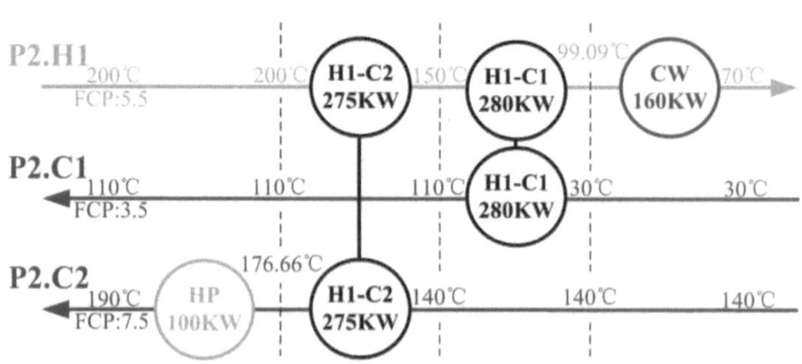

FIGURE 2.11 Optimal HEN design for plant P2. (Reproduced with permission from *Applied Energy* 2018, 211: 904–920. Copyright 2018 Elsevier.)

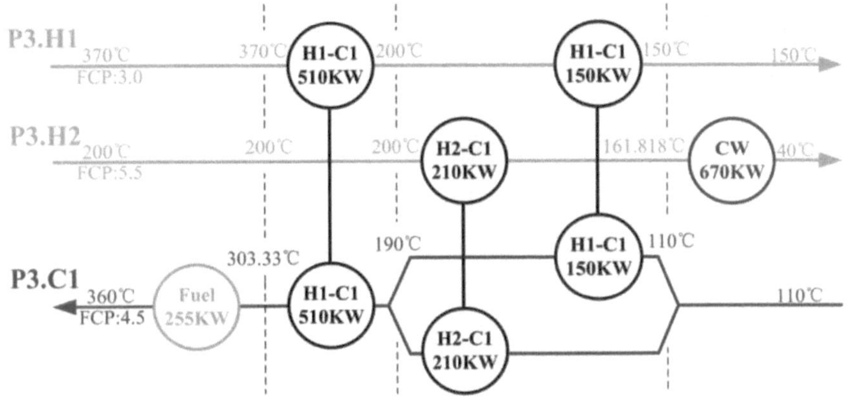

FIGURE 2.12 Optimal HEN design for plant P3. (Reproduced with permission from *Applied Energy* 2018, 211: 904–920. Copyright 2018 Elsevier.)

$$\left(tcs_{i,j,k} - t_{j,k+1}\right) rc_{i,j,k} F_j = q_{i,j,k} \tag{2.104}$$

where $i \in H$; $j \in C$; $k \in ST$; $rh_{i,j,k} \geq 0$ denotes the split ratio of a branch of hot stream i for a match with cold stream j in stage k, and $ths_{i,j,k}$ is the corresponding hot stream temperature at the exit of match (i, j) in stage k; $rc_{i,j,k} \geq 0$ denotes the split ratio of a branch of cold stream j for a match with hot stream i in stage k, and $tcs_{i,j,k}$ is the corresponding cold stream temperature at the exit of match (i, j) in stage k.

- **Sum of split ratios.** Since the split ratios are used to determine the flow rates of all branches of a process stream, the sum of these ratios should equal one. The corresponding equality constraints obviously must be included in the model, i.e.

$$rrh_{i,k} + \sum_{j \in C} rh_{i,j,k} = 1 \tag{2.105}$$

$$rrc_{j,k} + \sum_{i \in H} rc_{i,j,k} = 1 \tag{2.106}$$

where $i \in H$; $j \in C$; $k \in ST$;$rrh_{i,k} \geq 0$ denotes the split ratio of bypass on hot stream i in stage k;$rrc_{j,k} \geq 0$ denotes the split ratio of bypass on cold stream j in stage k.

- **Mixing-point temperatures.** Due to the removal of the isothermal mixing assumption, the mixing-point temperatures at the outlets of all stages must be determined on the basis of energy balances. Following are the corresponding constraints:

$$rrh_{i,k} t_{i,k} + \sum_{j \in C} rh_{i,j,k} ths_{i,j,k} = t_{i,k+1} \tag{2.107}$$

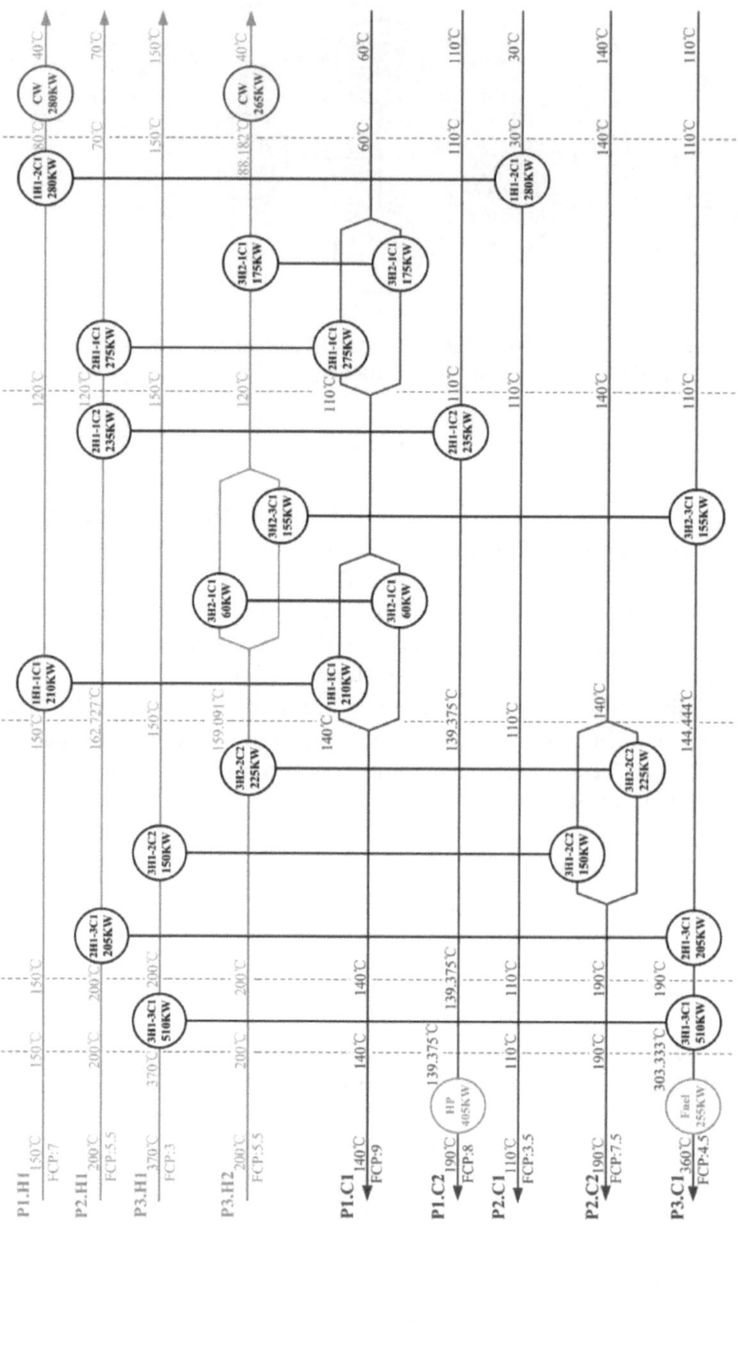

FIGURE 2.13 Optimal HEN design for plants P1, P2, and P3. (Reproduced with permission from *Applied Energy* 2018, 211: 904–920. Copyright 2018 Elsevier.)

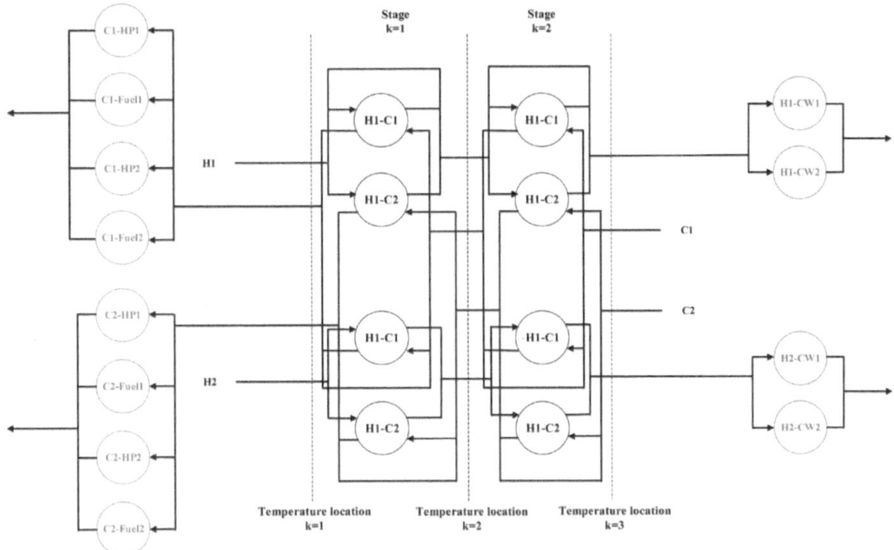

FIGURE 2.14 A relaxed superstructure for simultaneous synthesis of multi-plant HENs.

$$rrc_{j,k}t_{j_p,k+1} + \sum_{i \in H} rc_{i,j,k}tcs_{i,j,k} = t_{j,k} \tag{2.108}$$

where $i \in H$; $j \in C$; $k \in ST$.

2.3 CONCLUDING REMARKS

Without addressing the benefit allocation issues, the traditional sequential and simultaneous approaches for generating the HEN in a single plant can both be applied to produce the multi-plant HEN designs with minor modifications. These modifications are detailed in the aforementioned model formulations. Although it can be found in the examples presented in this chapter that a significant TAC saving can usually obtained via total-site heat integration, the subsequent allocation issues are still unsettled.

REFERENCES

Chen, J.J.J. 1987. Comments on improvements on a replacement for the logarithmic mean. *Chemical Engineering Science* 42: 2488–2489.

Floudas, C.A., Ciric, A.R., Grossmann, I.E. 1986. Automatic synthesis of optimum heat exchanger network configurations. *AIChE Journal* 32: 276–290.

Floudas, C.A., Grossmann, I.E. 1987. Automatic generation of multi-period heat exchanger network configurations. *Computers & Chemical Engineering* 11: 123–142.

Papoulias, S.A., Grossmann, I.E. 1983. A structural optimization approach in process synthesis – II: Heat recovery networks. *Computers & Chemical Engineering* 7: 707–721.

Yee, T.F., Grossmann, I.E. 1990. Simultaneous optimization models for heat integration—II. Heat exchanger network synthesis. *Computers & Chemical Engineering* 14: 1165–1184.
Yee, T.F., Grossmann, I.E., Kravanja, Z. 1990. Simultaneous optimization models for heat integration—I. Area and energy targeting and modeling of multi-stream exchangers. *Computers & Chemical Engineering* 14: 1151–1164.

3 Indirect HEN Designs for Batch Processes

3.1 INTRODUCTION

Batch processes have attracted increasing attention in the past decades because of the growing demand for low volume, high value-added products. The potential for energy savings in batch plants is thus appealing to researchers and engineers in the area of process systems engineering. The main challenge of heat integration in batch plants is the time dependence of the hot and cold process streams.

Heat integration in batch plants has direct and indirect heat exchange schemes. In the direct scheme, hot and cold process streams that occur at the same time are integrated to maximize heat recovery. In the indirect scheme, a recirculating heat transfer medium (HTM) is used to absorb surplus heat from hot streams and supply it to cold streams that come later in time. A number of tanks are used to temporarily store the hot and cold HTM to bypass the time constraint on heat exchange between time-dependent process streams. De Boer et al. (2006) evaluated the technical and economic feasibility of different heat storage methods in industrial processes. With technological advances, heat storage can play a significant role in shifting cooling and heating loads in the process industries by reducing the consumption of external utilities.

In batch plants, the unit processing times are determined by the production schedule. In most cases, the occurrences of hot and cold process streams do not fully coincide. This limits the potential of direct heat integration. To improve heat recovery, rescheduling of the process may be considered (Corominas et al., 1994). Alternatively, indirect heat integration can be implemented, which is much less schedule-sensitive with the use of an HTM for heat storage. Indirect heat exchange also allows greater operational flexibility compared to the direct way. Several design methods for indirect heat exchange systems have been proposed. Sadr-Kazemi and Polley (1996) presented a pinch analysis approach to the design of heat storage systems. A storage line between hot and cold composite curves is used to determine the tank temperatures first and then calculate the tank sizes. Krummenacher and Favrat (2001) followed this approach and addressed the problem of targeting the minimum number of heat storage units. Papageorgiou et al. (1994) proposed a mathematical programming framework to calculate the variation of mass and energy holdups of HTM over time under a known operating policy. Georgiadis and Papageorgiou (2001) extended the formulation of Papageorgiou et al. (1994) to incorporate fouling considerations during the heat integration of multipurpose batch plants.

Most early works on indirect heat integration in batch plants are based on pinch analysis or assume a known heat integration policy. In those works, the configuration of the indirect heat exchanger network (HEN) and the associated heat storage policy are usually determined sequentially. In this chapter, the general problem of indirect batch HEN synthesis and heat storage policy determination is addressed. A superstructure that considers all possible network configurations and the operating policy of HTM is proposed. Based on this superstructure, a mathematical programming model is developed for the synthesis of indirect batch HENs. This approach does not require any pinch-based heuristics. A literature example is then presented to demonstrate the application of the model.

3.2 PROBLEM STATEMENT

The problem of indirect batch HEN synthesis can be stated as follows.

- In a batch plant, there are a set of hot process streams $i \in \mathbf{HP}$ and a set of cold process streams $j \in \mathbf{CP}$. The start and end times of each stream are known a priori. A set of time periods $t \in \mathbf{TP}$ is then defined according to the timings of the process streams.
- There is also a set of tanks $k \in \mathbf{ST}$ for heat storage using an HTM.
- Given (1) the supply and target temperatures of the hot/cold process streams, (2) the start and end times of each stream, (3) the heat capacity flowrate or the latent/sensible heat of each stream, (4) the minimum temperature approach for heat transfer, and (5) the properties of the HTM (e.g., density and heat capacity), the objective is to determine the optimal heat integration strategy that minimizes the consumption of external utilities.
- The solution will provide (1) the number of heat storage tanks and their temperatures, (2) the heat exchange matches between the HTM and process streams, (3) the heat capacity flowrates of the HTM, (4) the amounts of heat transferred between process streams, the HTM, and external utilities, and (5) the outlet temperatures of hot and cold process streams at each match. With these details, the indirect batch HEN can be visualized.

3.3 CONCEPTUAL STRUCTURE OF HEAT STORAGE SYSTEMS

Figure 3.1 illustrates a heat storage system in a batch plant with two hot process streams and one cold process stream in different time periods. A recirculated HTM is used to absorb heat from the hot streams and the resulting hot HTM is temporarily stored in the hot tank. The stored heat is then transferred to the cold stream and the resulting cold HTM goes to the cold tank. In Figure 3.1, *absorption* refers to the process of transferring heat from hot process streams to the cold HTM and cold utilities in series or in parallel (Krummenacher and Favrat, 2001). On the other hand, *rejection* refers to the process of transferring heat from the hot HTM and external hot utilities to cold process streams. With the use of hot and cold tanks as well as an HTM for heat absorption and rejection, heat can be stored for later use, thereby relaxing the time constraint of batch process streams. This improves the potential for heat recovery and can significantly reduce the consumption of external utilities.

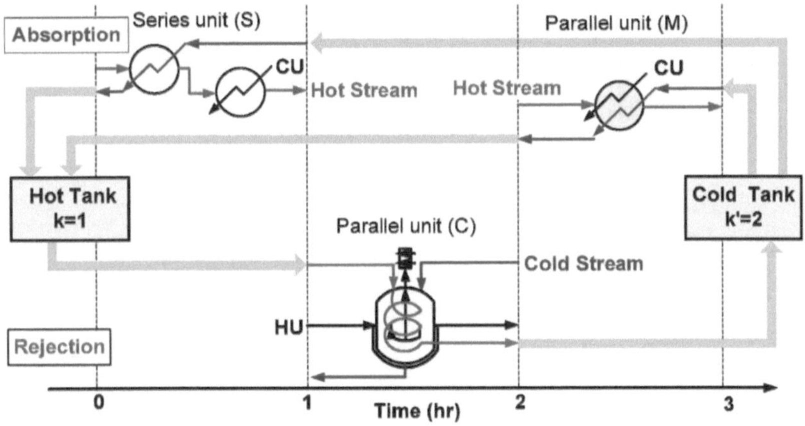

FIGURE 3.1 Conceptual structure of an indirect batch HEN with two hot streams, one cold stream, and one HTM. (Reprinted with permission from *Ind. Eng. Chem. Res.* 2008, 47, 14, 4817–4829. Copyright 2008 American Chemical Society.)

In an indirect batch HEN, heat exchange can be carried out in a series-type unit (S) or a parallel-type unit (P). In a series-type unit, heat is transferred from the hot stream to the cold HTM first in an exchanger and then to the cold utility in a downstream cooler. Alternatively, heat is transferred to the cold stream from the hot HTM first in an exchanger and then from the hot utility in a downstream heater. In a parallel-type unit, the process stream exchanges heat with the HTM and the utility simultaneously. Typical examples of a parallel unit include multi-stream heat exchangers (M) and jacket heat exchangers with internal coils (C).

The HTM absorbs and rejects sensible heat. Therefore, several factors need to be considered for HTM selection, such as the temperature range to maintain the liquid phase, heat capacity, density, volatility, corrosiveness, and heat losses. In particular, higher heat capacity and density of the HTM will lead to smaller storage tanks.

3.4 MODEL FORMULATION

To formulate the problem of synthesizing an indirect batch HEN, a superstructure that reflects all possible configurations of indirect heat integration is used. This superstructure consists of series-/parallel-type heat exchange units and heat storage tanks, as shown in Figure 3.2. The indices and sets, parameters, and variables used for model formulation are defined in the Nomenclature. In Figure 3.2(a), two binary variables ($za_{k'i}^{(T2H)}$ and $za_{ik}^{(H2T)}$) are defined to indicate whether or not there is an HTM stream from cold tank k' to hot tank k via the heat exchanger for absorbing heat from hot process stream i. Similarly, in Figure 3.2(b), two binary variables ($zr_{kj}^{(T2C)}$ and $zr_{k'j}^{(C2T)}$) are defined to indicate whether or not there is an HTM stream from hot tank k to cold tank k' via the heat exchanger for rejecting heat to cold process stream j. The time horizon is divided into a number of time periods $t \in \mathbf{TP}$, according to the

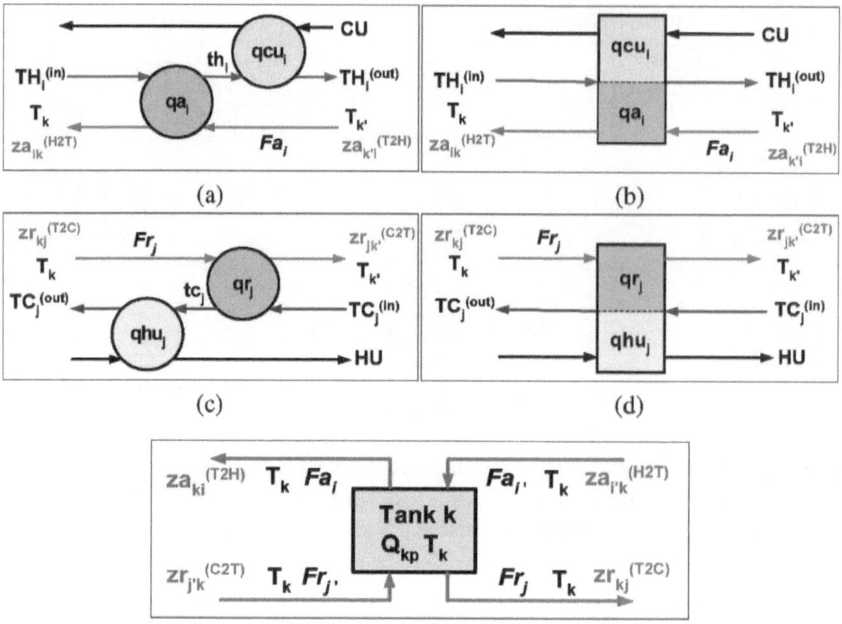

FIGURE 3.2 Schematics of heat exchange units: series-type (a) and parallel-type (b) units for heat absorption; series-type (c) and parallel-type (d) units for heat rejection; (e) an HTM tank. (Reprinted with permission from *Ind. Eng. Chem. Res.* 2008, 47, 14, 4817–4829. Copyright 2008 American Chemical Society.)

start and end times of the hot and cold process streams. With these definitions and the superstructure, the optimization model for indirect batch HEN synthesis can be formulated. To simplify the formulation, it is assumed that each HTM tank has a fixed temperature, at which the HTM enters and leaves the tank. This assumption can, however, be relaxed with a more sophisticated formulation.

3.4.1 Overall Heat Balance of the Recirculated HTM

An HTM is used to transfer heat from hot process streams to cold process streams indirectly using a heat storage system. After absorbing the heat from hot process streams, the hot HTM can be temporarily stored in a hot tank until cold process streams occur. After rejecting heat to cold streams, the cold HTM enters a cold tank and leaves when hot process streams occur. Equation (3.1) describes the overall heat balance of the HTM, which is recirculated between the cold and hot tanks in a closed loop.

$$\sum_{i \in \mathbf{HP}} qa_i \left(\sum_{p \in \mathbf{TP}} t_p Z_{ip}^{(\text{hot})} \right) = \sum_{j \in \mathbf{CP}} qr_j \left(\sum_{p \in \mathbf{TP}} t_p Z_{jp}^{(\text{cold})} \right) \tag{3.1}$$

where t_p is the length of time period p; $Z_{ip}^{(\text{hot})}$ and $Z_{ip}^{(\text{cold})}$ are binary parameters denoting the existence of hot stream i and cold stream j in time period p, respectively; qa_i and qr_j represent the amount of heat absorbed (by the cold HTM) from hot stream i and the amount of heat rejected (by the hot HTM) to cold stream j stream, respectively. The left-/right-hand sides of equation (3.1) denote the total energy occurring between the heat transfer medium and all hot/cold process streams.

3.4.2 Heat Balance for Series-Type Heat Exchange Units

Equation (3.2) describes the overall heat balance for hot stream i. The two terms on the left-hand side represent the latent heat and the sensible heat, respectively. The heat to be removed from hot stream i is absorbed by the HTM (qa_i) and the cold utility (qcu_i) in series. Likewise, equation (3.3) describes the overall heat balance for cold stream j. The heat to be supplied to cold stream j comes from the HTM (qr_j) first and then the hot utility (qhu_j).

$$\frac{LH_i}{\sum_{p\in \mathbf{TP}} t_p Z_{ip}^{(\text{hot})}} + \left(TH_i^{(\text{in})} - TH_i^{(\text{out})}\right) FH_i = qa_i + qcu_i \quad \forall i \in \mathbf{HPS} \tag{3.2}$$

$$\frac{LC_j}{\sum_{p\in \mathbf{TP}} t_p Z_{jp}^{(\text{cold})}} + \left(TC_j^{(\text{out})} - TC_j^{(\text{in})}\right) FC_j = qr_j + qhu_j \quad \forall j \in \mathbf{CPS} \tag{3.3}$$

Here it is assumed that for a hot or cold process stream with phase change, the heat absorbed or rejected by the HTM (qa_i or qr_j) has to be greater than the latent heat of the process stream if the match between the HTM and the process stream exists ($za_i = 1$ or $zr_j = 1$). Otherwise, the heat load (including the latent heat and sensible heat) is all removed or supplied by the utility ($za_i = 0$ or $zr_j = 0$). The corresponding heat balance for hot stream i is formulated in equations (3.4) and (3.5), where the interval temperature th_i between the heat exchanger and the cooler is calculated. If $za_i = 1$, the latent heat and part of the sensible heat of hot stream i are removed by the cold HTM. The remaining sensible heat is then removed by the cold utility. If $za_i = 0$ ($qa_i = 0$ and hence $th_i = TH_i^{(\text{in})}$ according to equation (3.4)), all the latent heat and sensible heat of hot stream i are absorbed by the cold utility. Similarly, equations (3.6) and (3.7) are formulated for cold stream j.

$$\frac{LH_i za_i}{\sum_{p\in \mathbf{TP}} t_p Z_{ip}^{(\text{hot})}} + \left(TH_i^{(\text{in})} - th_i\right) FH_i = qa_i \quad \forall i \in \mathbf{HPS} \tag{3.4}$$

$$\frac{LH_i \left(1 - za_i\right)}{\sum_{p\in \mathbf{TP}} t_p Z_{ip}^{(\text{hot})}} + \left(th_i - TH_i^{(\text{out})}\right) FH_i = qcu_i \quad \forall i \in \mathbf{HPS} \tag{3.5}$$

$$\frac{LC_j zr_i}{\sum_{p\in\mathbf{TP}} t_p Z_{jp}^{(\text{cold})}} + \left(TC_j^{(\text{out})} - TC_j^{(\text{in})}\right) FC_j = qr_j \quad \forall j \in \mathbf{CPS} \tag{3.6}$$

$$\frac{LC_j\left(1 - zr_i\right)}{\sum_{p\in\mathbf{TP}} t_p Z_{jp}^{(\text{cold})}} + \left(TC_j^{(\text{out})} - TC_j^{(\text{in})}\right) FC_j = qhu_j \quad \forall j \in \mathbf{CPS} \tag{3.7}$$

3.4.3 Heat Balance for Parallel-Type Heat Exchange Units

In parallel-type heat exchange units, the latent heat and sensible heat of a process stream are removed or supplied by the HTM and the utility in parallel, as described in equation (3.8) for hot stream i and in equation (3.9) for cold stream j.

$$\frac{LH_i + SH_i}{\sum_{p\in\mathbf{TP}} t_p Z_{ip}^{(\text{hot})}} = qa_i + qcu_i \quad \forall i \in \mathbf{HPP} \tag{3.8}$$

$$\frac{LC_j + SC_j}{\sum_{p\in\mathbf{TP}} t_p Z_{jp}^{(\text{hot})}} = qr_j + qhu_j \quad \forall j \in \mathbf{CPP} \tag{3.9}$$

3.4.4 Heat Balance for the HTM

Four more binary variables ($za_{k'i}^{(\text{T2H})}$, $za_{ik}^{(\text{H2T})}$, $zr_{kj}^{(\text{T2C})}$, and $zr_{jk'}^{(\text{C2T})}$) are defined to identify the matches between process streams and the HTM. The amounts of heat absorbed and rejected by HTM streams are then calculated using the following constraints.

$$\left(T_k - T_{k'}\right) Fa_i - qa_i \le \bar{q}\left(2 - za_{k'i}^{(\text{T2H})} - za_{ik}^{(\text{H2T})}\right) \quad \forall i \in \mathbf{HP}, k \in \mathbf{ST} \tag{3.10}$$

$$\left(T_k - T_{k'}\right) Fa_i - qa_i \ge -\bar{q}\left(2 - za_{k'i}^{(\text{T2H})} - za_{ik}^{(\text{H2T})}\right) \quad \forall i \in \mathbf{HP}, k \in \mathbf{ST} \tag{3.11}$$

$$\left(T_k - T_{k'}\right) Fr_j - qr_j \le \bar{q}\left(2 - zr_{kj}^{(\text{T2C})} - zr_{jk'}^{(\text{C2T})}\right) \quad \forall j \in \mathbf{CP}, k \in \mathbf{ST} \tag{3.12}$$

$$\left(T_k - T_{k'}\right) Fr_j - qr_j \ge -\bar{q}\left(2 - zr_{kj}^{(\text{T2C})} - zr_{jk'}^{(\text{C2T})}\right) \quad \forall j \in \mathbf{CP}, k \in \mathbf{ST} \tag{3.13}$$

where $za_{k'i}^{(\text{T2H})}$ indicates if there is a cold HTM stream from tank k' to absorb heat from hot stream i; $za_{ik}^{(\text{H2T})}$ indicates if there is a hot HTM stream to tank k after absorbing heat from hot stream i; $zr_{kj}^{(\text{T2C})}$ denotes if there is a hot HTM stream from tank k to reject heat to cold stream j; $zr_{jk'}^{(\text{C2T})}$ denotes if there is a cold HTM stream to tank k' after rejecting heat to cold stream j. It can be seen that the amount of heat absorbed or rejected is proportional to the temperature difference between the tanks $(T_k - T_{k'})$. These constraints are made redundant when there is not a match between the HTM and the process stream for heat absorption ($za_{k'i}^{(T2H)} = 0$ or $za_{ik}^{(H2T)} = 0$) or rejection ($zr_{kj}^{(T2C)} = 0$ or $zr_{jk'}^{(C2T)} = 0$).

3.4.5 Calculation of the Approach Temperature

A large enough temperature driving force is required to ensure feasible heat transfer between process streams and the HTM. Such a constraint will be made redundant if the match does not exist and the binary variable takes a value of zero. Note that for jacket exchangers with internal coils, the closest approach temperature of the process stream and the HTM is considered, as in the case of cocurrent heat exchangers.

For series-type units (S):

$$th_i - T_{k'} + \Gamma\left(1 - za_{k'i}^{(T2H)}\right) \geq \Delta T_{\min} \quad \forall i \in \mathbf{HPS}, k' \in \mathbf{ST} \tag{3.14}$$

$$TH_i^{(in)} - T_k + \Gamma\left(1 - za_{ik}^{(H2T)}\right) \geq \Delta T_{\min} \quad \forall i \in \mathbf{HPS}, k \in \mathbf{ST} \tag{3.15}$$

$$T_k - tc_j + \Gamma\left(1 - zr_{kj}^{(T2C)}\right) \geq \Delta T_{\min} \quad \forall j \in \mathbf{CPS}, k \in \mathbf{ST} \tag{3.16}$$

$$T_{k'} - TC_j^{(in)} + \Gamma\left(1 - zr_{jk'}^{(C2T)}\right) \geq \Delta T_{\min} \quad \forall j \in \mathbf{CPS}, k' \in \mathbf{ST} \tag{3.17}$$

For parallel-type units (M and C):

$$TH_i^{(out)} - T_{k'} + \Gamma\left(1 - za_{k'i}^{(T2H)}\right) \geq \Delta T_{\min} \quad \forall i \in \mathbf{HPP}, k' \in \mathbf{ST} \tag{3.18}$$

$$T_k - TC_j^{(out)} + \Gamma\left(1 - zr_{kj}^{(T2C)}\right) \geq \Delta T_{\min} \quad \forall j \in \mathbf{CPP}, k \in \mathbf{ST} \tag{3.19}$$

For multi-stream exchangers (M):

$$TH_i^{(in)} - T_k + \Gamma\left(1 - za_{ik}^{(H2T)}\right) \geq \Delta T_{\min} \quad \forall i \in \mathbf{HPM}, k \in \mathbf{ST} \tag{3.20}$$

$$T_{k'} - TC_j^{(in)} + \Gamma\left(1 - zr_{jk'}^{(C2T)}\right) \geq \Delta T_{\min} \quad \forall j \in \mathbf{CPM}, k' \in \mathbf{ST} \tag{3.21}$$

For jacket heat exchangers with internal coils (C):

$$TH_i^{(out)} - T_k + \Gamma\left(1 - za_{ik}^{(H2T)}\right) \geq \Delta T_{\min} \quad \forall i \in \mathbf{HPC}, k \in \mathbf{ST} \tag{3.22}$$

$$T_{k'} - TC_j^{(\text{out})} + \Gamma\left(1 - zr_{jk'}^{(\text{C2T})}\right) \geq \Delta T_{\min} \quad \forall j \in \mathbf{CPC}, k' \in \mathbf{ST} \tag{3.23}$$

3.4.6 Maximum Number of Tanks

The maximum number of tanks (MNT) for HTM recirculation is specified in equation (3.24). With the assumption of constant tank temperature, at least two tanks are needed for heat storage.

$$2 \leq \sum_{k \in \mathbf{ST}} zt_k \leq MNT \tag{3.24}$$

where zt_k is a binary variable indicating the use of tank k.

3.4.7 Temperature Trend of HTM Tanks

It has been assumed that the tanks are arranged in descending order of temperature. To ensure this, the following constraints are included. For cooling hot stream i using the cold HTM, the cumulative sum of $za_{li}^{(\text{T2H})}$ values (from $l = 1$ to k) should be less than or equal to that of $za_{il}^{(\text{H2T})}$ values (from $l = 1$ to k), as formulated in equation (3.25). For heating cold stream j using the hot HTM, the cumulative sum of $zr_{lj}^{(\text{T2C})}$ values should be greater than or equal to that of $zr_{jl}^{(\text{C2T})}$ values, as formulated in equation (3.26). Such restrictions force the tank at a lower temperature to have a larger index number and the tank at a higher temperature to have a smaller index number. Table 3.1 shows an example with four tanks. A cold HTM stream from tank 3 absorbs heat from hot stream i and the resulting hot HTM stream enters tank 1 ($za_{3i}^{(\text{T2H})} = 1$; $za_{i1}^{(\text{H2T})} = 1$). A hot HTM stream from tank 1 then rejects heat to cold stream j and the resulting cold HTM stream enters tank 3 ($zr_{1i}^{(\text{T2C})} = 1$; $zr_{j3}^{(\text{C2T})} = 1$). If

TABLE 3.1
Example of Tank Ordering

k	1	2	3	4	k	1	2	3	4
$za_{li}^{(\text{T2H})}$	0	0	1	0	$\sum_{l=1}^{k} za_{li}^{(\text{T2H})}$	0	0	1	1
$za_{il}^{(\text{H2T})}$	1	0	0	0	$\sum_{l=1}^{k} za_{il}^{(\text{H2T})}$	1	1	1	1
$zr_{lj}^{(\text{T2C})}$	1	0	0	0	$\sum_{l=1}^{k} zr_{lj}^{(\text{T2C})}$	1	1	1	1
$zr_{jl}^{(\text{C2T})}$	0	0	1	0	$\sum_{l=1}^{k} zr_{jl}^{(\text{C2T})}$	0	0	1	1

the hot and cold tanks are numbered correctly, the cumulative sums of the binary variables in Table 3.1 satisfy the constraints in equations (3.25) and (3.26), and vice versa.

$$\sum_{l=1}^{k} za_{li}^{(\mathrm{T2H})} \le \sum_{l=1}^{k} za_{il}^{(\mathrm{H2T})} \quad \forall i \in \mathbf{HP}, k \in \mathbf{ST} \tag{3.25}$$

$$\sum_{l=1}^{k} zr_{lj}^{(\mathrm{T2C})} \ge \sum_{l=1}^{k} zr_{jl}^{(\mathrm{C2T})} \quad \forall j \in \mathbf{CP}, k \in \mathbf{ST} \tag{3.26}$$

In addition, there is a minimum temperature difference (DT) between the adjacent tanks k and $k+1$, as given in equation (3.27).

$$T_{k+1} + \Gamma zt_k \ge T_k \ge T_{k+1} + DTzt_k \quad \forall k \in \mathbf{ST}^{-} \tag{3.27}$$

Equation (3.28) is used to set the lower and upper limits to the tank temperature, which is also limited by the freezing and boiling points of the HTM.

$$\underline{T} \le T_k \le \overline{T} \quad \forall k \in \mathbf{ST} \tag{3.28}$$

3.4.8 HTM Levels in Storage Tanks

The HTM level in tank k at the start of time period $p+1$ ($Q_{k,p+1}$) is equal to that at the start of time period p (Q_{kp}) plus the difference between the inlet and outlet flowrates during time period p, as given by equation (3.29). For cyclic operation, the end of a cycle corresponds to the start of the next cycle, as formulated in equation (3.30). The tank size is then determined by the maximum HTM level during the cycle time.

$$\begin{aligned} Q_{k,p+1} = Q_{kp} &+ \left[\sum_{i \in \mathbf{HP}} \left(za_{ik}^{(\mathrm{H2T})} - za_{ki}^{(\mathrm{T2H})}\right) \frac{Fa_i}{Cp^{(\mathrm{HTM})}} Z_{ip}^{(\mathrm{hot})} t_p\right] \\ &+ \left[\sum_{j \in \mathbf{CP}} \left(zr_{jk}^{(\mathrm{C2T})} - zr_{kj}^{(\mathrm{T2C})}\right) \frac{Fr_j}{Cp^{(\mathrm{HTM})}} Z_{jp}^{(\mathrm{cold})} t_p\right] \quad \forall k \in \mathbf{ST}, p \in \mathbf{TP}^{-} \end{aligned} \tag{3.29}$$

$$\begin{aligned} Q_{k,1} = Q_{k,N_p} &+ \left[\sum_{i \in \mathbf{HP}} \left(za_{ik}^{(\mathrm{H2T})} - za_{ki}^{(\mathrm{T2H})}\right) \frac{Fa_i}{Cp^{(\mathrm{HTM})}} Z_{i,N_p}^{(\mathrm{hot})} t_{N_p}\right] \\ &+ \left[\sum_{j \in \mathbf{CP}} \left(zr_{jk}^{(\mathrm{C2T})} - zr_{kj}^{(\mathrm{T2C})}\right) \frac{Fr_j}{Cp^{(\mathrm{HTM})}} Z_{j,N_p}^{(\mathrm{cold})} t_{N_p}\right] \quad \forall k \in \mathbf{ST} \end{aligned} \tag{3.30}$$

3.4.9 Logical Constraints

Equation (3.31) ensures that the required HTM tanks are numbered successively.

$$zt_k \geq zt_{k+1} \quad \forall k \in \mathbf{ST}^- \tag{3.31}$$

Equations (3.32)–(3.35) correlate the binary variables and the corresponding continuous variables, while setting the lower ($\underline{q}$) and upper limits ($\overline{q}$) on the amount of heat transferred.

$$\underline{q}za_i \leq qa_i \leq \overline{q}za_i \quad \forall i \in \mathbf{HP} \tag{3.32}$$

$$\underline{q}zcu_i \leq qcu_i \leq \overline{q}zcu_i \quad \forall i \in \mathbf{HP} \tag{3.33}$$

$$\underline{q}zr_j \leq qr_j \leq \overline{q}zr_j \quad \forall j \in \mathbf{CP} \tag{3.34}$$

$$\underline{q}zhu_j \leq qhu_j \leq \overline{q}zhu_j \quad \forall j \in \mathbf{CP} \tag{3.35}$$

where zcu_i denotes if cold utility is used for hot stream i; zhu_j denotes if hot utility is used for cold stream j. The heat capacity flowrate of an HTM stream for heat absorption or rejection can also be limited using equations (3.36) and (3.37).

$$\underline{F}za_i \leq Fa_i \leq \overline{F}za_i \quad \forall i \in \mathbf{HP} \tag{3.36}$$

$$\underline{F}zr_j \leq Fr_j \leq \overline{F}zr_j \quad \forall j \in \mathbf{CP} \tag{3.37}$$

To have a simpler indirect batch HEN configuration, each process stream is allowed to use only one HTM stream for heat absorption or rejection, as given in equations (3.38) and (3.39).

$$\sum_{k \in \mathbf{ST}} za_{ki}^{(\mathrm{T2H})} = za_i = \sum_{k \in \mathbf{ST}} za_{ik}^{(\mathrm{H2T})} \quad \forall i \in \mathbf{HP} \tag{3.38}$$

$$\sum_{k \in \mathbf{ST}} zr_{kj}^{(\mathrm{T2C})} = zr_j = \sum_{k \in \mathbf{ST}} zr_{jk}^{(\mathrm{C2T})} \quad \forall j \in \mathbf{CP} \tag{3.39}$$

An existing HTM tank has at least one inlet stream and at least one outlet stream, as stated in equations (3.40) and (3.41).

$$\sum_{i \in \mathbf{HP}} za_{ik}^{(\mathrm{H2T})} + \sum_{j \in \mathbf{CP}} zr_{jk}^{(\mathrm{C2T})} \geq zt_k \quad \forall k \in \mathbf{ST} \tag{3.40}$$

$$\sum_{i \in \mathbf{HP}} za_{ki}^{(\mathrm{T2H})} + \sum_{j \in \mathbf{CP}} zr_{kj}^{(\mathrm{T2C})} \geq zt_k \quad \forall k \in \mathbf{ST} \tag{3.41}$$

In addition, an HTM stream cannot return to the same tank after exchanging heat with a process stream, as given in equations (3.42) and (3.43).

$$za_{ki}^{(\mathrm{T2H})} + za_{ik}^{(\mathrm{H2T})} \le zt_k \quad \forall i \in \mathbf{HP}, k \in \mathbf{ST} \tag{3.42}$$

$$zr_{kj}^{(\mathrm{T2C})} + zr_{jk}^{(\mathrm{C2T})} \le zt_k \quad \forall i \in \mathbf{HP}, k \in \mathbf{ST} \tag{3.43}$$

3.4.10 Objective Functions

A relevant objective of indirect batch HEN design is to minimize the total annual cost, which consists of the costs of external utilities, heat exchange units, storage tanks, and the HTM. However, the objective function adopted here is to minimize the total consumption of external hot and cold utilities, as given in equation (3.44), in order to compare the proposed mixed-integer nonlinear programming (MINLP) approach with the pinch design method.

$$\min J_1 = \sum_{i \in \mathbf{HP}} \sum_{p \in \mathbf{TP}} Z_{ip}^{(\mathrm{hot})} t_p qcu_i + \sum_{j \in \mathbf{CP}} \sum_{p \in \mathbf{TP}} Z_{jp}^{(\mathrm{cold})} t_p qhu_j \tag{3.44}$$

After determining the minimum utility consumption, a second objective may be set to determine the minimum tank capacity for heat storage, as formulated in equation (3.45).

$$\min J_2 = \sum_{k \in \mathbf{ST}} V_k \tag{3.45}$$

Two additional constraints are needed with this objective. Equation (3.46) indicates the required tank volume, while equation (3.47) limits the utility consumption to the minimum.

$$V_k \ge Q_{kp} / \rho^{(\mathrm{HTM})} \quad \forall k \in \mathbf{ST}, p \in \mathbf{TP} \tag{3.46}$$

$$\sum_{i \in \mathbf{HP}} \sum_{p \in \mathbf{TP}} Z_{ip}^{(\mathrm{hot})} t_p qcu_i + \sum_{j \in \mathbf{CP}} \sum_{p \in \mathbf{TP}} Z_{jp}^{(\mathrm{cold})} t_p qhu_j \le J_1^* \tag{3.47}$$

The MINLP models for the design of indirect batch HENs (equation (44) subject to equations (3.1)–(3.43) for utility minimization and equation (3.45) subject to equations (3.1)–(3.43), (3.46), and (3.47) for storage minimization) are implemented and solved in GAMS using solver BARON.

3.5 ILLUSTRATIVE EXAMPLES

Two examples are presented to demonstrate the effectiveness of the proposed MINLP model for indirect batch heat integration problems. The first example is a simple one used to compare the results obtained using the proposed approach and the pinch design method. The second example considers a single-product batch plant.

3.5.1 Example 1

Several methods for designing an indirect batch heat integration system have been developed in the literature. Sadr-Kazemi and Polley (1996) proposed the use of a storage line between the hot and cold composite curves to determine the tank temperatures and the tank sizes.

A simple problem with two hot and two cold streams is considered first, as shown in Table 3.2. Four time periods are defined according to the start and end times of the streams, as shown in Table 3.3. Applying the pinch targeting method of Sadr-Kazemi and Polley (1996), the hot and cold composite curves and the HTM storage line are shown in Figure 3.3, where the storage temperatures are determined to be 140°C

TABLE 3.2
Stream Data for the Simple Example

Hot stream	$TH_i^{(in)}$ (°C)	$TH_i^{(out)}$ (°C)	$t^{(s)}$ (h)	$t^{(f)}$ (h)	LH_i (MJ)	SH_i (MJ)	FH_i (MJ/h/°C)	Type
H1	170	60	0.3	0.9	0	330	5	S
H2	150	30	0	0.5	0	180	3	S
Cold stream	$TC_j^{(in)}$ (°C)	$TC_j^{(out)}$ (°C)	$t^{(s)}$ (h)	$t^{(f)}$ (h)	LH_j (MJ)	SH_j (MJ)	FC_j (MJ/h/°C)	**Type**
C1	20	135	0.3	0.5	0	230	10	S
C2	80	140	0.5	1	0	240	8	S

ΔT_{min} = 10°C; DT = 10°C.
HTM: glycerol (heat capacity = 0.0024 MJ/kg/°C; density = 1.2578 kg/L) operating in 18–290°C.

TABLE 3.3
Existence of Streams for the Simple Example

Period	1	2	3	4
t_p (h)	0.3	0.2	0.4	0.1
Hot stream		**Value of $Z_{ip}^{(hot)}$**		
H1	0	1	1	0
H2	1	1	0	0
Cold stream		**Value of $Z_{jp}^{(cold)}$**		
C1	0	1	0	0
C2	0	0	1	1

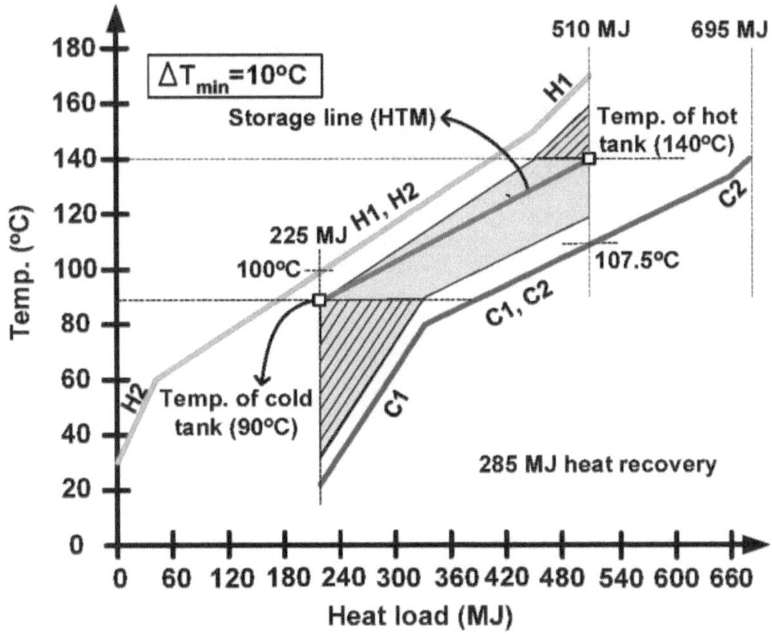

FIGURE 3.3 Pinch analysis for the simple example. (Reprinted with permission from *Ind. Eng. Chem. Res.* 2008, 47, 14, 4817–4829. Copyright 2008 American Chemical Society.)

and 90°C for the specified minimum approach temperature of 10°C. Figure 3.3 also shows that the maximum heat recovery is 285 MJ/cycle and the minimum hot and cold utility requirements are 185 MJ/cycle and 225 MJ/cycle, respectively. There may be multiple network configurations that achieve these targets. According to the information in Figure 3.3, an obvious solution is to heat both cold streams up to 107.5°C by the hot HTM (140°C) and to cool both hot streams down to 100°C by the cold HTM (90°C). The heaters and coolers are then used for the hot and cold streams to reach their target temperatures. The corresponding network is shown in Figure 3.4(a). The 90°C HTM from the cold tank absorbs 210 MJ (= 350 MJ/h × 0.6 h) of heat from H1 in 0.3–0.9 h and 75 MJ (= 150 MJ/h × 0.5 h) from H2 in 0–0.5 h, before entering the hot tank at 140°C. Two coolers are then used to remove the remaining 120 MJ (= 200 MJ/h × 0.6 h) of heat from H1 and 105 MJ (= 210 MJ/h × 0.5 h) from H2. The heat absorbed from the hot streams (285 MJ in total) is then transferred from the hot HTM to the cold streams, with 175 MJ (= 875 MJ/h × 0.2 h) to C1 in 0.3–0.5 h and 110 MJ (= 220 MJ/h × 0.5 h) to C2 in 0.5–1 h. Two heaters are then used to supply a further 55 MJ (= 275 MJ/h × 0.2 h) of heat to C1 and 130 MJ (= 260 MJ/h × 0.5 h) to C2. With the heat recovery of 285 MJ/cycle, the hot and cold utility requirements are 185 and 225 MJ/cycle, respectively. The HTM levels in the hot and cold tanks are shown in Figure 3.5(a) and (b) The required volume of both tanks is determined to be 497 L.

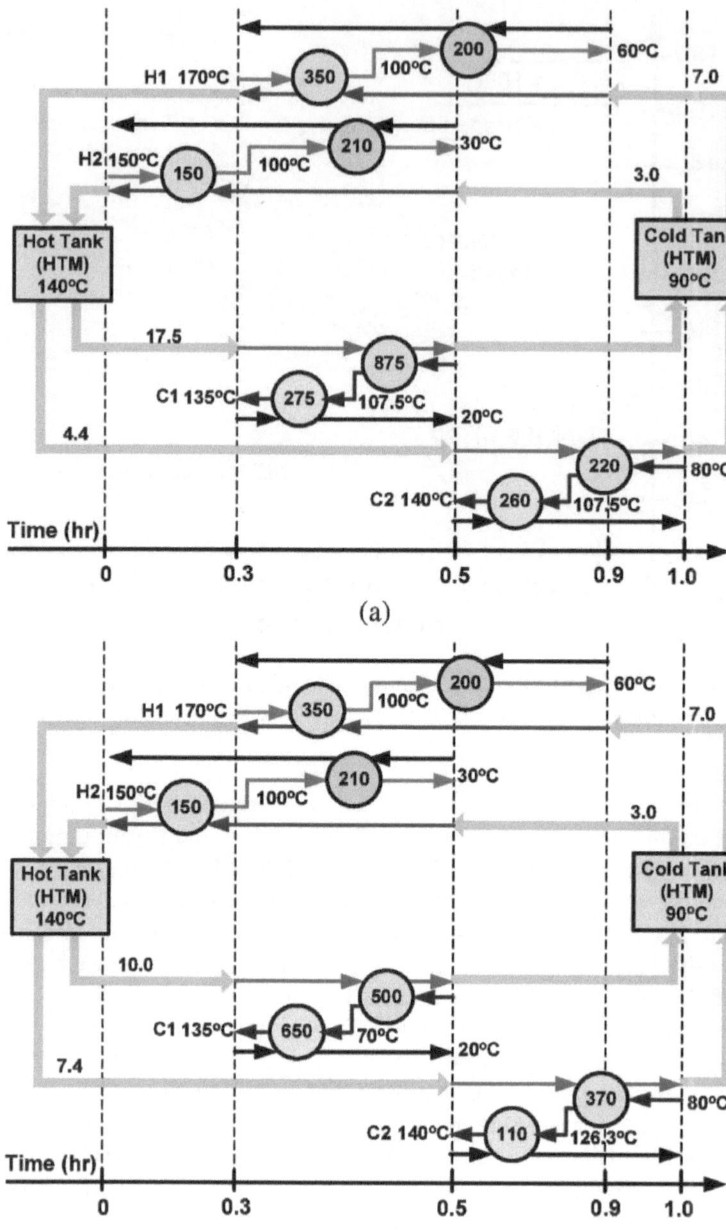

FIGURE 3.4 Indirect batch HENs obtained using (a) the pinch design method and (b) the proposed MINLP approach. (Reprinted with permission from *Ind. Eng. Chem. Res.* 2008, 47, 14, 4817–4829. Copyright 2008 American Chemical Society.)

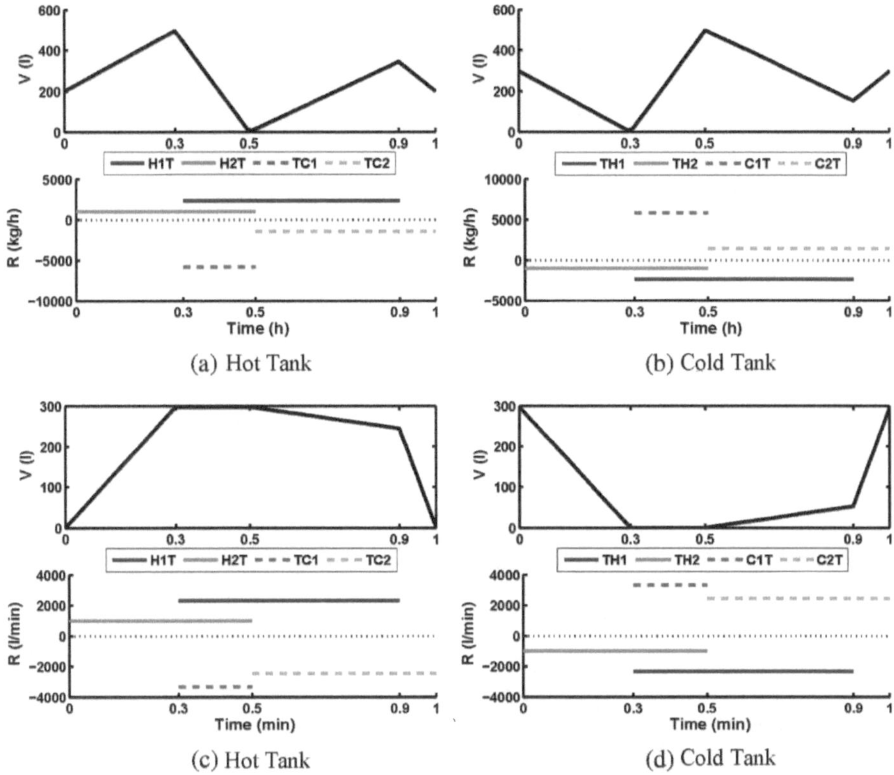

FIGURE 3.5 HTM levels in the hot and cold tanks and the corresponding HTM flowrates in the designs obtained by the pinch method (a, b) and the proposed MINLP approach (c, d). (Reprinted with permission from *Ind. Eng. Chem. Res.* 2008, 47, 14, 4817–4829. Copyright 2008 American Chemical Society.)

The resulting network configuration determined by the proposed MINLP approach is shown in Figure 3.4(b). This design is very similar to that in Figure 3.4(a) and achieves the same energy targets. The difference is found in the allocation of recovered heat to cold streams. C1 receives 100 MJ (= 500 MJ/h × 0.2 h) and C2 receives 185 MJ (= 370 MJ/h × 0.5 h) from the hot HTM. Two heaters are then used to supply a further 130 MJ (= 650 MJ/h × 0.2 h) of heat to C1 and 55 MJ (= 110 MJ/h × 0.5 h) to C2. This different heat allocation results in different heat capacity flowrates of the hot HTM in 0.3–1 h, and hence smaller tank sizes. The HTM levels in the hot and cold tanks are shown in Figure 3.5(c) and (d). The required volume of both tanks is determined to be 298 L, which is much smaller than that determined using the pinch design method.

It is worth noting that although the pinch method and the MINLP approach result in the same utility consumption for this simple example, the tank size determined by the proposed MINLP approach is 40% smaller than that by the pinch method. This shows the significance of adding a second objective to determine the minimum tank size to achieve the minimum utility consumption. Note that the pinch method is applicable only to flowing streams using series-type units. The MINLP approach, however, is capable of handling various types of streams and heat exchange units. The applicability of the MINLP approach is further demonstrated in the next example.

3.5.2 Example 2

This example considers a single-product batch plant with two batch reactors and a batch distillation column, as shown in Figure 3.6 (Gremouti, 1991). The process involves the following tasks. Feed A is fed into reactor R1 and heated from 10 to 60°C before the exothermic reaction starts. Without further heating, the R1 product leaves at 100°C and enters column D1, which operates at 120°C. The condenser of column D1 operates at 110°C. The condensed distillate is cooled from 110 to 50°C and sent to an overhead receiver (M) before being charged to reactor R2 together with feed B (available at 15°C). The mixture is heated to 95°C before the reaction starts, which is also exothermic. The temperature in R2 increases to 135°C, where the heat of reaction is removed using a reflux condenser. On completion of the reaction, the R2 product is cooled from 140 to 35°C before downstream treatment.

The tasks performed by the main process units of the batch plant (R1, D1, and R2) are represented in the Gantt chart in Figure 3.7. The batch processing time is

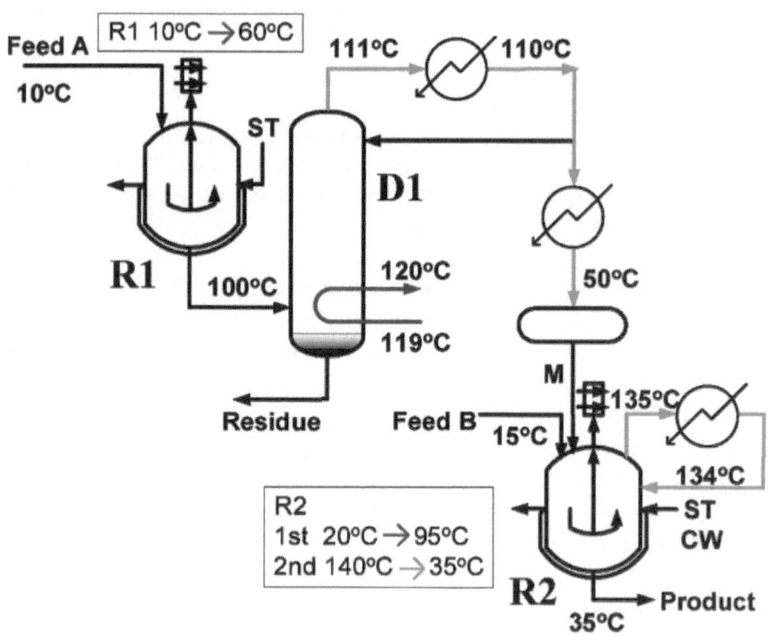

FIGURE 3.6 Flowsheet for the single-product batch process (Gremouti, 1991). (Reprinted with permission from *Ind. Eng. Chem. Res.* 2008, 47, 14, 4817–4829. Copyright 2008 American Chemical Society.)

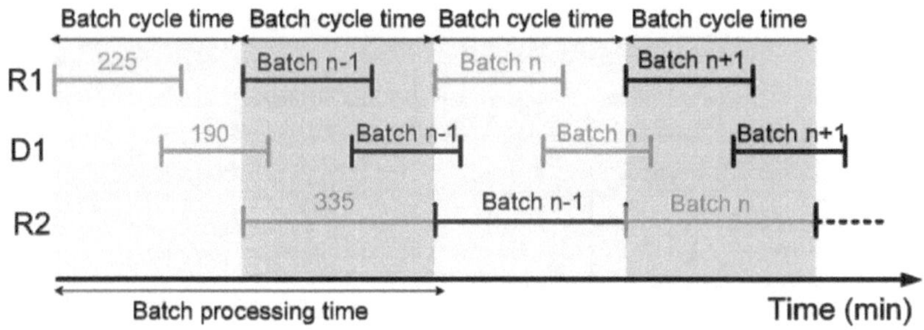

FIGURE 3.7 Gantt chart for the single-product batch process. (Reprinted with permission from *Ind. Eng. Chem. Res.* 2008, 47, 14, 4817–4829. Copyright 2008 American Chemical Society.)

690 min, while the batch cycle time is 335 min, as determined by the unit with the longest processing time (R2).

The hot and cold streams within the batch cycle time are shown in Figure 3.8. Seven time periods are defined according to the start and end times of these streams as shown in Table 3.4. The stream data are given in Table 3.5. Before exploring heat integration opportunities, all heating and cooling demands of the cold and hot streams are satisfied by external steam and cooling water. The total utility consumption was 21225.1 MJ/cycle ($=\sum_i(LH_i+SH_i)+\sum_i(LC_j+SC_j)$).

In this example, the hot and cold streams use different types of heat exchange units, as indicated in Table 3.5. Two HTM tanks can be used for heat storage. An indirect batch HEN with the minimum utility consumption is obtained as shown in Figure 3.9. It can be seen that only H3, C1, and C3 are involved in the indirect heat integration. The heating demand of C2 and the cooling demands of H1, H2, and H4 are satisfied only by external utilities. The process flowsheet with two storage tanks is shown in Figure 3.10.

As shown in Figures 3.9 and 3.10, the hot HTM from tank 1 (125°C) rejects heat to C1 and C3 in 30–70 min and 70–135 min, respectively. The cooled HTM (105°C) enters tank 2 for temporary storage. The heat capacity flowrates of the hot HTM are 1.38 MJ/min/°C for C1 and 2.48 MJ/min/°C for C3. In 135–225 min, the cold HTM from tank 2 (105°C) absorbs heat from H3, and the heated HTM (125°C) is stored in tank 1. Such indirect heat integration with two HTM tanks can significantly reduce the total utility consumption from 21,225.1 to 12,590.1 MJ/cycle. This corresponds to a 40.7% reduction. Figure 3.11 shows the HTM levels in the hot and cold tanks and the corresponding HTM flowrates. For tank 1 (see Figure 3.11(a)), there is an inlet HTM flowrate from the heat exchange match with H3 (H3T) and two outlet HTM flowrates to the matches with C1 (TC1) and C3 (TC3). The required volume of both tanks is determined to be 71,512 L.

From the heat integration network in Figures 3.9 and 3.10, it can be observed that the possible temperature of tank 1 ranges between 105 and 125°C, as limited by the supply temperature of H3 (135°C – ΔT_{min}) and the target temperatures of

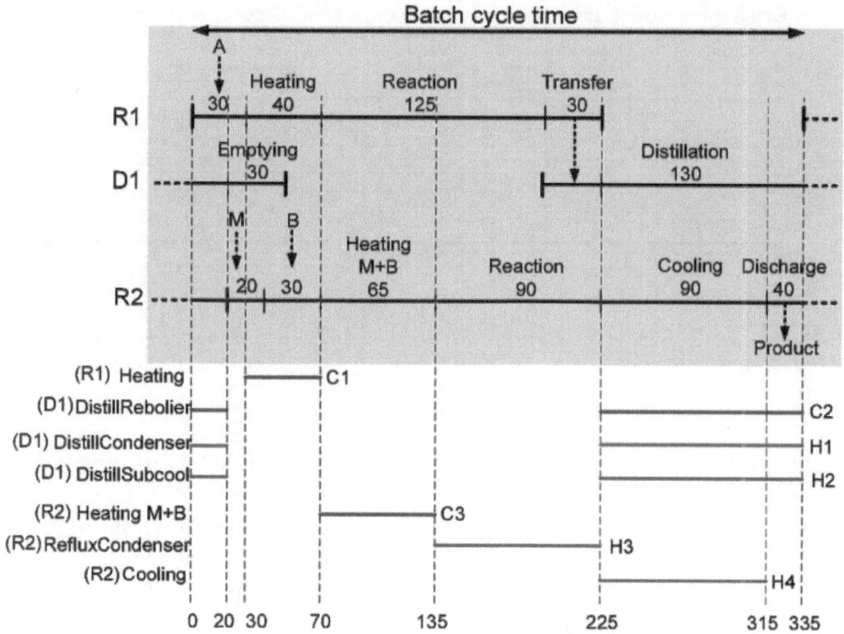

FIGURE 3.8 Gantt chart for the single-product batch process within a batch cycle. (Reprinted with permission from *Ind. Eng. Chem. Res.* 2008, 47, 14, 4817–4829. Copyright 2008 American Chemical Society.)

TABLE 3.4
Existence of Streams the Single-Product Batch Process

Period	**1**	**2**	**3**	**4**	**5**	**6**	**7**
t_p (min)	20	10	40	65	90	90	20
Hot stream			**Value of $Z_{ip}^{(hot)}$**				
H1	1	0	0	0	0	1	1
H2	1	0	0	0	0	1	1
H3	0	0	0	0	1	0	0
H4	0	0	0	0	0	1	0
Cold stream			**Value of $Z_{jp}^{(cold)}$**				
C1	0	0	1	0	0	0	0
C2	1	0	0	0	0	1	1
C3	0	0	0	1	0	0	0

TABLE 3.5
Stream Data for the Single-Product Batch Process

Hot stream	$TH_i^{(in)}$ (°C)	$TH_i^{(out)}$ (°C)	$t^{(s)}$ (min)	$t^{(f)}$ (min)	LH_i (MJ)	SH_i (MJ)	FH_i (MJ/min/°C)	Type
H1	111	110	0	20	3149.9	0	-	M
			225	335				
H2	110	50	0	20	0	358.8	0.046	S
			225	335				
H3	135	134	135	225	4955.4	0	-	M
H4	140	35	225	315	0	4914	-	C

Cold stream	$TC_j^{(in)}$ (°C)	$TC_j^{(out)}$ (°C)	$t^{(s)}$ (min)	$t^{(f)}$ (min)	LH_j (MJ)	SH_j (MJ)	FC_j (MJ/min/°C)	Type
C1	10	60	30	70	0	1100	-	C
C2	119	120	0	20	3529.5	0	-	M
			225	335				
C3	20	95	70	135	0	3217.5	-	C

ΔT_{min} = 10°C; DT = 10°C.
HTM: glycerol (heat capacity = 0.0024 MJ/kg/°C; density = 1.2578 kg/L) operating in 18–290°C.

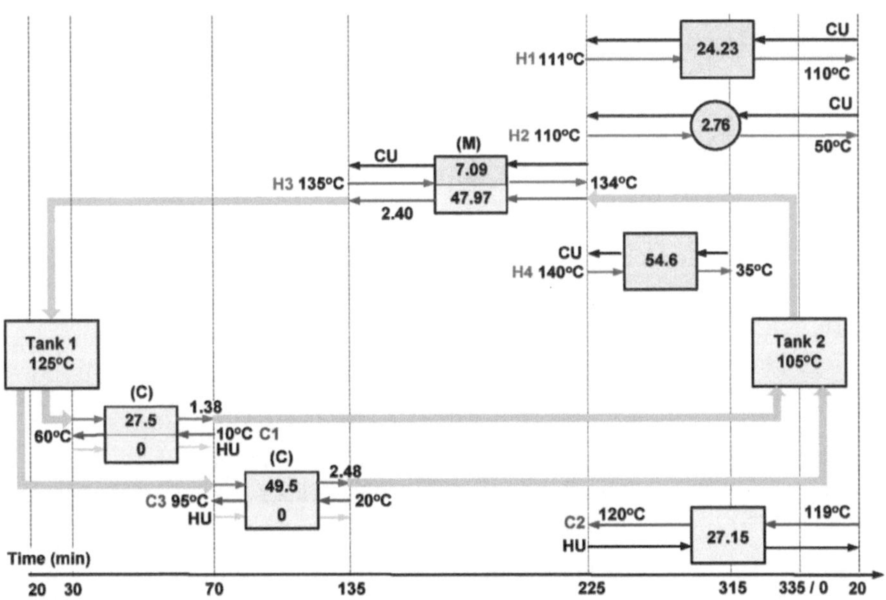

FIGURE 3.9 Indirect heat integration of the single-product batch plant with two HTM tanks. (Reprinted with permission from *Ind. Eng. Chem. Res.* 2008, 47, 14, 4817–4829. Copyright 2008 American Chemical Society.)

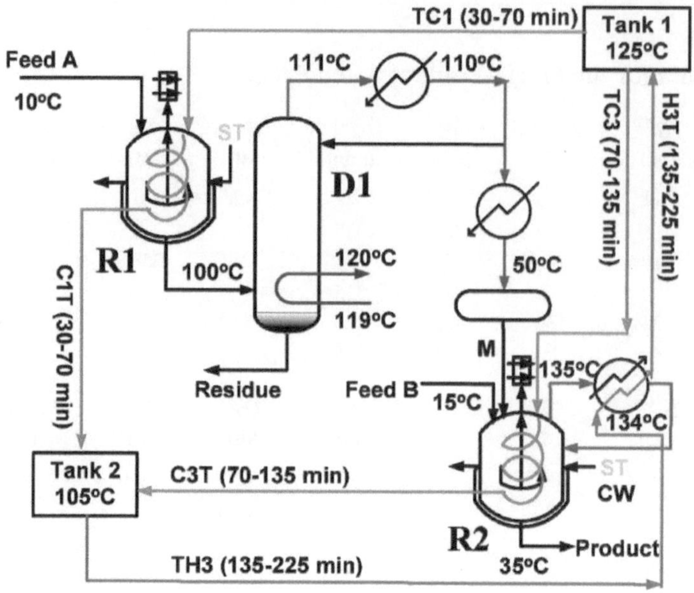

FIGURE 3.10 Flowsheet for the heat-integrated single-product batch plant with two HTM tanks. (Reprinted with permission from *Ind. Eng. Chem. Res.* 2008, 47, 14, 4817–4829. Copyright 2008 American Chemical Society.)

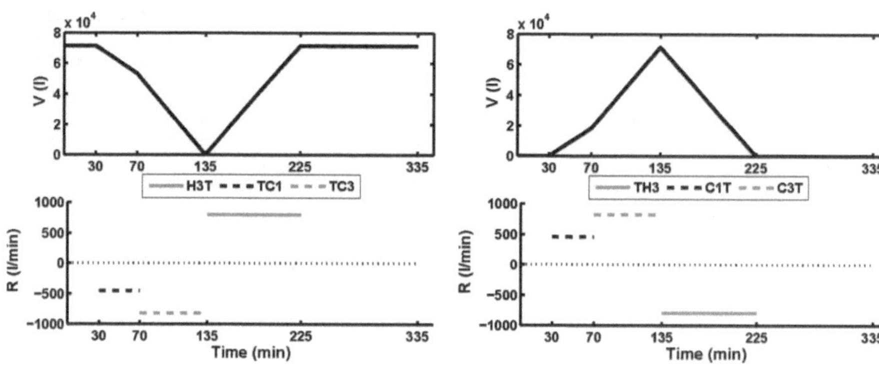

(a) Tank 1 (b) Tank 2

FIGURE 3.11 HTM levels in the two tanks and the HTM flowrates. (Reprinted with permission from *Ind. Eng. Chem. Res.* 2008, 47, 14, 4817–4829. Copyright 2008 American Chemical Society.)

C1 (60°C + ΔT_{min}) and C3 (95°C + ΔT_{min}). Similarly, the possible temperature of tank 2 (105–124°C) should be greater than the target temperatures of C1 (60°C + ΔT_{min}) and C3 (95°C + ΔT_{min}) and lower than the target temperature of H3 (134°C − ΔT_{min}). The greater the temperature difference between the two tanks, the smaller the HTM flowrate and the tank size. Therefore, the optimal temperatures of tanks

1 and 2 were determined to be 125 and 105°C, respectively. In the case of two HTM tanks, the temperature of tank 2 is limited by the target temperature of C3 and could not be any lower. It would then be interesting to examine the effect of adding additional HTM tanks and allowing the tank temperatures to vary over time on the design of indirect batch HENs (Chen and Ciou, 2008, 2009).

3.6 SUMMARY

The main challenge of batch heat integration is the time dependence of process streams. The design of indirect batch HENs with heat storage has been addressed in this chapter. In the heat storage system, a recirculating HTM is used to absorb heat from hot process streams and reject it to cold process streams that occur in later time periods, thereby relaxing the time constraint. A superstructure is presented to facilitate the modeling of time-dependent batch process streams for indirect heat integration. Based on this superstructure, the model for indirect batch HEN synthesis is formulated as an MINLP, with the objective of minimizing the consumption of external utilities. Two illustrative examples were used to demonstrate the applicability of the proposed model. In addition to the minimum utility consumption, the minimum heat storage tank capacity can be determined by setting a second objective. Although this chapter has focused on heat integration in single batch plants, the proposed indirect heat integration scheme and the formulation is applicable to multi-plant cases. In fact, indirect heat integration is less schedule-sensitive and allows better operability for multi-plant batch HENs.

NOMENCLATURE

Indices and Sets

$i \in \mathbf{HP} = \{1, \cdots, N_H\}$	hot process streams
$i \in \mathbf{HPC}$	hot process streams using jacket exchangers with internal coils
$i \in \mathbf{HPM}$	hot process streams using multi-stream exchangers
$i \in \mathbf{HPP}$	hot process streams using parallel units (= $\mathbf{HPM} \cup \mathbf{HPC}$)
$i \in \mathbf{HPS}$	hot process streams using series units
$j \in \mathbf{CP} = \{1, \cdots, N_C\}$	cold process streams
$j \in \mathbf{CPC}$	cold process streams using jacket exchangers with internal coils
$j \in \mathbf{CPM}$	cold process streams using multi-stream exchangers
$j \in \mathbf{CPP}$	cold process streams using parallel units (= $\mathbf{CPM} \cup \mathbf{CPC}$)
$j \in \mathbf{CPS}$	cold process streams using series units
$k \in \mathbf{ST} = \{1, \cdots, N_T\}$	storage tanks
$k \in \mathbf{ST}^- = \{1, \cdots, N_T - 1\}$	storage tanks
$t \in \mathbf{TP} = \{1, \cdots, N_P\}$	storage tanks
$t \in \mathbf{TP}^- = \{1, \cdots, N_P - 1\}$	storage tanks

Parameters

$C_p^{(HTM)}$	heat capacity of the HTM
DT	minimum temperature difference between a pair of HTM tanks
$\underline{F}$	lower limit to the heat capacity flowrate
$\overline{F}$	upper limit to the heat capacity flowrate
FC_j	heat capacity flowrate of cold stream j in a series unit
FH_i	heat capacity flowrate of hot stream i in a series unit
$\underline{q}$	lower limit to the amount of heat transferred
$\overline{q}$	upper limit to the amount of heat transferred
LC_j	latent heat of cold stream j
LH_i	latent heat of hot stream i
MNT	maximum number of tanks
SC_j	sensible heat of cold stream j
SH_i	sensible heat of hot stream i
$\underline{T}$	lower limit to the tank temperature
$\overline{T}$	upper limit to the tank temperature
t_p	length of time period p
$TC_j^{(in)}$	inlet/supply temperature of cold stream j
$TC_j^{(out)}$	outlet/target temperature of cold stream j
$TH_i^{(in)}$	inlet/supply temperature of hot stream i
$TH_i^{(out)}$	outlet/target temperature of hot stream i
$Z_{ip}^{(hot)} = 0/1,$	denoting the existence of hot stream i in time period p
$Z_{jp}^{(cold)} = 0/1,$	denoting the existence of cold stream j in time period p
$\rho^{(HTM)}$	density of the HTM
Γ	arbitrary large number
ΔT_{min}	minimum temperature difference for a heat exchange match

Variables

Fa_i	heat capacity flowrate of the HTM stream for absorbing heat from hot stream i
Fr_j	heat capacity flowrate of the HTM stream for rejecting heat to cold stream j
Q_{kp}	HTM level in tank k at the start of time period p
qa_i	amount of heat transferred from hot stream i to the HTM stream
qcu_i	amount of heat removed from hot stream i by cold utility
qhu_j	amount of heat supplied to cold stream j by hot utility
qr_j	amount of heat transferred from the HTM stream to cold stream j
tc_j	outlet temperature of cold stream j after exchanging heat with the HTM in a series unit
th_i	outlet temperature of hot stream i after exchanging heat with the HTM in a series unit
T_k	temperature of tank k
V_k	volume of tank k

$za_i = 0/1$, indicating if the HTM is used for absorbing heat from hot stream i

$za_{ik}^{(H2T)} = 0/1$, indicating if there is an HTM stream from the match with hot stream i to tank k

$za_{ki}^{(T2H)} = 0/1$, indicating if there is an HTM stream from tank k to the match with hot stream i

$zcu_i = 0/1$, indicating if cold utility is used to remove heat from hot stream i

$zhu_j = 0/1$, indicating if hot utility is used to supply heat to cold stream j

$zr_j = 0/1$, indicating if the HTM is used for rejecting heat to cold stream j

$zr_{jk}^{(C2T)} = 0/1$, indicating if there is an HTM stream from the match with cold stream j to tank k

$zr_{kj}^{(T2C)} = 0/1$, indicating if there is an HTM stream from tank k to the match with cold stream j

$zt_k = 0/1$, indicating the existence of tank k

REFERENCES

Chen, C.-L., Ciou, Y.-J. 2008. Design and optimization of indirect energy storage systems for batch process plants. *Industrial and Engineering Chemistry Research* 47(14): 4817–4829.

Chen, C.-L., Ciou, Y.-J. 2009. Design of indirect heat recovery systems with variable-temperature storage for batch plants. *Industrial and Engineering Chemistry Research* 48(9): 4375–4387.

Corominas, J., Espuña, A., Puigjaner, L. 1994. Method to incorporate energy integration considerations in multiproduct batch processes. *Computers and Chemical Engineering* 18(11–12): 1043–1055.

de Boer, R., Semeding, S.F., Bach, P.W. 2006. Heat storage systems for use in an industrial batch plant (Result of) A case study. In: Proceedings of the Tenth International Conference on Thermal Energy Storage, Ecostock, New Jersey, 31 May–2 June, 2006.

Georgiadis, M.C., Papageorgiou, L.G. 2001. Optimal scheduling of heat- integrated multipurpose plants under fouling conditions. *Applied Thermal Engineering* 21(16): 1675–1697.

Gremouti, I.D. 1991. Integration of batch processes for energy savings and debottlenecking. Master's thesis, Department of Chemical Engineering, University of Manchester Institute of Science and Technology (UMIST), UK.

Krummenacher, P., Favrat, D. 2001. Indirect and mixed direct-indirect heat integration of batch processes based on Pinch Analysis. *International Journal of Applied Thermodynamics* 4(3): 135–143.

Papageorgiou, L.G., Shah, N., Pantelides, C.C. 1994. Optimal scheduling of heat-integrated multipurpose plants. *Industrial and Engineering Chemistry Research* 33(12): 3168–3186.

Sadr-Kazemi, N., Polley, G.T. 1996. Design of energy storage systems for batch process plants. *Chemical Engineering Research and Design* 74(5): 584–596.

4 Benefit Allocation Methods for Interplant Heat Integration Based on Non-Cooperative Games

As mentioned in the previous chapters, the total operating cost of almost every chemical plant can be largely attributed to the needs for heating and cooling utilities. The heat exchanger network (HEN) embedded in a chemical process is usually configured for the purpose of minimizing the total utility consumption rates. A HEN design is traditionally produced with either a simultaneous optimization strategy (Yee et al., 1990; Yee and Grossmann, 1990) or a stepwise procedure for determining the minimum utility consumption rates and the minimum match number first (Papoulias and Grossmann, 1983) and then the optimal network structure (Floudas et al., 1986). The former usually yields a better trade-off between utility and capital costs, but the computational effort required for solving the corresponding mixed-integer nonlinear programming (MINLP) model can be overwhelming. On the other hand, although only suboptimal solutions can be obtained in the latter case, implementing a stepwise method is easier. For this reason, it is often advantageous to adopt a sequential approach to synthesize the inner-plant heat-exchange networks in three steps. In the first two steps, a linear program (LP) and a mixed-integer linear program (MILP) are solved respectively to determine the minimum total utility cost and to identify the minimum number of matches and their heat duties. A nonlinear programming (NLP) model is then solved in the final step for generating the cost-optimal network.

Driven by the belief that significant extra improvement can be achieved by expanding the feasible region of any optimization problem, a number of studies have been carried out to develop various interplant heat integration schemes, e.g., see Bagajewicz and Rodera (2002), Kralj (2008), and Liew et al. (2017). The available synthesis methods for total-site heat integration (TSHI) can be classified into three kinds: the insight-based pinch analysis (Hackl and Harvey, 2015), the model-based methods (Chang et al., 2017), and the hybrid methods (Liew et al., 2017), while the required interplant energy flows may be either realized with direct heat exchanges between process streams or facilitated indirectly with extraneous fluids (Wang et al., 2015; Chang et al., 2017).

The main advantages of insight-based pinch analysis can be attributed to its target-setting strategy and flexible design steps. Matsuda et al. (2009) applied the area-wide pinch technology which incorporated the R-curve analysis and site-source-sink-profile analysis to TSHI of Kashima industrial area. For fluctuating renewable energy supply, Liew et al. (2014a) proposed the graphical targeting procedures based on the time slices to handle the energy supply/demand variability in TSHI. In addition, a retrofit framework was proposed by the same research group (Liew et al., 2014b), and the framework showed that energy retrofit projects should be approached from the total-site context first. Furthermore, Tarighaleslami et al. (2017) developed a new improved TSHI method in order to address the non-isothermal utilities targeting issues.

On the other hand, the model-based methods are more rigorous and thus better equipped to identify the true optimum. Zhang et al. (2016) proposed to use a superstructure for building an MINLP model to synthesize multi-plant HEN designs. Chang et al. (2017) presented a simultaneous optimization methodology for interplant heat integration using the intermediate fluid circle(s), while Wang et al. (2015) adopted a hybrid approach for the same problems. The performances for heat integration across plant boundaries using direct, indirect, and combined methods were analyzed and compared through composite curves, while the mathematical programming models were adopted to determine the optimal conditions of direct and/or indirect options.

As indicated in Cheng et al. (2014), the traditional model-based interplant heat integration arrangements were often not implementable in practice due to the fact that the profit margin might not be acceptable to every participating party. This drawback can be primarily attributed to the conventional HEN design objective, i.e., minimization of overall energy cost. Thus, the key to a successful interplant heat integration scheme should be to allow every plant to maximize its own benefit while striving for the largest possible overall saving at the same time. To address this benefit distribution issue, a non-cooperative game-based sequential optimization strategy was first developed by Cheng et al. (2014) to generate "fair" interplant integration schemes via direct heat exchanges between the hot and cold process streams across plant boundaries. In addition to a lighter computation load, this approach is justified by the fact that the game theoretic models can be more naturally incorporated into a step-by-step design practice when the same type of decision variables are evaluated one-at-a-time on a consistent basis.

Other than the benefit allocation issues, it is clearly of equal importance to examine alternative means for facilitating the desired energy flows among plants. As mentioned before, these flows can be materialized via heat exchange(s) either directly between hot and cold process streams located in different plants or indirectly between the process streams and an intermediate fluid (or the heating and cooling utilities). Although the direct heat exchanges are thermodynamically more efficient than their indirect counterparts, the resulting highly coupled interplant HEN may pose a control problem in the industrial environment. On the other hand, since the indirect heat integration is facilitated with the auxiliary streams (i.e., steam, cooling water, and/or hot oil) that do not take part in any production process, a greater degree of operational flexibility can be achieved and, thus, should be regarded as a more practical alternative. To improve the practical feasibility of interplant heat integration projects, the aforementioned sequential optimization approach has been modified in a later study (Chang et al., 2018) by replacing the direct interplant heat-transfer options with indirect ones. Extensive case studies were performed to demonstrate the feasibility of the modified procedures and to compare the pros and cons of different indirect heat-exchange alternatives.

As mentioned before, if the participating plants of a TSHI plan are owned by different companies, then the benefit distribution issues may be assumed to be identical to those in a non-cooperative game. The corresponding interplant heat integration schemes can then be synthesized directly and indirectly according to the following procedures:

4.1 AN ILLUSTRATIVE EXAMPLE

A simple example is utilized throughout this chapter for illustrating the implementation steps of various direct and indirect multi-plant HEN synthesis strategies. The process and utility data in Table 4.1 and Table 4.2 are taken from Liew et al. (2017).

TABLE 4.1
Process Data Used in Illustrative Example

Plant	Stream	T_{in} (°C)	T_{out} (°C)	F_{cp} (kW/°C)
P1	H1	150	40	7
P1	C1	60	140	9
P1	C2	110	190	8
P2	H1	200	70	5.5
P2	C1	30	110	3.5
P2	C2	140	190	7.5
P3	H1	370	150	3.0
P3	H2	200	40	5.5
P3	C1	110	360	4.5

TABLE 4.2
Utility Data Used in Illustrative Example

Plant	Utility	Temperature (°C)	Unit Cost (USD/kW·yr)	Upper Bound (kW)
P1	Cooling water	25	10	5000
P1	HPS (240 psig)	200	90	5000
P1	Fuel	500	80	5000
P2	Cooling water	25	22.5	5000
P2	HPS (240 psig)	200	30	5000
P2	Fuel	500	120	5000
P3	Cooling water	25	30	5000
P3	HPS (240 psig)	200	60	5000
P3	Fuel	500	40	5000

All single-plant heat-flow cascades and the corresponding pinch points, which can be independently determined without considering interplant heat integration, have already been shown in Figure 2.2 in Chapter 2. The minimum consumption rates of hot utilities in plants P1–P3 can be found to be 800 kW, 100 kW, and 255 kW respectively, while those of the cold utilities are 210 kW, 160 kW, and 670 kW respectively. The corresponding annual energy costs of these plants should be 66,100 USD/yr, 6600 USD/yr, and 30,300 USD/yr, respectively.

4.2 DIRECT INTEGRATION

As mentioned previously, the traditional sequential approach for single-plant HEN synthesis is suitable for the present applications. Specifically, a series of four optimization problems are solved consecutively as follows:

1. The minimum acceptable total utility cost of the entire industrial site can be first determined with a linear program, which should be formulated by modifying the conventional transshipment model (Papoulias and Grossmann, 1983).
2. By incorporating the constraints of minimum acceptable overall utility cost (obtained in Step 1) and also Nash equilibrium in a nonlinear program, the heat flows between every pair of plants on site and also their fair trade prices can be calculated accordingly.
3. By fixing the interplant heat-flow patterns determined in Step 2, the minimum total number of both inner- and interplant matches and their heat duties can be determined with an extended version of the conventional MILP model (Papoulias and Grossmann, 1983).
4. By following the approach suggested by Floudas et al. (1986) the superstructure of total-site HEN designs (in which all possible flow configurations are embedded) can be built to facilitate the matches identified in Step 3. A nonlinear programming model (Floudas et al., 1986; Floudas and Grossmann, 1987) can then be constructed for generating the optimal HEN configuration that maximizes the individual TAC saving of every plant.

4.2.1 MINIMUM ACCEPTABLE SITE-WIDE UTILITY COST

As mentioned before, the minimum acceptable total utility cost of the industrial park is determined on the basis of a modified version of the traditional transshipment model. To construct such a model, the entire temperature range is partitioned according to the inlet and outlet temperatures of *all* process streams on site (Papoulias and Grossmann, 1983). The heat flows into and out of every temperature interval in *each* plant are depicted in Figure 4.1, and the corresponding linear program (LP) can be formulated as follows:

$$\overline{ZT} = \min \sum_{p=1}^{P} \hat{Z}_p^U \tag{4.1}$$

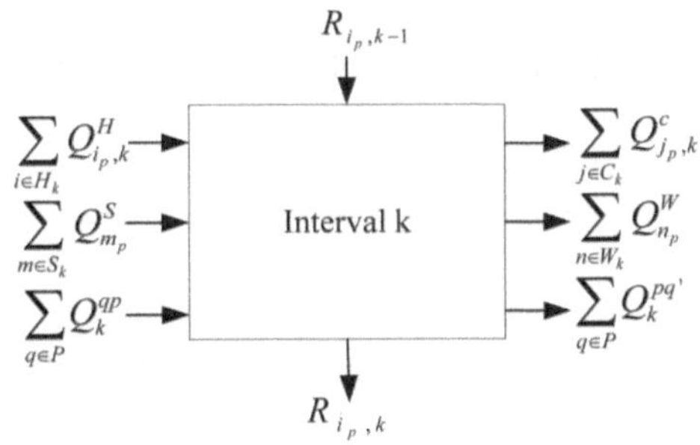

FIGURE 4.1 Direct heat flows around interval k in plant p. (Reproduced with permission from *Chem. Eng. Sci.* 2014, 118: 60–73. Copyright 2014 Elsevier.)

subject to

$$R_k^p - R_{k-1}^p + \sum_{n_p \in W_k^p} Q_{n_p}^W - \sum_{m_p \in S_k^p} Q_{m_p}^S + \sum_{\substack{q'=1 \\ q' \neq p}}^{P} Q_k^{p,q'} - \sum_{\substack{q=1 \\ q \neq p}}^{P} Q_k^{q,p} = \Delta H_k^p \tag{4.2}$$

$$\Delta H_k^p = \sum_{i_p \in H_k^p} Q_{i_p,k}^H - \sum_{j_p \in C_k^p} Q_{j_p,k}^C = \left(\sum_{i_p \in H_k^p} F_{i_p}^H - \sum_{j_p \in C_k^p} F_{j_p}^C \right) \Delta T_k \tag{4.3}$$

$$\hat{Z}_p^U = \sum_{m_p \in S^p} c_{m_p} Q_{m_p}^S + \sum_{n_p \in W^p} c_{n_p} Q_{n_p}^W \tag{4.4}$$

$$\bar{Z}_p^U - \hat{Z}_p^U \geq 0 \tag{4.5}$$

$$R_0^p = R_K^p = 0 \tag{4.6}$$

$$R_1^p, R_2^p, \cdots, R_{K-1}^p, Q_k^{q,p}, Q_k^{p,q'} \geq 0 \tag{4.7}$$

where $k\left(=1,2,\ldots,K\right)$ is the numerical label used for identifying a temperature interval, and p, q, and $q'\left(=1,2,\ldots,P\right)$ are those for the plants; $\bar{Z}_p^U$ is the lower bound of utility cost of plant p obtained via inner-plant heat integration only; $\hat{Z}_p^U$ is the minimum utility cost of plant p obtained via both inner- and interplant heat exchanges *without* treating the latter as energy trades. Note that the energy balance around the temperature interval in Figure 4.1 is described mathematically with equations (4.2) and (4.3). Finally, note that equations (4.1)–(4.4), (4.6), and (4.7) are essentially reduced to the conventional transshipment formulation if P=1 and the extra constraint, i.e., equation (4.5), is imposed primarily to ensure the individual cost saving achieved by interplant heat integration is acceptable (i.e., nonnegative).

For the illustrative example, the integrated heat-flow cascade can be obtained by solving the model described above (see Figure 2.3 in Chapter 2). The minimum consumption rates of heating utilities in plants P1–P3 were found to be 0, 220, and 444 kW respectively, while those of the cooling utilities were 485, 0, and 60 kW. The corresponding utility cost savings of these three plants can then be calculated accordingly, i.e., 61,250, 0, and 10,900 USD/yr. Notice from Figure 2.3 that, although the hot utility of plant P2 is cheaper than those in the other plants, its consumption rate cannot exceed 220 kW due to the inequality constraints in equation (4.5).

4.2.2 Feasible Interplant Heat Flows and Their Fair Trade Prices

The feasible interplant heat flows and their fair trade prices can be determined simultaneously in the second step with a nonlinear program formulated by imposing Nash equilibrium constraints in the transshipment model. This approach takes into account not only the fairness in energy trade but also heat-transfer efficiency, while only financial arrangements can be considered with other available allocation methods, e.g., Hiete et al. (2012). To facilitate clear illustration, let us consider the key components one at a time in this NLP model, i.e., the energy balances, the payoff matrices, the game strategies, the equilibrium constraints, the objective function, and the upper limits of utility cost and consumption rates.

- *Energy balances*:
  The energy balances around every temperature interval in each plant should be treated as constraints in the present model, i.e., equations (4.2), (4.3), (4.6), and (4.7).
- *Payoffs*:
  In the proposed multi-player game, every player (say plant p) can select one or more strategy from four alternatives, i.e., exporting/importing heat at a temperature above or below the pinch (denoted with U and L respectively). The structure of the payoff matrix for plant p, i.e., $\mathbf{A}_p\left(=\left[\mathbf{A}_{p,q_1} \quad \cdots \quad \mathbf{A}_{p,q_N}\right]\right)$, can be expressed as

$$\begin{bmatrix} P_F^{pU,q_1U} & P_F^{pU,q_1L} & NA & NA & & P_F^{pU,q_NU} & P_F^{pU,q_NL} & NA & NA \\ P_F^{pL,q_1U} & P_F^{pL,q_1L} & NA & NA & & P_F^{pL,q_NU} & P_F^{pL,q_NL} & NA & NA \\ NA & NA & P_F^{q_1U,pU} & P_F^{q_1L,pU} & \cdots & NA & NA & P_F^{q_NU,pU} & P_F^{q_NL,pU} \\ NA & NA & P_F^{q_1U,pL} & P_F^{q_1L,pL} & & NA & NA & P_F^{q_NU,pL} & P_F^{q_NL,pL} \end{bmatrix}$$

  where $N = P - 1$, $q_i \in \{1,2,\cdots,p-1,p+1,\ldots,P\}$, and i=1,2, …, N. Notice that, in this case, plant p and plant q_i are treated respectively as the row and column players of sub-matrix $\mathbf{A}_{p,q_i}$ and

$$\mathbf{A}_{p,q_i} = \begin{bmatrix} P_F^{pU,q_iU} & P_F^{pU,q_iL} & NA & NA \\ P_F^{pL,q_iU} & P_F^{pL,q_iL} & NA & NA \\ NA & NA & P_F^{q_iU,pU} & P_F^{q_iL,pU} \\ NA & NA & P_F^{q_iU,pL} & P_F^{q_iL,pL} \end{bmatrix} \tag{4.8}$$

Notice also that *NA* indicates that the corresponding heat exchange is forbidden and the remaining payoff values (of plant *p*) are calculated according to the following formulas:

$$P_F^{pU,q_iU} = -C_p^{HU} - C_{trd}^{pU,q_iU} \tag{4.9}$$

$$P_F^{pU,q_iL} = -C_p^{HU} - C_{trd}^{pU,q_iL} \tag{4.10}$$

$$P_F^{pL,q_iU} = +C_p^{CU} - C_{trd}^{pL,q_iU} \tag{4.11}$$

$$P_F^{pL,q_iL} = +C_p^{CU} - C_{trd}^{pL,q_iL} \tag{4.12}$$

$$P_F^{q_iU,pU} = +C_p^{HU} + C_{trd}^{q_iU,pU} \tag{4.13}$$

$$P_F^{q_iL,pU} = +C_p^{HU} + C_{trd}^{q_iL,pU} \tag{4.14}$$

$$P_F^{q_iU,pL} = -C_p^{CU} + C_{trd}^{q_iU,pL} \tag{4.15}$$

$$P_F^{q_iL,pL} = -C_p^{CU} + C_{trd}^{q_iL,pL} \tag{4.16}$$

On the left side of each equation above, the two superscripts of payoff denote the origin and destination of the corresponding heat flow respectively. On the right sides of these equations, C_p^{HU} and C_p^{CU} denote the unit costs of hot and cold utilities of plant *p*, respectively, and their values should always be positive and *a priori* given. The second right-side terms represent the unknown unit trade prices of the corresponding heat flows between plant *p* and plant q_i. As mentioned before, each interplant heat flow can be identified according to the corresponding superscripts. For examples, C_{trd}^{pL,q_iU} denotes the unit trade price for the heat flow from below the pinch in plant *p* to above the pinch in plant q_i, and $C_{trd}^{q_iU,pL}$ denotes that from above the pinch in plant q_i to below the pinch in plant *p*. To facilitate consistent model formulation, a positive cash flow is chosen in this study to coincide with the heat flow. A fee should be paid by the source of heat flow to the corresponding sink if the trade price is positive, and vice versa if a negative price is called for. If all trade prices are zero, notice that essentially half of the above heat exchanges are counterproductive. Specifically, the payoffs in equations (4.9), (4.10), (4.15), and (4.16) should all be negative without the second terms on the right sides and, thus, extra costs are incurred in these scenarios. On the other hand, the other four heat transfers should all result in positive payoffs and extra savings under the condition that no fee can be assessed to plant *p*.

Introducing a nonzero trade price obviously adjusts the payoffs of the two plants involved in interplant heat exchanges. This practice is adopted in here only for the purpose of

(a) Shifting the extra saving of one party partially to make up for a portion of the extra cost incurred to the other, and/or

(b) Redistributing the extra savings (or costs) if both benefited (or suffered) from such a heat transfer.

These purposes can be achieved by imposing the constraints below:

$$-\max\left(C_p^{HU}, C_{q_i}^{HU}\right) \le C_{trd}^{pU,q_iU} \le -\min\left(C_p^{HU}, C_{q_i}^{HU}\right) \tag{4.17}$$

$$-C_p^{HU} \le C_{trd}^{pU,q_iL} \le +C_{q_i}^{CU} \tag{4.18}$$

$$-C_{q_i}^{HU} \le C_{trd}^{pL,q_iU} \le +C_p^{CU} \tag{4.19}$$

$$+\min\left(C_p^{CU}, C_{q_i}^{CU}\right) \le C_{trd}^{pL,q_iL} \le +\max\left(C_p^{CU}, C_{q_i}^{CU}\right) \tag{4.20}$$

$$-\max\left(C_{q_i}^{HU}, C_p^{HU}\right) \le C_{trd}^{q_iU,pU} \le -\min\left(C_{q_i}^{HU}, C_p^{HU}\right) \tag{4.21}$$

$$-C_{q_i}^{HU} \le C_{trd}^{q_iU,pL} \le +C_p^{CU} \tag{4.22}$$

$$-C_p^{HU} \le C_{trd}^{q_iL,pU} \le +C_{q_i}^{CU} \tag{4.23}$$

$$+\min\left(C_{q_i}^{CU}, C_p^{CU}\right) \le C_{trd}^{q_iL,pL} \le +\max\left(C_{q_i}^{CU}, C_p^{CU}\right) \tag{4.24}$$

For brevity, let us consider only the aforementioned first and third constraints as examples. Notice that transferring a unit of heat from above the pinch in plant p to above the pinch in plant q_i inevitably results in an extra cost of C_p^{HU} for the former and a saving of $C_{q_i}^{HU}$ for the latter. On the basis of equation (4.17), plant q_i should pay plant p a fee of $-C_{trd}^{pU,q_iU}$ per unit of transferred heat so as to make the payoff of plant p in equation (4.9) less negative if $C_p^{HU} \ge C_{q_i}^{HU}$ or positive if otherwise. On the other hand, by transferring a unit of heat from below the pinch in plant p to above the pinch in plant q_i, both parties can achieve savings. In this scenario, the overall saving per unit of transferred heat is $C_p^{CU} + C_{q_i}^{HU}$ and it is redistributed on the basis of equation (4.19). It can also be observed from equation (4.11) that the allowed payoff of plant p is bounded between zero and this upper limit.

- *Strategy allocation*:
  For plant p, the weights placed upon the aforementioned four row strategies can be expressed as

$$w_p^{U,D} = \frac{1}{Q_p^E} \sum_{k \in K_p^U} \sum_{\substack{q=1 \\ q \ne p}}^{P} Q_k^{p,q} \tag{4.25}$$

$$w_p^{L,D} = \frac{1}{Q_p^E} \sum_{k \in K_p^L} \sum_{\substack{q=1 \\ q \neq p}}^{P} Q_k^{p,q} \tag{4.26}$$

$$w_p^{U,A} = \frac{1}{Q_p^E} \sum_{k \in K_p^U} \sum_{\substack{q=1 \\ q \neq p}}^{P} Q_k^{q,p} \tag{4.27}$$

$$w_p^{L,A} = \frac{1}{Q_p^E} \sum_{k \in K_p^L} \sum_{\substack{q=1 \\ q \neq p}}^{P} Q_k^{q,p} \tag{4.28}$$

$$Q_p^E = \sum_{k \in K} \sum_{\substack{q=1 \\ q \neq p}}^{P} \left(Q_k^{p,q} + Q_k^{q,p} \right) \tag{4.29}$$

where $Q_k^{p,q}$ and $Q_k^{q,p}$ respectively denote the heat flow transferred from interval k in plant p to interval k in plant q and vice versa; Q_p^E is the total amount of heat exchanged externally by plant p; K_p^U and K_p^L denote sets of intervals above and below the pinch point of plant p. It is also clear that $K_p^U \cap K_p^L = \varnothing$, $K_p^U \cup K_p^L = K$, and $w_p^{U,D} + w_p^{L,D} + w_p^{U,A} + w_p^{L,A} = 1$.

- *Nash constraints*:
  The Nash equilibrium constraints (Quintas, 1989) can be expressed as

$$\mathbf{x}_p^T \sum_{\substack{q=1 \\ q \neq p}}^{P} \mathbf{A}_{p,q} \mathbf{x}_q = \alpha_p \tag{4.30}$$

$$\sum_{\substack{q=1 \\ q \neq p}}^{P} \mathbf{A}_{p,q} \mathbf{x}_q \leq \alpha_p \mathbf{j}_p \tag{4.31}$$

$$\mathbf{x}_p^T \mathbf{j}_p = 1 \tag{4.32}$$

where $\mathbf{j}_p = \begin{bmatrix} 1 & 1 & 1 & 1 \end{bmatrix}^T$; $\mathbf{x}_p = \begin{bmatrix} w_p^{U,D} & w_p^{L,D} & w_p^{U,A} & w_p^{L,A} \end{bmatrix}^T$ is the strategy vector of plant p; $\mathbf{x}_q = \begin{bmatrix} w_q^{U,D} & w_q^{L,D} & w_q^{U,A} & w_q^{L,A} \end{bmatrix}^T$ denotes the strategy vector of plant q; $\mathbf{A}_{p,q}$ is a sub-matrix of the payoff matrix $\mathbf{A}_p$ in which the payoff values between plant p and plant q are specified; α_p denotes the average payoff value of plant p.

The above constraints are incorporated into the proposed NLP model to ensure that the heat exchanges in the interplant integration scheme are acceptable to all involved parties. Since no plant can gain any expected payoff if one or more party in the game chooses to deviate from the Nash equilibrium, the corresponding set of strategies adopted by each plant should be the best against those of the other parties.

- *Objective function*:
The objective function of the maximization problem in Step 2 is formulated as

$$\max \prod_{p=1}^{P} S_p^U \tag{4.33}$$

where $S_p^U \geq 0$ denotes the utility cost saving achieved by plant p after interplant integration with energy trades, and its value can be calculated with the following formula

$$S_p^U = \bar{Z}_p^U - \tilde{Z}_p^U + pf_p \tag{4.34}$$

where $\bar{Z}_p^U$ denotes the minimum utility costs of plant p obtained with only inner-plant heat integration; $\tilde{Z}_p^U$ denotes the total utility cost of plant p obtained via both inner- and interplant heat integration with *nonzero* trade prices; pf_p represents the revenue gained by plant p via interplant energy trades, i.e.

$$\begin{aligned} pf_p = & -\sum_{\substack{q'=1 \\ q' \neq p}}^{P} \left(C_{trd}^{pU,q'U} \sum_{k \in K_p^U \cap K_{q'}^U} Q_k^{p,q'} + C_{trd}^{pU,q'L} \sum_{k \in K_p^U \cap K_{q'}^L} Q_k^{p,q'} \right) \\ & -\sum_{\substack{q'=1 \\ q' \neq p}}^{P} \left(C_{trd}^{pL,q'U} \sum_{k \in K_p^L \cap K_{q'}^U} Q_k^{p,q'} + C_{trd}^{pL,q'L} \sum_{k \in K_p^L \cap K_{q'}^L} Q_k^{p,q'} \right) \\ & +\sum_{\substack{q=1 \\ q \neq p}}^{P} \left(C_{trd}^{qU,pU} \sum_{k \in K_p^U \cap K_q^U} Q_k^{q,p} + C_{trd}^{qU,pL} \sum_{k \in K_p^L \cap K_q^U} Q_k^{q,p} \right) \\ & +\sum_{\substack{q'=1 \\ q' \neq p}}^{P} \left(C_{trd}^{qL,pU} \sum_{k \in K_p^U \cap K_q^L} Q_k^{q,p} + C_{trd}^{qL,pL} \sum_{k \in K_p^L \cap K_q^L} Q_k^{q,p} \right) \end{aligned} \tag{4.35}$$

Notice that $\tilde{Z}_p^U$ in equation (4.34) and $\hat{Z}_p^U$ in the previous model are constrained in the same way, i.e., via equations (4.4) and (4.5), and S_p^U must be nonnegative because, if otherwise, there are really no incentives for plant p to take part in the interplant heat integration scheme.

- *Upper limits of utility cost and consumption rates*:
  To produce a reasonable pricing structure, the minimum total utility cost obtained in the previous step should be treated as an upper bound in the present step, i.e.,

$$\sum_{p=1}^{P} \tilde{Z}_p^{U} \leq \overline{ZT} \tag{4.36}$$

Finally, it is sometimes also reasonable to assume that the supply rate of each utility generated on site is bounded by a given upper limit.

By solving the above model for the illustrative example, one can obtain the following strategy vectors and the corresponding payoff matrices as follows:

$$\mathbf{x}_1 = \begin{bmatrix} 0 \\ 0 \\ 1 \\ 0 \end{bmatrix}; \mathbf{x}_2 = \begin{bmatrix} 0.415 \\ 0.361 \\ 0 \\ 0.224 \end{bmatrix}; \mathbf{x}_3 = \begin{bmatrix} 0 \\ 0 \\ 1 \\ 0 \end{bmatrix};$$

$$\mathbf{A}_1 = \begin{bmatrix} \mathbf{A}_{12} & \mathbf{A}_{13} \end{bmatrix} = \begin{bmatrix} \begin{bmatrix} -60 & 0 & NA & NA \\ 40 & 0 & NA & NA \\ NA & NA & 52.2 & 105 \\ NA & NA & 0 & 12.5 \end{bmatrix} & \begin{bmatrix} 0 & 0 & NA & NA \\ 70 & 0 & NA & NA \\ NA & NA & 0 & 51.7 \\ NA & NA & -70 & 0 \end{bmatrix} \end{bmatrix}$$

$$\mathbf{A}_2 = \begin{bmatrix} \mathbf{A}_{21} & \mathbf{A}_{23} \end{bmatrix} = \begin{bmatrix} \begin{bmatrix} 7.5 & -40 & NA & NA \\ 7.5 & 0 & NA & NA \\ NA & NA & 0 & 0 \\ NA & NA & -112.5 & -12.5 \end{bmatrix} & \begin{bmatrix} 0 & -10.7 & NA & NA \\ 82.5 & 0 & NA & NA \\ NA & NA & -30 & 0 \\ NA & NA & -82.5 & 7.5 \end{bmatrix} \end{bmatrix}$$

$$\mathbf{A}_3 = \begin{bmatrix} \mathbf{A}_{31} & \mathbf{A}_{32} \end{bmatrix} = \begin{bmatrix} \begin{bmatrix} 30 & 0 & NA & NA \\ 68.3 & 20 & NA & NA \\ NA & NA & -30 & 0 \\ NA & NA & -120 & -20 \end{bmatrix} & \begin{bmatrix} 0 & 0 & NA & NA \\ 60 & 0 & NA & NA \\ NA & NA & 30 & 0 \\ NA & NA & -49.3 & -7.5 \end{bmatrix} \end{bmatrix}$$

The resulting unit trade prices can be found in Table 4.3. The average payoffs received by the plants can be respectively determined to be 111.4, 7.5, and 68.3 USD/yr, which indicate that plant P1 benefits the most from interplant heat integration. The required utility costs of every plant before and after integration are presented in Table 4.4, and the corresponding cost savings are also listed in the same table. The total revenue received by each plant via energy trades and the resulting net saving in utility cost can be found in Table 4.5. To provide further insights into the optimal integration scheme, the utility consumption rates of each plant and the interplant heat flows are also presented in Table 4.6 and Table 4.7 respectively. Notice that these data will also be used in the next step.

TABLE 4.3
Unit Trade Prices (USD/kW·yr) for Direct Interplant Heat Flows in Illustrative Example

$C_{trd}^{1U,2U}$	−30	$C_{trd}^{2U,1U}$	−37.5	$C_{trd}^{3U,1U}$	−90
$C_{trd}^{1U,2L}$	−90	$C_{trd}^{2U,1L}$	10	$C_{trd}^{3U,1L}$	−60
$C_{trd}^{1U,3U}$	−90	$C_{trd}^{2U,3U}$	−30	$C_{trd}^{3U,2U}$	−60
$C_{trd}^{1U,3L}$	−90	$C_{trd}^{2U,3L}$	−19.3	$C_{trd}^{3U,2L}$	−60
$C_{trd}^{1L,2U}$	−30	$C_{trd}^{2L,1U}$	15	$C_{trd}^{3L,1U}$	−38.3
$C_{trd}^{1L,2L}$	10	$C_{trd}^{2L,1L}$	22.5	$C_{trd}^{3L,1L}$	10
$C_{trd}^{1L,3U}$	−60	$C_{trd}^{2L,3U}$	−60	$C_{trd}^{3L,2U}$	−30
$C_{trd}^{1L,3L}$	10	$C_{trd}^{2L,3L}$	22.5	$C_{trd}^{3L,2L}$	30

TABLE 4.4
Utility Cost Analysis of Direct Synthesis Strategy in Illustrative Example

Plants	Utilities	Costs before Integration (USD/yr)	Costs after Integration (USD/yr)	Cost Savings (USD/yr)
P1	Fuel	64,000	0	64,000
P1	CW	2100	4850	−2750
P2	Fuel	0	0	0
P2	Steam	3000	12,150	−9150
P2	CW	3600	1350	2250
P3	Fuel	10,200	10,200	0
P3	CW	20,100	0	20,100

TABLE 4.5
Utility Cost Savings Achieved with Direct Synthesis Strategy in Illustrative Example

Plant	Trade Revenue (USD/yr)	Net Saving (USD/yr)
P1	−26,781	34,469
P2	12,413	5513
P3	14,368	34,468

4.2.3 Minimum Number of Matches and Their Heat Duties

In the third step of the proposed procedure, the minimum number of exchangers and the corresponding heat duties are determined by solving a modified version of the conventional MILP model (Floudas and Ciric, 1989). This model can be formulated according to the generalized heat-flow pattern in Figure 4.2, which is in essence a

TABLE 4.6
Utility Consumption Rates Needed for Direct Interplant Heat Integration with Energy Trades in Illustrative Example

Plant	Fuel (kW)	Steam (kW)	CW (kW)
P1	0	0	485
P2	0	405	60
P3	255	0	0

TABLE 4.7
Interplant Heat Flows Facilitated with Direct Heat Exchanges with Energy Trades in Illustrative Example

Interval	P2→P1 (kW)	P3→P1 (kW)	P3→P2 (kW)
1	0	0	0
2	305	95	0
3	165	135	0
4	100	275	0
5	0	0	165

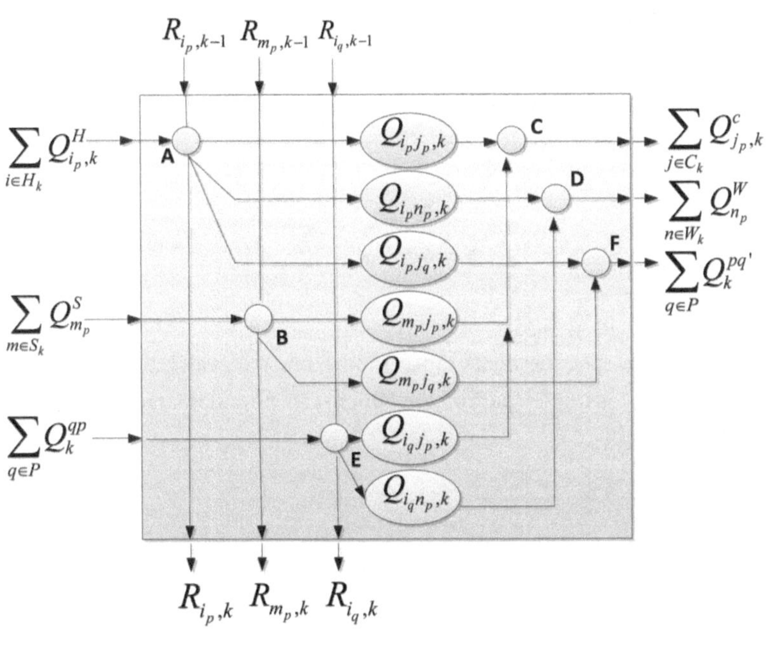

FIGURE 4.2 Generalized heat-flow pattern around and within interval k in plant p for direct interplant heat integration. (Reproduced with permission from *Chem. Eng. Sci.* 2014, 118: 60–73. Copyright 2014 Elsevier.)

zoomed-in picture of Figure 4.1. The detailed energy balances associated with the corresponding temperature interval can be established in a straightforward fashion by considering the input and output heat flows around every node in Figure 4.2. Specifically, these model constraints are listed exhaustively in the sequel:

Node A:

$$R_{i_p,k} - R_{i_p,k-1} + \sum_{j_p \in C_k^P} Q_{i_p,j_p,k} + \sum_{n_p \in W_k^P} Q_{i_p,n_p,k} + \sum_{\substack{q'=1 \\ q' \neq p}}^{P} \sum_{j_{q'} \in C_k^{q'} \cup W_k^{q'}} Q_{i_p,j_{q'},k} = Q_{i_p,k}^H \tag{4.37}$$

where $i_p \in H_k^P$. Note that the heat transported to interval k from hot stream i_p in plant p, i.e., $Q_{i_p,k}^H$, is a model parameter which can be calculated according to the given stream data.

Node B:

$$R_{m_p,k} - R_{m_p,k-1} + \sum_{j_p \in C_k^P} Q_{m_p,j_p,k} + \sum_{\substack{q'=1 \\ q' \neq p}}^{P} \sum_{j_{q'} \in C_k^{q'}} Q_{m_p,j_{q'},k} = Q_{m_p}^S \tag{4.38}$$

where $m_p \in S_k^P$. It should be noted that the consumption rate of hot utility m_p (i.e., $Q_{m_p}^S$) in interval k is a given parameter in the present model, and its value can be determined in Step 2 with a nonlinear program.

Node C:

$$\sum_{i_p \in H_k^P} Q_{i_p,j_p,k} + \sum_{m_p \in S_k^P} Q_{m_p,j_p,k} + \sum_{\substack{q=1 \\ q \neq p}}^{P} \sum_{i_q \in H_k^q \cup S_k^q} Q_{i_q,j_p,k} = Q_{j_p,k}^C \tag{4.39}$$

where $j_p \in C_k^P$. Note that the heat transported from interval k to cold stream j_p in plant p, i.e., $Q_{j_p,k}^C$, is a model parameter which can be calculated according to the given stream data.

Node D:

$$\sum_{i_p \in H_k^P} Q_{i_p,n_p,k} + \sum_{\substack{q=1 \\ q \neq p}}^{P} \sum_{i_q \in H_k^q \cup S_k^q} Q_{i_q,n_p,k} = Q_{n_p}^W \tag{4.40}$$

where $n_p \in W_k^P$. Note that the consumption rate of cold utility n_p (i.e., $Q_{n_p}^W$) in interval k is a given parameter which can be determined in Step 2.

Node E:

$$\sum_{i_q \in H_k^q} \left(\sum_{j_p \in C_k^p} Q_{i_q,j_p,k} + \sum_{n_p \in W_k^p} Q_{i_q,n_p,k} \right) + \sum_{m_q \in S_k^q} \sum_{j_p \in C_k^p} Q_{m_q,j_p,k} = Q_k^{q,p} \tag{4.41}$$

It should be emphasized that the interplant heat flow rate on the right side of this equation is a given parameter which must be determined in Step 2, and $q=1, 2, \ldots, p-1, p+1, \ldots, P$.

Node F:

$$\sum_{j_{q'} \in C_k^{q'}} \left(\sum_{i_p \in H_k^p} Q_{i_p,j_{q'},k} + \sum_{m_p \in S_k^p} Q_{m_p,j_{q'},k} \right) + \sum_{n_{q'} \in W_k^{q'}} \sum_{i_p \in H_k^p} Q_{i_p,n_{q'},k} = Q_k^{p,q'} \tag{4.42}$$

where $q' = 1,2,\ldots,p-1,p+1,\ldots,P$ and the interplant heat flow on the right side, i.e., $Q_k^{p,q'}$, is a given parameter which must be determined in Step 2.

Let us next define a set of binary variables:

$$z_{i_p,j_q} = \begin{cases} 1, \text{ if there is heat exchange between } i_p \text{ and } j_q \\ 0, \text{ otherwise} \end{cases} \tag{4.43}$$

where $i_p \in H_k^p \cup S_k^p$, $j_q \in C_k^q \cup W_k^q$, and $p,q = 1,2,\ldots,P$. The following inequality constraints can then be imposed accordingly:

$$\sum_{k \in K} Q_{i_p,j_q,k} - z_{i_p,j_q} \Gamma_{i_p,j_q} \le 0 \tag{4.44}$$

where Γ_{i_p,j_q} is the maximum heat exchange rate between hot stream i_p in plant p and cold stream j_q in plant q, and $Q_{i_p,j_q,k}$ is the rate of this heat exchange within interval k.

The objective function of the proposed MILP model can be expressed as:

$$\begin{aligned} \overline{NU} = \min \Bigg(& \sum_{p=1}^{P} \sum_{i_p \in H^p} \sum_{j_p \in C^p} z_{i_p,j_p} + \sum_{p=1}^{P} \sum_{i_p \in S^p} \sum_{j_p \in C^p} z_{i_p,j_p} + \sum_{p=1}^{P} \sum_{i_p \in H^p} \sum_{j_p \in W^p} z_{i_p,j_p} \\ & + \sum_{p=1}^{P} \sum_{\substack{q=1 \\ q \ne p}}^{P} \sum_{i_p \in H^p} \sum_{j_q \in C^q} z_{i_p,j_q} + \sum_{p=1}^{P} \sum_{\substack{q=1 \\ q \ne p}}^{P} \sum_{i_p \in S^p} \sum_{j_q \in C^q} z_{i_p,j_q} + \sum_{p=1}^{P} \sum_{\substack{q=1 \\ q \ne p}}^{P} \sum_{i_p \in H^p} \sum_{j_q \in W^q} z_{i_p,j_q} \Bigg) \end{aligned} \tag{4.45}$$

After constructing the MILP model for Example 4.1 and carrying out the corresponding optimization run, the minimum unit number can be found to be 14 and the optimal matches are shown in Table 4.8.

TABLE 4.8
Optimal Inner- and Interplant Matches Obtained with Direct Synthesis Strategy in Illustrative Example

Match #	Hot Stream	Cold Stream	Heat Duty (kW)
1	P1_H1	P1_C1	285
2	P1_H1	P1_CW	485
3	P2_H1	P1_C1	160
4	P2_H1	P1_C2	380
5	P2_H1	P2_C1	175
6	P3_H1	P3_C1	660
7	P3_H2	P1_C1	275
8	P3_H2	P1_C2	230
9	P3_H2	P2_C1	105
10	P3_H2	P3_C1	210
11	P3_H2	P2_CW	60
12	P2_HP	P1_C2	30
13	P2_HP	P2_C2	375
14	P2_Fuel	P3_C1	255

4.2.4 Optimal Network Configuration

Since only the matches and their heat duties are fixed in Step 3, the network configuration and specifications of each embedded exchanger must be obtained for calculating the total investment cost of a HEN design. This final design has been traditionally generated with a superstructure-based NLP model (Floudas et al., 1986). Essentially the same synthesis strategy can be taken to build the model constraints, while a different objective function is adopted to facilitate reasonable allocation of TAC savings. Specifically, this last step tries to maximize the product of overall savings of all plants, i.e.,

$$\max \prod_{p=1}^{P} S_p^T \tag{4.46}$$

and each individual TAC saving can be determined according to the following formula:

$$S_p^T = S_p^U + Af\left(\bar{Z}_p^C - \tilde{Z}_p^C - \sum_{\substack{q'=1 \\ q'\neq p}}^{P} SC_{p,q'}^P - \sum_{\substack{q=1 \\ q\neq p}}^{P} SC_{q,p}^P \right) \tag{4.47}$$

where S_p^U denotes the total utility cost saving achieved by plant p (which can be calculated in Step 2); Af is a constant annualization factor; $\bar{Z}_p^C$ is the minimum total capital cost of HEN in plant p *without* interplant heat integration (which must be determined independently with a traditional sequential design procedure); $\tilde{Z}_p^C$ is the total capital cost for all inner-plant matches in plant p (which can be determined in Step 3); $SC_{p,q'}^p$ is the total capital cost shared by plant p to facilitate interplant heat exchanges between the hot streams in plant p and the cold streams in plant q'; $SC_{q,p}^p$ is the total capital cost shared by plant p to facilitate interplant heat exchanges between the hot streams in plant q and the cold streams in plant p. Notice that only the last three costs are adjustable in the present model and they can be evaluated with the following formulas:

$$\tilde{Z}_p^C = \sum_{i_p \in H^p \cup S^p} \sum_{j_p \in C^p \cup W^p} z_{i_p,j_p} c_{i_p,j_p} \left\{ \frac{Q_{i_p,j_p}}{U_{i_p,j_p} \left[\theta^1_{i_p,j_p} \theta^2_{i_p,j_p} \left(\theta^1_{i_p,j_p} + \theta^2_{i_p,j_p} \right) / 2 \right]^{\frac{1}{3}}} \right\}^{\beta} \tag{4.48}$$

$$SC_{p,q'} = \sum_{i_p \in H^p \cup S^p} \sum_{j_{q'} \in C^{q'} \cup W^{q'}} z_{i_p,j_{q'}} \gamma^p_{i_p,j_{q'}} c_{i_p,j_{q'}} \left\{ \frac{Q_{i_p,j_{q'}}}{U_{i_p,j_{q'}} \left[\theta^1_{i_p,j_{q'}} \theta^2_{i_p,j_{q'}} \left(\theta^1_{i_p,j_{q'}} + \theta^2_{i_p,j_{q'}} \right) / 2 \right]^{\frac{1}{3}}} \right\}^{\beta} \tag{4.49}$$

$$SC_{q,p} = \sum_{i_q \in H^q \cup S^q} \sum_{j_p \in C^p \cup W^p} z_{i_q,j_p} \gamma^p_{i_q,j_p} c_{i_q,j_p} \left\{ \frac{Q_{i_q,j_p}}{U_{i_q,j_p} \left[\theta^1_{i_q,j_p} \theta^2_{i_q,j_p} \left(\theta^1_{i_q,j_p} + \theta^2_{i_q,j_p} \right) / 2 \right]^{\frac{1}{3}}} \right\}^{\beta} \tag{4.50}$$

Each term in the above cost model is associated with a distinct match, which is denoted in the sequel as (i, j) for illustration conciseness; $Q_{i,j}$, $c_{i,j}$, $U_{i,j}$, $\theta^1_{i,j}$, and $\theta^2_{i,j}$ respectively denote the heat duty, cost coefficient, the overall heat transfer coefficient, and the temperature differences at the hot and cold ends of the corresponding heat exchanger; $\gamma^p_{i,j}$ represents the proportion of capital cost that is paid by plant p for interplant match (i, j). Obviously, the capital cost proportions must be constrained in the NLP model according to the following equations:

$$\gamma^p_{i_p,j_{q'}} + \gamma^{q'}_{i_p,j_{q'}} = 1 \tag{4.51}$$

$$\gamma^q_{i_q,j_p} + \gamma^p_{i_q,j_p} = 1 \tag{4.52}$$

A superstructure has been constructed according to Table 4.8 for each process stream in Example 4.1, and the material and energy balances in this structure were then formulated as equality constraints accordingly. The model parameters chosen for Example 4.1 are as follows: Af=0.1349, β=0.83, $c_{i,j} = 670\ \text{USD/m}^{1.66}$, $\Delta T_{min} = 5°C$, and $U_{i,j} = 1\ \text{W/m}^2\text{K}$. From Table 4.3, it can be observed that the net utility cost savings achieved by P1, P2, and P3 are 34,469 USD/yr (S_1^U), 5513 USD/yr (S_2^U), and 34,468 USD/yr (S_3^U), respectively. By applying the traditional sequential design strategy to generate the HENs for the three plants individually, the optimal annualized capital costs (i.e., $Af \times \overline{Z}_p^C$ and $p=1,2,3$) were found to be 5891 USD/yr, 5861 USD/yr, and 7865 USD/yr, respectively. The resulting optimal interplant HEN design is presented in Figure 4.3. The capital costs of all interplant exchangers in this design can be found in Table 4.9. Note that, in this work, the capital cost of every interplant unit is shared by the two parties involved in the corresponding heat exchange. The optimal proportions of their payments are shown in Table 4.10. Based on the above data, the individual TAC savings of the three plants can be determined to be 30,099, 8039, and 30,099 USD/yr, respectively. A more detailed economic analysis can also be found in Table 4.11. It can be observed that, although the interplant heat integration scheme results in an increase in the capital cost, the reduction in the utility cost is more than enough to justify the extra investment. The proposed optimization procedure also ensures fair distribution of financial benefits among all participating members. Finally, it should be noted that the additional energy saving achieved with interplant integration also implies that the corresponding CO_2 emission rate is much less.

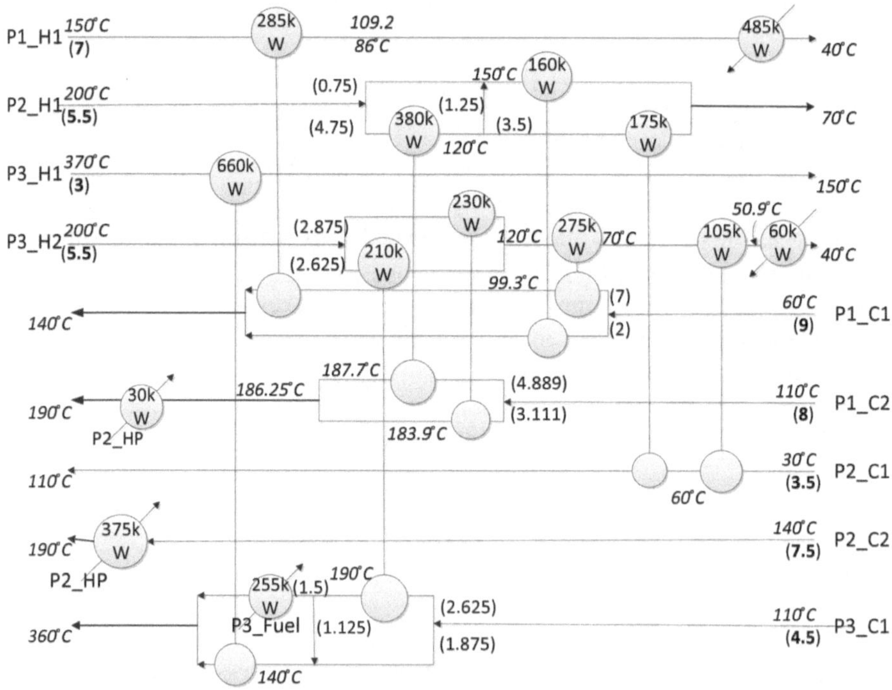

FIGURE 4.3 Optimal multi-plant HEN design obtained with direct synthesis strategy in illustrative example. (Reproduced with permission from *Chem. Eng. Sci.* 2014, 118: 60–73. Copyright 2014 Elsevier.)

TABLE 4.9
Capital Costs of Interplant Heat Exchangers in Optimal HEN Design of Illustrative Example

Hot Stream	Cold Stream	Area (m^2)	Capital Cost (USD)
P2_H1	P1_C1	16	13,291
P2_H1	P1_C2	34.25	19,185
P2_HP	P1_C2	1.375	7473
P3_H2	P1_C1	18.698	14,215
P3_H2	P1_C2	17.970	13,968
P3_H2	P2_C1	7.102	10,010
P3_H2	P2_CW	3.372	8438

TABLE 4.10
Optimal Allocation of Capital Costs of Direct Interplant Heat Exchangers in Illustrative Example

Hot Stream	Cold Stream	Plant 1	Plant 2	Plant 3
P2_H1	P1_C1	1	0	–
P2_H1	P1_C2	1	0	–
P2_HP	P1_C2	1	0	–
P3_H2	P1_C1	0.411	–	0.589
P3_H2	P1_C2	0.054	–	0.946
P3_H2	P2_C1	–	0	1
P3_H2	P2_CW	–	0	1

TABLE 4.11
Comparison of the Utility Cost Savings, Capital Cost Savings, and TAC Savings Achieved by Direct Synthesis Strategy in Illustrative Example

Plant	Utility Cost Saving (USD/yr)	Capital Cost Saving (USD/yr)	TAC Saving (USD/yr)
P1	34,469	–4370	30,099
P2	5513	2526	8039
P3	34,468	–4370	30,099

4.2.5 An Additional Example of the Direct Interplant HEN Synthesis Strategy

The vinyl chloride monomer (VCM) is traditionally produced with ethylene, chlorine, and oxygen. In this example, let us assume that three different firms are interested in a joint venture to build manufacturing facilities and produce VCM on an industrial park. Due to the unique technical expertise and financial constraint of each company, this process is divided into three separate plants (see Figure 4.4) for the participating parties to invest and run independently. The process data used in this example (see Table 4.12) were taken from Lakshmanan et al. (1999). Since the utilities are assumed to be provided by the same systems on site, their prices are identical for all plants (see Table 4.13).

To facilitate realistic cost estimation, the following models have been utilized to replace equations (4.48)–(4.50):

$$\tilde{Z}_p^C = \frac{M\&S}{280} \left[101.3\left(2.29 + F_c\right)\left(\sum_{i_p \in H^p \cup S_H^p} \sum_{j_p \in C^p \cup W^p} z_{i_p,j_p} A_{i_p,j_p}^{0.65} \right) + 5070\left(1.23 + F_c^{fuel}\right)\left(\sum_{i_p \in S_{fuel}^p} \sum_{j_p \in C^p} z_{i_p,j_p} Q_{i_p,j_p}^{0.85} \right) \right] \tag{4.53}$$

$$SC_{p,q'} = \frac{M\&S}{280} \left[101.3\left(2.29 + F_c\right)\left(\sum_{i_p \in H^p \cup S_H^p} \sum_{j_{q'} \in C^{q'} \cup W^{q'}} z_{i_p,j_{q'}} \gamma_{i_p,j_{q'}}^p A_{i_p,j_{q'}}^{0.65} \right) + 5070\left(1.23 + F_c^{fuel}\right)\left(\sum_{i_p \in S_{fuel}^p} \sum_{j_{q'} \in C^{q'}} z_{i_p,j_{q'}} \gamma_{i_p,j_{q'}}^p Q_{i_p,j_{q'}}^{0.85} \right) \right] \tag{4.54}$$

$$SC_{q,p} = \frac{M\&S}{280} \left[101.3\left(2.29 + F_c\right)\left(\sum_{i_q \in H^q \cup S_H^q} \sum_{j_p \in C^p \cup W^p} z_{i_q,j_p} \gamma_{i_q,j_p}^p A_{i_q,j_p}^{0.65} \right) + 5070\left(1.23 + F_c^{fuel}\right)\left(\sum_{i_q \in S_{fuel}^q} \sum_{j_p \in C^p} z_{i_q,j_p} \gamma_{i_q,j_p}^p Q_{i_q,j_p}^{0.85} \right) \right] \tag{4.55}$$

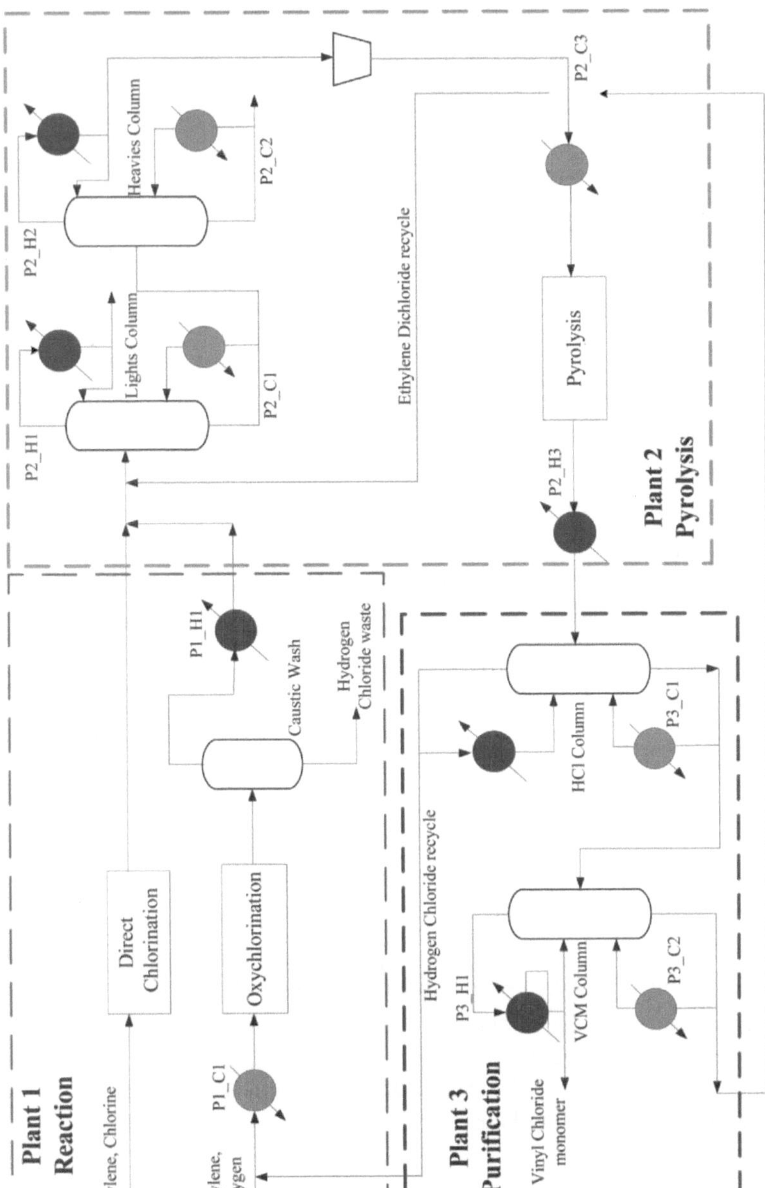

FIGURE 4.4 The VCM process. (Reproduced with permission from *Chem. Eng. Sci.* 2014, 118: 60–73. Copyright 2014 Elsevier.)

TABLE 4.12
Process Data Used in VCM Example

Plant	Stream	T_{in} (°C)	T_{out} (°C)	F_{cp} (kW/°C)
P1	H1	192.6	57.3	1.37
P1	C1	25.0	193.1	1.19
P2	H1	120.9	119.9	608.14
P2	H2	147.0	146.0	1600.24
P2	H3	499.6	57.4	6.46
P2	C1	207.9	208.9	12.32
P2	C2	158.5	159.5	2560.38
P2	C3	199.8	498.8	0.06
P3	H1	36.1	35.1	667.81
P3	C1	86.6	87.6	2699.9
P3	C2	157.3	158.3	1392.21

TABLE 4.13
Utility Data Used in VCM Example

Plant	Utility	Temperature (°C)	Unit Cost (USD/kW·yr)	Maximum Level (kW)
P1	Steam (240 psig)	200	150	5000
P1	Cooling water	20	60	1000
P2	Steam (240 psig)	200	150	5000
P2	Fuel oil	600	130	5000
P2	Cooling water	20	60	1000
P3	Steam (240 psig)	200	150	5000
P3	Cooling water	20	60	1000

In the above models, S_H^p denotes the set of regular hot utilities in plant p and S_{fuel}^p denotes the set of fuels in plant p. Note that $S_H^p \cap S_{fuel}^p = \varnothing$ and $S_H^p \cup S_{fuel}^p = S^p$. Note also that the Marshall and Swift (M&S) index is chosen to be 914 (at 1989), because this year's utility costs are also adopted in the present example. Finally, F_c and F_c^{fuel} respectively denote the correction factors for heat exchangers and furnaces. Notice that the following formulas for computing the heat-transfer areas of two types of heat exchangers, i.e., the floating-head and kettle reboilers, have been adopted in the VCM example:

- Floating head (F_c=1):

$$A_{i,j} = \frac{Q_{i,j}}{U_{i,j}\left[\theta_{i,j}^1\theta_{i,j}^2\left(\theta_{i,j}^1+\theta_{i,j}^2\right)/2\right]^{\frac{1}{3}}} \tag{4.56}$$

- Kettle (F_c=1.36):

$$A_{i,j} = \frac{Q_{i,j}}{U_{i,j}\left[2\left(\theta_{i,j}^{1}\theta_{i,j}^{2}\right)^{1/2}/3+\left(\theta_{i,j}^{1}+\theta_{i,j}^{2}\right)/6\right]} \tag{4.57}$$

The correction factor F_c^{fuel} and all overall heat-transfer coefficients are chosen to be 1 in the present example for the sake of simplicity, while the inlet and outlet temperature difference of the cold utility stream of every cooler is set to be 5°C.

Case 1

In this first case, we would like to find out whether it is necessary to involve all three companies in the interplant heat integration scheme. After implementing the aforementioned sequential optimization strategy, the network structure in Figure 4.5 can be obtained. The required utility consumption rates before and after interplant heat integration are summarized in Table 4.14, while a comparison of utility cost savings, capital cost savings, and TAC savings can be found in Table 4.15. Since the utility cost saving of plant P1 is nil and the capital cost saving is only 107 USD/yr, the

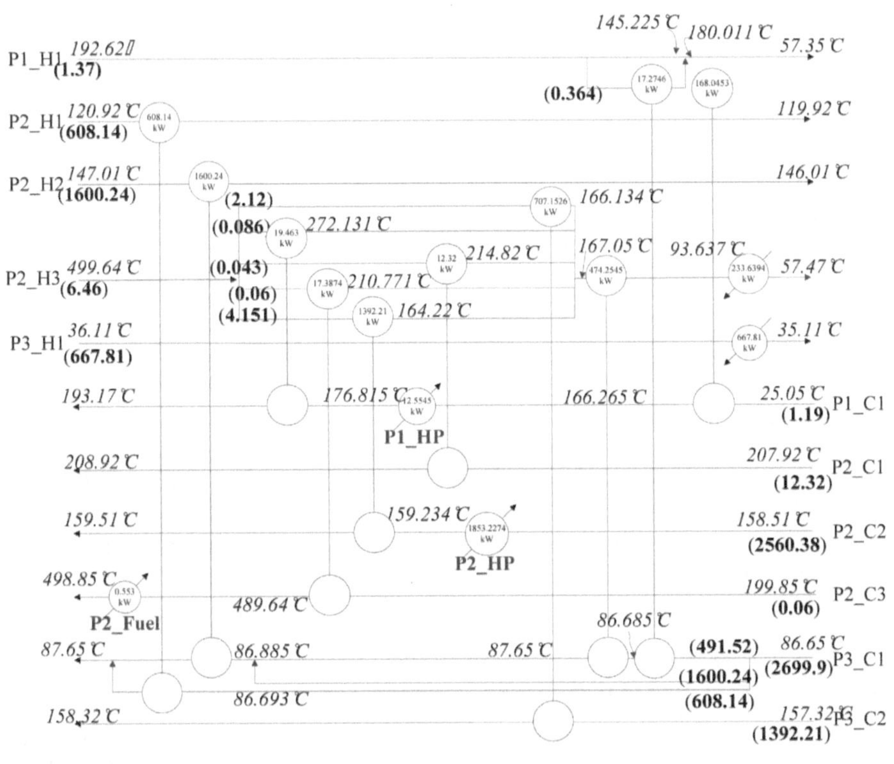

FIGURE 4.5 Optimal interplant HEN design in Case 1 of VCM example. (Reproduced with permission from *Chem. Eng. Sci.* 2014, 118: 60–73. Copyright 2014 Elsevier.)

TABLE 4.14
Utility Consumption Rates before and after Interplant Heat Integration in Case 1 of VCM Example

	Before Integration (kW)			After Integration (kW)		
Plant	**HP**	**Fuel**	**CW**	**HP**	**Fuel**	**CW**
P1	14.7	–	0	12.6	–	0
P2	451.0	0.553	2925.7	185.2	0.553	233.6
P3	4092.1	–	667.8	0	–	667.8

TABLE 4.15
A Comparison of Utility Cost Savings, Capital Cost Savings, and TAC Savings Achieved in Case 1 of VCM Example

Plant	Total Utility Saving (USD/yr)	Capital Cost Saving (USD/yr)	TAC Saving (USD/yr)
P1	0.002	107	107
P2	282,932	–1480	281,452
P3	282,400	1094	283,494

decision-maker of plant P1 may feel that the incentives are not enough. Therefore, it is necessary to evaluate the feasibility of interplant heat integration between P2 and P3 only.

Case 2

Before interplant heat integration, the minimum consumption rate of fuel in plant P2 is 0.553 kW and those of hot utilities in plant P2 and P3 are 450.988 kW and 4092.110 kW, while the cold utility consumption rates of P2 and P3 are 2925.698 kW and 667.81 kW. On the basis of these results, the interplant HEN structure in Figure 4.6 can be synthesized according to the proposed procedure. The corresponding economic analyses for these two plants are presented in Table 4.16 and Table 4.17, and their TACs can be found to be 280,621 USD/yr and 280,622 USD/yr. By comparing these values with those achieved in the aforementioned three-plant integration scheme, it can be concluded that the TACs of plants P2 and P3 are virtually unchanged and, therefore, the participation of plant P1 is in fact unnecessary.

4.3 INDIRECT INTEGRATION

Basically the four-step procedure described previously in Section 4.2 is also followed in the sequel to produce the indirect heat integration schemes. Two types of auxiliary streams, i.e., utilities and intermediate fluids, can be adopted to facilitate interplant

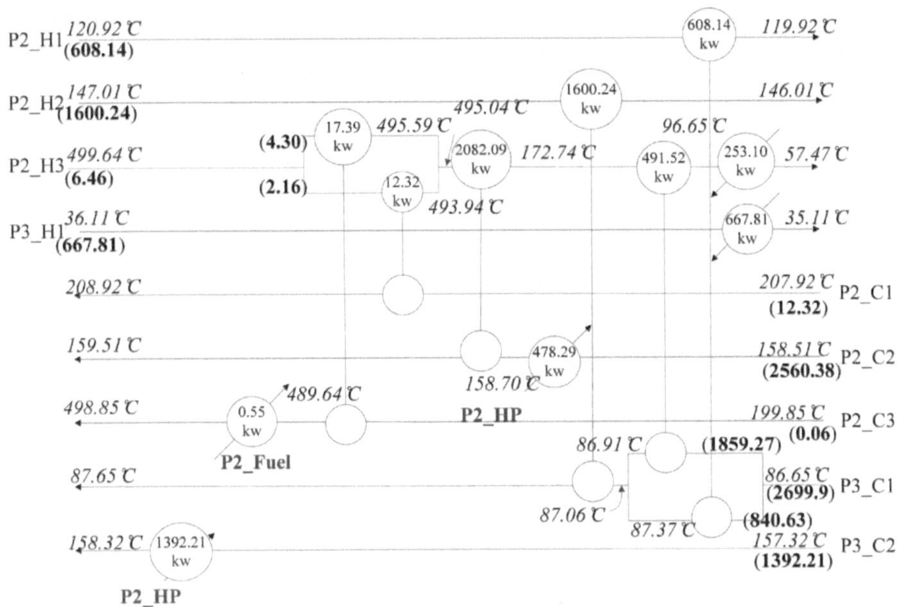

FIGURE 4.6 Optimal interplant HEN design in Case 2 of VCM example. (Reproduced with permission from *Chem. Eng. Sci.* 2014, 118: 60–73. Copyright 2014 Elsevier.)

TABLE 4.16
Utility Consumption Rates before and after Interplant Heat Integration in Case 2 of VCM Example

	Before Integration (kW)			After Integration (kW)		
Plant	**HP**	**Fuel**	**CW**	**HP**	**Fuel**	**CW**
P2	451.0	0.553	2925.7	1870.5	0.553	253.1
P3	4092.1	–	667.8	0	–	667.8

TABLE 4.17
A Comparison of Utility Cost Savings, Capital Cost Savings, and TAC Savings Achieved in Case 2 of VCM Example

Plant	Total Utility Saving (USD/yr)	Capital Cost Saving (USD/yr)	TAC Saving (USD/yr)
P2	280,623	−2	280,621
P3	280,623	−1	280,622

heat flows and the corresponding indirect HEN synthesis methods are referred to as strategies I and II, respectively. The needed mathematical programming models are presented below in two separate sections.

4.4 USING UTILITIES AS AUXILIARY STREAMS

This section details the four steps of indirect heat integration strategy by using utilities as the auxiliary streams.

4.4.1 Step 1 of Indirect Strategy I

This step is used for determining the minimum acceptable total utility cost of the entire industrial park with a linear program (LP) formulated according to the given process data. If the interplant heat flows are facilitated with the heating and cooling utilities only, the transshipment model described in Section 4.2.1 must be further constrained to forbid any interplant heat exchange between hot and cold process streams. To facilitate model construction, the entire temperature range should be first partitioned into K intervals according to the initial and target temperatures of all hot and cold streams. Starting from the high temperature end, these intervals are labeled sequentially as k=1,2, ..., K. Let us assume that a total of P plants take part in the interplant heat integration project and they are labeled as p=1,2, ..., P. Figure 4.7 shows the interior and exterior heat-flow patterns of interval k in plant p ($p \neq q$ and $p \neq q'$), and the modified transshipment model can be formulated accordingly as follows:

$$\min \sum_{p=1}^{P} Z'_p \tag{4.58}$$

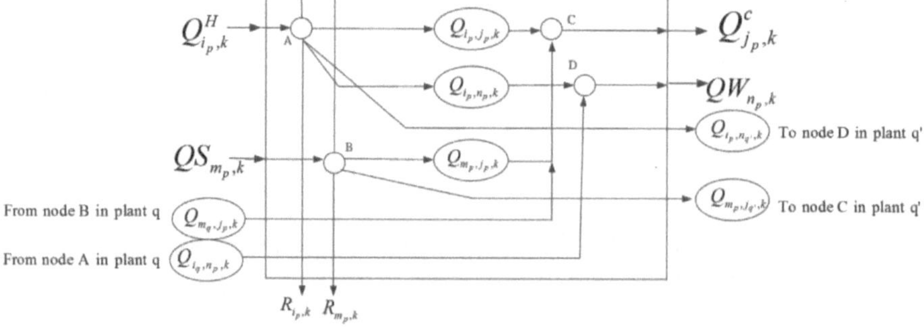

FIGURE 4.7 Interior and exterior heat-flow patterns of interval k in plant p with utility-facilitated interplant heat exchanges. (Reproduced with permission from *Energy* 2018, 148: 90–111. Copyright 2018 Elsevier.)

s.t.

$$R_{i_p,k} - R_{i_p,k-1} + \sum_{j_p \in C_k^p} Q_{i_p,j_p,k} + \sum_{n_p \in W_k^p} Q_{i_p,n_p,k} + \sum_{\substack{q'=1 \\ q' \neq p}}^{P} \sum_{n_{q'} \in W_k^{q'}} Q_{i_p,n_{q'},k} = Q_{i_p,k}^H$$

$$i_p \in \widetilde{H}_k^P \tag{4.59}$$

$$R_{m_p,k} - R_{m_p,k-1} + \sum_{j_p \in C_k^p} Q_{m_p,j_p,k} + \sum_{\substack{q'=1 \\ q' \neq p}}^{P} \sum_{j_{q'} \in C_k^{q'}} Q_{m_p,j_{q'},k} = QS_{m_p,k}, \quad m_p \in \tilde{S}_k^p \tag{4.60}$$

$$\sum_{i_p \in \tilde{H}_k^p} Q_{i_p,j_p,k} + \sum_{m_p \in \tilde{S}_k^p} Q_{m_p,j_p,k} + \sum_{\substack{q=1 \\ q \neq p}}^{P} \sum_{m_q \in \tilde{S}_k^q} Q_{m_q,j_p,k} = Q_{j_p,k}^C, j_p \in C_k^p \tag{4.61}$$

$$\sum_{i_p \in \tilde{H}_k^p} Q_{i_p,n_p,k} + \sum_{\substack{q=1 \\ q \neq p}}^{P} \sum_{i_q \in \tilde{H}_k^q} Q_{i_q,n_p,k} = QW_{n_p,k}, n_p \in W_k^p \tag{4.62}$$

$$QW_{n_p} = \sum_{k=1}^{K} QW_{n_p,k}, \quad n_p \in W^p \tag{4.63}$$

$$QS_{m_p} = \sum_{k=1}^{K} QS_{m_p,k}, \quad m_p \in S^p \tag{4.64}$$

$$Z_p' = \sum_{m_p \in S^p} c_{m_p} QS_{m_p} + \sum_{n_p \in W^p} c_{n_p} QW_{n_p} \tag{4.65}$$

$$R_{i_p,0} = R_{m_p,0} = R_{i_p,K} = R_{m_p,K} = 0 \tag{4.66}$$

where $\tilde{H}_k^p$ denotes the set of hot process streams in or above interval k of plant p; $\tilde{S}_k^p$ denotes the set of hot utility streams in or above interval k of plant p; $Q_{i_p,k}^H$ is a model parameter which is used to represent the heat released by hot stream i_p in temperature interval k; $Q_{j_p,k}^C$ is another model parameter which is used to represent the heat absorbed by cold stream j_p in temperature interval k; c_{m_p} and c_{n_p} are model parameters used to denote the unit costs of hot utility m_p and cold utility n_p, respectively, in plant p.

4.4.2 Step 2 of Indirect Strategy I

This step is followed to build a nonlinear program (NLP) by incorporating the constraints of minimum acceptable overall utility cost obtained in Step 1 and also the Nash equilibrium, and then to solve this NLP for identifying the heat flows between every pair of plants on site and their fair trade prices.

Specifically, the proposed NLP formulation is outlined below:

- ***Payoff matrices and strategy vectors***

  The payoff matrices (denoted as **A***p* and $p = 1,2, \ldots, P$) in the present applications are essentially the same as those used in Section 4.1.1, while the strategy vectors are more constrained. For illustration clarity, the general structure of the payoff matrices is repeated below:

$$\mathbf{A}_p = \left[\mathbf{A}_{p,q_1} \middle| \mathbf{A}_{p,q_2} \cdots \middle| \mathbf{A}_{q_N} \right] \tag{4.67}$$

where $N = P - 1$; $q_i \in \{1,2,\cdots,p-1,p+1,\cdots,P\}$ and $i = 1,2, \ldots, N$. Note that each submatrix of $\mathbf{A}_p$ is exactly the same as that in equation (4.8). Note again that each element in this submatrix is used to represent the payoff received by plant p per unit heat transferred from plant p to plant q_i or vice versa, while the direction of every feasible heat flow and its source and sink temperatures in relation to their respective pinch points are denoted in the superscript. For example, P_F^{pU,q_iL} represents the payoff for transferring a unit of heat from above the pinch in plant p to below the pinch in plant q_i. Note also that the entry *NA* means that the corresponding heat transfer is not allowed since in this case both plants are chosen to be the sources (or sinks). For all these heat flows, the payoffs can be computed according to the formulas given in equations (4.9)–(4.16). Note that the trade prices are treated as variables in this work. A positive price reveals that cash and heat flows are moving in the same direction, while a negative one denotes otherwise. Since any interplant heat flow simultaneously alters the utility consumption rates at the source and sink ends, the trade price of this exchange must be bounded according to equations (4.17)–(4.24). The more detailed analysis of these inequality constraints can be found in Section 4.1.1.

The game strategies are concerned only with the utility-facilitated interplant heat exchanges and the corresponding heat transfer rates can be calculated as follows:

$$QHU_k^{q,p} = \sum_{m_q \in \tilde{S}_k^q} \sum_{j_p \in C_k^p} Q_{m_q,j_p,k} \tag{4.68}$$

$$QCU_k^{q,p} = \sum_{i_q \in \tilde{H}_k^q} \sum_{n_p \in W_k^p} Q_{i_q,n_p,k} \tag{4.69}$$

$$QHU_k^{p,q} = \sum_{m_p \in \tilde{S}_k^p} \sum_{j_q \in C_k^q} Q_{m_p,j_q,k} \tag{4.70}$$

$$QCU_k^{p,q} = \sum_{i_p \in \tilde{H}_k^p} \sum_{n_q \in W_k^q} Q_{i_p,n_q,k} \tag{4.71}$$

where $QHU_k^{q,p}$ denotes the total heat-exchange rate between heating utilities in plant q and cold streams in interval k of plant p; $QCU_k^{q,p}$ denotes the total heat-exchange rate between hot streams in interval k of plant q and cold utilities in plant p; $QHU_k^{p,q}$ denotes the total heat-exchange rate between heating utilities in plant p and cold streams in interval k of plant q; $QCU_k^{p,q}$ denotes the total heat-exchange rate between hot streams in interval k of plant p and cold utilities in plant q. In order to keep the total utility cost at the minimum level determined in Step 1, these utility-facilitated interplant heat flows should remain unchanged in the present step. As a result, it is only necessary to determine their trade prices.

Four different types of heat exchanges may be selected by plant p and each can be uniquely characterized on the basis of pinch location and the corresponding interplant heat-flow direction. A game strategy is taken as the ratio between the total amount of a particular type of heat exchanges and that of all possible heat flows in and out of plant p. Specifically, equations (4.25)–(4.28) should now be rewritten as

$$\hat{w}_p^{U,D} = \frac{1}{\hat{Q}_p^E} \sum_{k \in K_p^U} \sum_{\substack{q=1 \\ q \neq p}}^{P} \left(QCU_k^{p,q} + QHU_k^{p,q} \right) \tag{4.72}$$

$$\hat{w}_p^{L,D} = \frac{1}{\hat{Q}_p^E} \sum_{k \in K_p^L} \sum_{\substack{q=1 \\ q \neq p}}^{P} \left(QCU_k^{p,q} + QHU_k^{p,q} \right) \tag{4.73}$$

$$\hat{w}_p^{U,A} = \frac{1}{\hat{Q}_p^E} \sum_{k \in K_p^U} \sum_{\substack{q=1 \\ q \neq p}}^{P} \left(QCU_k^{q,p} + QHU_k^{q,p} \right) \tag{4.74}$$

$$\hat{w}_p^{L,A} = \frac{1}{\hat{Q}_p^E} \sum_{k \in K_p^L} \sum_{\substack{q=1 \\ q \neq p}}^{P} \left(QCU_k^{q,p} + QHU_k^{q,p} \right) \tag{4.75}$$

where K_p^U and K_p^L denote the sets of temperature intervals above and below the pinch in plant p, respectively, and $K_p^U \cap K_p^L = \varnothing$. Instead of equation (4.29), the total volume of energy traffic in and out of plant p should be expressed as

$$\hat{Q}_p^E = \sum_{k=1}^{K} \sum_{\substack{q=1 \\ q \neq p}}^{P} \left(QHU_k^{p,q} + QCU_k^{p,q} + QHU_k^{q,p} + QCU_k^{q,p} \right) \tag{4.76}$$

Thus, the strategy vector of plant p can be constructed according to equations (4.72)–(4.75) as

$$\mathbf{x}_p = \left[\hat{w}_p^{U,D} \quad \hat{w}_p^{L,D} \quad \hat{w}_p^{U,A} \quad \hat{w}_p^{L,A} \right]^T \tag{4.77}$$

- ***Trade prices under Nash equilibrium constraints***
  As mentioned before, the multi-player Nash equilibrium constraints were formulated explicitly by Quintas (1989) and they have already been presented previously in equations (4.30)–(4.32). Note that $\mathbf{x}_p$ and $\mathbf{x}_q$ in the present application should be the strategy vectors of plant p and plant q determined according to equations (4.72)–(4.77).
- ***Objective function***
  The following objective function is maximized in an NLP model for setting the proper trade prices in equations (4.17)–(4.24):

$$\max \prod_{p=1}^{P} \left(S_p^U \right)^{\varpi_p} \tag{4.78}$$

where ϖ_p and S_p^U respectively denote the negotiation power and the total saving of utility cost of plant p. They can be computed as follows:

$$\varpi_p = Z'_p / \bar{Z}_p \tag{4.79}$$

$$S_p^U = \bar{Z}_p - Z'_p + pf_p \tag{4.80}$$

where $\bar{Z}_p$ is the minimum total utility cost of a standalone HEN in plant p; Z'_p denotes the minimum total utility cost of plant p achieved with interplant heat integration; pf_p is the total revenue received by plant p via energy trades. The minimum total utility cost of plant p in an interplant heat integration scheme (Z'_p) can be determined by solving equations (4.58)–(4.66), while in a standalone HEN this cost ($\bar{Z}_p$) can be calculated with the conventional transshipment model (Papoulias and Grossmann, 1983). Note that the negotiation power of plant p is clearly weakened/strengthened by lowering/raising ϖ_p. In other words, if Step 1 results in significant differences in the utility cost savings, they are moderated in the present step by stipulating proper energy trade prices to maximize the objective function in equation (4.78). Finally, the total trade revenue of plant p can be computed according to the following formula:

$$
\begin{aligned}
pf_p = &-\sum_{\substack{q'=1\\q'\neq p}}^{P}\left(C_{trd}^{pU,q'U}\sum_{k\in K_p^U\cap K_{q'}^U}\left(QCU_k^{p,q'}+QHU_k^{p,q'}\right)+C_{trd}^{pU,q'L}\sum_{k\in K_p^U\cap K_{q'}^L}\left(QCU_k^{p,q'}+QHU_k^{p,q'}\right)\right)\\
&-\sum_{\substack{q'=1\\q'\neq p}}^{P}\left(C_{trd}^{pL,q'U}\sum_{k\in K_p^L\cap K_{q'}^U}\left(QCU_k^{p,q'}+QHU_k^{p,q'}\right)+C_{trd}^{pL,q'L}\sum_{k\in K_p^L\cap K_{q'}^L}\left(QCU_k^{p,q'}+QHU_k^{p,q'}\right)\right)\\
&+\sum_{\substack{q=1\\q\neq p}}^{P}\left(C_{trd}^{qU,pU}\sum_{k\in K_p^U\cap K_q^U}\left(QCU_k^{q,p}+QHU_k^{q,p}\right)+C_{trd}^{qU,pL}\sum_{k\in K_p^L\cap K_q^U}\left(QCU_k^{q,p}+QHU_k^{q,p}\right)\right)\\
&+\sum_{\substack{q'=1\\q'\neq p}}^{P}\left(C_{trd}^{qL,pU}\sum_{k\in K_p^U\cap K_q^L}\left(QCU_k^{q,p}+QHU_k^{q,p}\right)+C_{trd}^{qL,pL}\sum_{k\in K_p^L\cap K_q^L}\left(QCU_k^{q,p}+QHU_k^{q,p}\right)\right)
\end{aligned}
\tag{4.81}
$$

Notice that the highly nonlinear function in equation (4.78) can be rewritten in the following form to somewhat reduce the computational effort:

$$\max\sum_{p=1}^{P}\varpi_p \ln S_p^U \tag{4.82}$$

Although the objective function and the majority of the equality and inequality constraints of this optimization problem have already been presented above in equations (4.9)–(4.24) and (4.59)–(4.81), it is still necessary to impose a lower bound on the utility cost saving of each plant and set the upper limits of the heating and cooling utility consumption rates in order to make sure that the optimum solution is acceptable to all participating members and also energy efficient. Specifically, these additional inequalities are summarized as follows:

$$S_p^U \geq 0 \tag{4.83}$$

$$\overline{QS}_{m_p} \geq QS_{m_p},\quad m_p \in S^P \tag{4.84}$$

$$\overline{QW}_{n_p} \geq QW_{n_p},\quad n_p \in W^P \tag{4.85}$$

where the upper limits $\overline{QS}_{m_p}$ and $\overline{QW}_{n_p}$ can be obtained in Step 1 according to equations (4.63) and (4.64).

4.4.3 Step 3 of Indirect Strategy I

This step is used to determine the minimum total number of both inner- and interplant matches and the corresponding heat duties with a mixed-integer linear programming (MILP) model in which the interplant heat-flow patterns are fixed according to those obtained in Step 2. The optimization results obtained in the previous two steps can be used for building a MILP model to identify the optimal inner- and interplant matches and their heat duties. Specifically, the energy balances given in equations (4.59)–(4.62) and (4.68)–(4.71) should all be included in this model as equality constraints. Since the consumption rates of heating and cooling utilities on the right sides of equations (4.59) and (4.62) have already been determined in Step 1, $QS_{m_p,k}$ and $QW_{n_p,k}$ are treated as given parameters in the present step. Similarly, since the consumption rates of heating and cooling utilities on the left sides of equations (4.68)–(4.71) have already been determined in the previous steps and these utilities are used only for the interplant heat exchanges, $QHU_k^{p,q}$, $QCU_k^{p,q}$, $QHU_k^{q,p}$, and $QCU_k^{q,p}$ should also be considered as given parameters in the present model. The objective function should be

$$\min Z = \sum_{p=1}^{P}\left(\sum_{i_p \in H^p}\sum_{j_p \in C^p} z_{i_p,j_p} + \sum_{m_p \in S^p}\sum_{j_p \in C^p} z_{m_p,j_p} + \sum_{i_p \in H^p}\sum_{n_p \in W^p} z_{i_p,n_p}\right) + \sum_{p=1}^{P}\sum_{\substack{q=1 \\ q\neq p}}^{P}\left(\sum_{m_p \in S^p}\sum_{j_q \in C^q} z_{m_p,j_q} + \sum_{i_p \in H^p}\sum_{n_q \in W^q} z_{i_p,n_q}\right) \quad (4.86)$$

In addition to the equality constraints mentioned above, the following inequality constraints should also be imposed:

$$Q_{i_p,j_q} - z_{i_p,j_q} Q_{i_p,j_q}^U \le 0 \quad (4.87)$$

where $p,q = 1,2,\ldots,P; i_p \in S^p \cup H^p; j_q \in W^q \cup C^q; z_{i_p,j_q}$ is a binary variable defined as follows

$$z_{i_p,j_q} = \begin{cases} 1 & \text{if hot stream } i_p \text{ exchange heat with cold stream } j_q \\ 0 & \text{otherwise} \end{cases} \quad (4.88)$$

where $Q_{i_p,j_q} = \sum_{k \in K} Q_{i_p,j_q,k}$ denotes the heat duty of match (i_p, j_q) and the model parameter Q_{i_p,j_q}^U is its upper bound. It should be noted that, although there are actually seven different types of matches in Figure 4.1, Q_{i_p,j_q} should be viewed as a generalized notation for representation of the total amount of heat exchanged between hot stream (or utility) i_p and cold stream (or utility) j_q.

4.4.4 Step 4 of Indirect Strategy I

This step is taken first to construct a superstructure to facilitate all matches identified in Step 3, then to formulate a nonlinear programming (NLP) model accordingly, and finally to solve this NLP model for simultaneously generating the HEN configuration and distributing TAC savings among all plants fairly. Since only the matches and their duties are fixed in Step 3, it is necessary to further synthesize the network structure and produce the design specifications of each exchanger. A superstructure-based NLP model can be used for this purpose, e.g., see Floudas et al. (1986), and essentially an identical approach is adopted here to build the model constraints. On the other hand, a modified objective function is utilized in this NLP model to facilitate reasonable distribution of the TAC savings of all individual plants (S_p^T), i.e.,

$$\max \prod_{p=1}^{P} \left(S_p^T\right)^{\omega_p} \tag{4.89a}$$

or

$$\max \sum_{p=1}^{P} \omega_p \ln S_p^T \tag{4.89b}$$

where the negotiation power of plant p for the present step, i.e., ω_p, can be determined according to all utility cost savings (S_p^U) as follows

$$\omega_p = \frac{1}{S_p^U}\left(\sum_{p=1}^{P} \frac{1}{S_p^U}\right)^{-1} \tag{4.90}$$

Notice that, with this formulation, the negotiation power of each plant is inversely proportional to its utility cost saving obtained in Step 2. In other words, the plant with a large utility cost saving may be required to shoulder a large portion of the extra capital investment. On the other hand, the TAC saving of plant p can be computed with the following formula

$$S_p^T = S_p^U + Af\left(CL_p^{\min} - TC_p^{\text{inner}} - \sum_{\substack{q'=1 \\ q'\neq p}}^{P} TRD_{p,q'} - \sum_{\substack{q=1 \\ q\neq p}}^{P} TRD_{q,p} \right) \tag{4.91}$$

where Af is the annualization factor; $CL_p^{\min}$ denotes the minimum capital cost of a standalone HEN design in plant p, which is treated as a given model parameter in this study; TC_p^{inner} denotes the total capital cost of all inner-plant matches of plant p in an interplant heat integration scheme; $TRD_{p,q'}$ represents the capital cost of interplant match $\left(i_p, j_{q'}\right)$ shared by plant p and this match is either between a hot stream in plant p and a cold utility in plant q' or between a hot utility in plant p and a cold stream in plant q'; $TRD_{q,p}$ also represents the capital cost of interplant match $\left(i_q, j_p\right)$ shared by plant p and this match is either between a hot stream in plant q and a cold utility in plant p or between a hot utility in plant q and a cold stream in plant p.

The minimum capital cost of a standalone HEN design (CL_p^I) should be computed in advance by following the existing methods, while the other three aforementioned capital costs can be expressed mathematically as follows:

$$TC_p^{\text{inner}} = \sum_{i_p \in H^p \cup S^p} \sum_{j_p \in C^p \cup W^p} z_{i_p,j_p} c_{i_p,j_p} \left\{ \frac{Q_{i_p,j_p}}{U_{i_p,j_p} \left[\theta^1_{i_p,j_p} \theta^2_{i_p,j_p} \left(\theta^1_{i_p,j_p} + \theta^2_{i_p,j_p} \right) / 2 \right]^{\frac{1}{3}}} \right\}^{\beta} \tag{4.92}$$

$$\begin{aligned} TRD_{p,q'} = {} & \sum_{i_p \in S^p} \sum_{j_{q'} \in C^{q'}} z_{i_p,j_{q'}} c_{i_p,j_{q'}} \gamma^p_{i_p,j_{q'}} \left\{ \frac{Q_{i_p,j_{q'}}}{U_{i_p,j_{q'}} \left[\theta^1_{i_p,j_{q'}} \theta^2_{i_p,j_{q'}} \left(\theta^1_{i_p,j_{q'}} + \theta^2_{i_p,j_{q'}} \right) / 2 \right]^{\frac{1}{3}}} \right\}^{\beta} \\ & + \sum_{i_p \in H^p} \sum_{j_{q'} \in W^{q'}} z_{i_p,j_{q'}} c_{i_p,j_{q'}} \gamma^p_{i_p,j_{q'}} \left\{ \frac{Q_{i_p,j_{q'}}}{U_{i_p,j_{q'}} \left[\theta^1_{i_p,j_{q'}} \theta^2_{i_p,j_{q'}} \left(\theta^1_{i_p,j_{q'}} + \theta^2_{i_p,j_{q'}} \right) / 2 \right]^{\frac{1}{3}}} \right\}^{\beta} \end{aligned} \tag{4.93}$$

$$\begin{aligned} TRD_{q,p} = {} & \sum_{i_q \in S^q} \sum_{j_p \in C^p} z_{i_q,j_p} c_{i_q,j_p} \gamma^p_{i_q,j_p} \left\{ \frac{Q_{i_q,j_p}}{U_{i_q,j_p} \left[\theta^1_{i_q,j_p} \theta^2_{i_q,j_p} \left(\theta^1_{i_q,j_p} + \theta^2_{i_q,j_p} \right) / 2 \right]^{\frac{1}{3}}} \right\}^{\beta} \\ & + \sum_{i_q \in H^q} \sum_{j_p \in W^p} z_{i_q,j_p} c_{i_q,j_p} \gamma^p_{i_q,j_p} \left\{ \frac{Q_{i_q,j_p}}{U_{i_q,j_p} \left[\theta^1_{i_q,j_p} \theta^2_{i_q,j_p} \left(\theta^1_{i_q,j_p} + \theta^2_{i_q,j_p} \right) / 2 \right]^{\frac{1}{3}}} \right\}^{\beta} \end{aligned} \tag{4.94}$$

where $\gamma^p_{i_p,j_{q'}}$ and $\gamma^{q'}_{i_p,j_{q'}}$ denote the fractions of capital cost of match $\left(i_p, j_{q'}\right)$ shared by plant p and plant q' respectively; $\gamma^q_{i_q,j_p}$ and $\gamma^p_{i_q,j_p}$ denote the fractions of capital cost of match $\left(i_q, j_p\right)$ shared by plant q and plant p respectively. Thus, the following equality constraints should also be imposed under the conditions that $p \neq q \neq q'$:

$$\gamma^p_{i_p,j_{q'}} + \gamma^{q'}_{i_p,j_{q'}} = 1 \tag{4.95}$$

$$\gamma^q_{i_q,j_p} + \gamma^p_{i_q,j_p} = 1 \tag{4.96}$$

Finally, by following the conventional approach (Floudas et al., 1986), a superstructure and the corresponding material and energy balances can be constructed on the basis of the optimization results obtained in Step 3. These constraints should also be included in the NLP model used in the present step for maximizing the objective function defined in equation (4.89) or (4.90).

4.4.5 Application of Indirect Strategy I for Multi-Plant HEN Synthesis

Let us again consider the illustrative example described in Section 4.1. Based on a minimum temperature approach of 10 K, the three standalone heat-flow cascades in Figure 2.2 in Chapter 2 can be constructed and the corresponding minimum utility costs for P1, P2, and P3 found to be $66,100\left(\bar{Z}_{p1}\right)$, $6,600\left(\bar{Z}_{p2}\right)$, and $30,300\left(\bar{Z}_{p3}\right)$ USD/yr, respectively.

- ***Step 1***
  After solving equations (4.58)–(4.66), one can produce the heat-flow cascades in Figure 4.8. Note that the lowest-cost heating and cooling utilities are used in all plants to minimize the overall utility cost. The corresponding cost savings and the negotiation powers gained by implementing this step are summarized in Table 4.18.

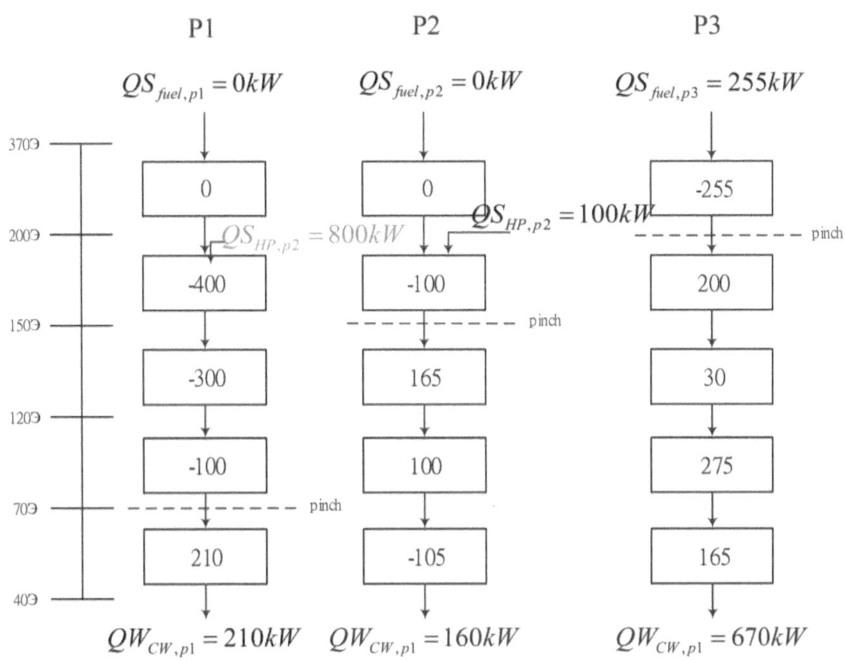

FIGURE 4.8 Integrated heat-flow cascade built at the first step of indirect strategy I in the illustrative example. (Adapted with permission from *Energy* 2018, 148: 90–111. Copyright 2018 Elsevier.)

TABLE 4.18
Utility Cost Savings and Negotiation Powers Established at the First Step of Indirect Strategy I in the Illustrative Example

Plant #	Utility Type	Utility Costs without Integration (USD/yr)	Utility Costs after Step 1 (USD/yr)	Total Utility Cost Savings after Step 1 (USD/yr)	Negotiation Powers Gained after Step 1 (ϖ_p)
P1	Fuel oil	64,000	0	55,700	0.157
	Cooling water	2100	10,400		
P2	High-pres. steam	3000	27,000	−20,400	4.091
	Cooling water	3600	0		
P3	Fuel oil	10,200	10,200	20,100	0.337
	Cooling water	20,100	0		

- ***Step 2***
  This step yields the following fair trade prices of all interplant heat exchanges and the corresponding strategy vectors.

$$C_{trd}^{1U,2U} = -30,\ C_{trd}^{1U,2L} = \ 0,\ C_{trd}^{1U,3U} = -90,\ C_{trd}^{1U,3L} = -90,$$

$$C_{trd}^{1L,2U} = -30,\ C_{trd}^{1L,2L} = +10,\ C_{trd}^{1L,3U} = -60,\ C_{trd}^{1L,3L} = +30,$$

$$C_{trd}^{2U,1U} = -40,\ C_{trd}^{2U,1L} = -30,\ C_{trd}^{2U,3U} = -30,\ C_{trd}^{2U,3L} = -30,$$

$$C_{trd}^{2L,1U} = -90,\ C_{trd}^{2L,1L} = +10,\ C_{trd}^{2L,3U} = -60,\ C_{trd}^{2L,3L} = +22.5,$$

$$C_{trd}^{3U,1U} = -90,\ C_{trd}^{3U,1L} = +10,\ C_{trd}^{3U,2U} = -30,\ C_{trd}^{3U,2L} = +22.5,$$

$$C_{trd}^{3L,1U} = -90,\ C_{trd}^{3L,1L} = +10,\ C_{trd}^{3L,2U} = \ 0,\ C_{trd}^{3L,2L} = +30.$$

$$\begin{bmatrix} \hat{w}_1^{U,D} \\ \hat{w}_1^{L,D} \\ \hat{w}_1^{U,A} \\ \hat{w}_1^{L,A} \end{bmatrix} = \begin{bmatrix} 0 \\ 0 \\ 0.491 \\ 0.509 \end{bmatrix}, \begin{bmatrix} \hat{w}_2^{U,D} \\ \hat{w}_2^{L,D} \\ \hat{w}_2^{U,A} \\ \hat{w}_2^{L,A} \end{bmatrix} = \begin{bmatrix} 0.833 \\ 0.167 \\ 0 \\ 0 \end{bmatrix}, \begin{bmatrix} \hat{w}_3^{U,D} \\ \hat{w}_3^{L,D} \\ \hat{w}_3^{U,A} \\ \hat{w}_3^{L,A} \end{bmatrix} = \begin{bmatrix} 0 \\ 1 \\ 0 \\ 0 \end{bmatrix}.$$

  The resulting utility cost savings and the negotiation powers gained after implementing this step are summarized in Table 4.19.
- ***Step 3***
  The optimal matches obtained in this step are listed in Table 4.20. In this table, a binary number is given in each cell to denote whether or not a corresponding match should be assigned in the HEN design. A value of 1 indicates that this match is chosen and its heat duty (kW) is specified in a

TABLE 4.19
Net Utility Cost Savings and Negotiation Powers Established at the Second Step of Indirect Strategy I in the Illustrative Example

Plant #	Total Utility Cost Savings after Step 1 (USD/yr)	Total Payoffs from Energy Trades in Step 2 (USD/yr)	Net Utility Cost Savings after Step 2 (USD/yr)	Negotiation Powers Gained after Step 2 (ω_p)
P1	55,700	−43,700	12,000	0.436
P2	−20,400	50,400	30,000	0.174
P3	20,100	−6700	13,400	0.39

TABLE 4.20
Optimal Matches Obtained at the Third Step of Indirect Strategy I in the Illustrative Example

Hot \ Cold	P1_C1	P1_C2	P2_C1	P2_C2	P3_C1	P1_CW
P1_H1	1 (560)	0	0	0	0	1 (210)
P2_H1	0	0	1 (280)	1 (275)	0	1 (160)
P3_H1	0	0	0	0	1 (510)	1 (150)
P3_H2	0	0	0	0	1 (360)	1 (520)
P2_HP	1 (160)	1 (640)	0	1 (100)	0	0
P3_Fuel	0	0	0	0	1 (255)	0

parenthesis in the same cell, while 0 means the corresponding exchanger should be excluded. From Table 4.20, one can see that there are seven inner-plant heat exchangers and five interplant heat-transfer units.

- ***Step 4***
  On the basis of the optimal matches listed in Table 4.20, one could construct a superstructure for each process stream in the three plants under consideration. The objective function used in the present step, i.e., equations (4.89)–(4.96), can then be formulated according to the negotiation powers given in Table 4.19, while the other model constraints can be obtained by applying the basic principles of material and energy balances to the superstructures. The annualization factor in equation (4.91), i.e., Af, was set at 0.1349. All

cost coefficients in equations (4.92)–(4.94) were chosen to be 670 USD/$m^{1.66}$ and $\beta=0.83$, while all overall heat-transfer coefficients were taken to be 1 W/m^2·K. Finally, the temperature rise of cooling water in every cooler was set to be 5°C. By solving the corresponding NLP program, one can produce the interplant HEN design presented in Figure 4.9 and the final economic assessment in Table 4.21. This interplant heat integration project should be quite feasible because all TAC savings are positive and reasonably distributed.

4.5 USING INTERMEDIATE FLUIDS AS AUXILIARY STREAMS

This section details the four steps of indirect heat integration strategy by using intermediate fluids as the auxiliary streams, i.e., indirect strategy II. The mathematical programming models presented here can be adopted to identify the proper interplant heat integration schemes by using an intermediate fluid or "hot" oil as the indirect heat-transfer medium. To simplify model formulations, let us assume that this fluid can only be treated as either a cold or a hot process stream in each plant that joins the integration project and, also, its temperature ranges in all plants must be identical. To facilitate clear explanation, let us use two label sets to classify these plants accordingly, i.e., sets *PC* and *PH*. In particular, they represent two groups of plants ($PC \cap PH = \varnothing$) and the intermediate fluid is used as a cold stream in the former set and a hot stream in the latter. The optimization steps for the present applications are applied to achieve essentially the same objectives as those given in the previous section, while the model solved in each step should be modified according to the aforementioned assumptions.

4.5.1 Step 1 of Indirect Strategy II

This step is again used to calculate the minimum total utility cost on the basis of a modified version of the conventional transshipment model (Papoulias and Grossmann, 1983). Figure 4.10 shows the interior heat-flow pattern of interval k in plant p if the intermediate fluid is used as a cold stream, while Figure 4.11 is its counterpart under the condition that this fluid is a hot stream. The optimization goal here is the same as that of Step 1 in strategy I, i.e., to minimize the total utility cost according to equation (4.58), while the equality constraints of the corresponding MINLP model can be obtained primarily on the basis of energy balances.

For $p \in PC$, the energy balances can be established according to Figure 4.12 as follows:

$$\text{A.} \quad R_{i_p,k} - R_{i_p,k-1} + \sum_{j_p \in C_k^P} Q_{i_p,j_p,k} + \sum_{n_p \in W_k^P} Q_{i_p,n_p,k} + QF_{i_p,k}^{C} = Q_{i_p,k}^{H}, i_p \in \tilde{H}_k^P; \tag{4.97}$$

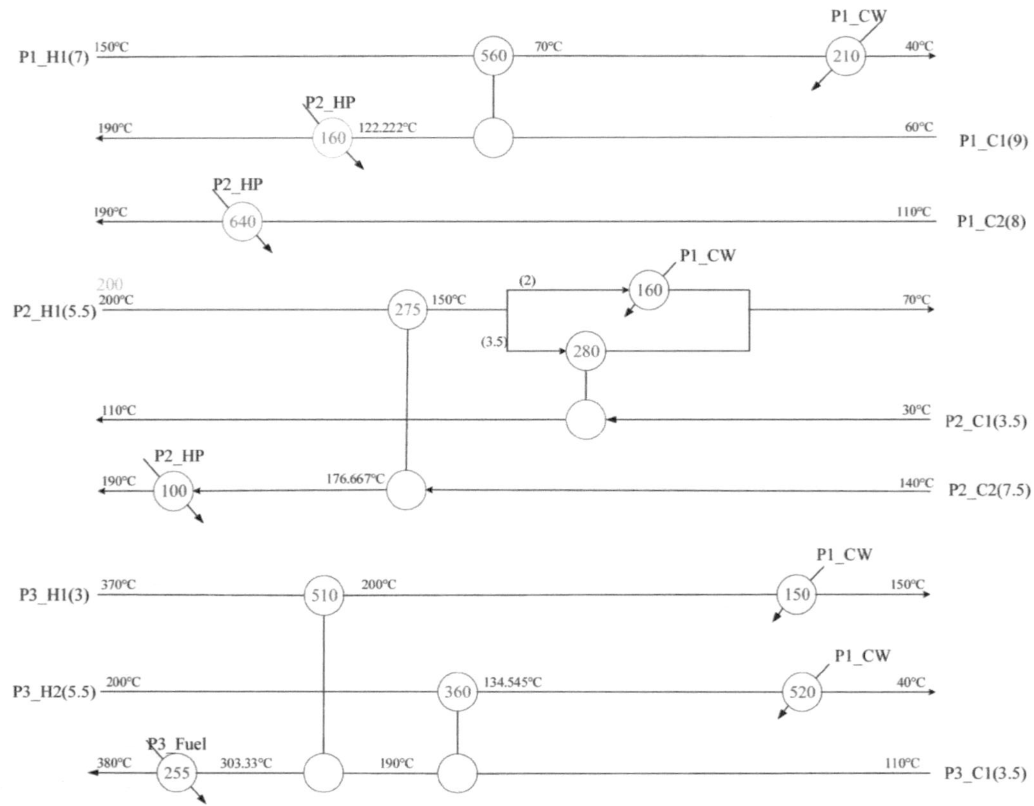

FIGURE 4.9 Interplant HEN design produced at the fourth step of indirect strategy I in the illustrative example. (Reproduced with permission from *Energy* 2018, 148: 90–111. Copyright 2018 Elsevier.)

TABLE 4.21
Total Annual Cost Savings Achieved with Indirect Strategy I in the Illustrative Example

Plant #	Net Utility Cost Savings after Step 2 (USD/yr)	Total Capital Costs without Integration (USD/yr)	Total Capital Costs after Step 4 (USD/yr)	Total Capital Cost Savings after Step 4 (USD/yr)	TAC Savings after Step 4 (USD/yr)
P1	12,000	5891	7548	−1657	10,343
P2	30,000	5861	7251	−1390	28,610
P3	13,400	7865	7725	140	13,540

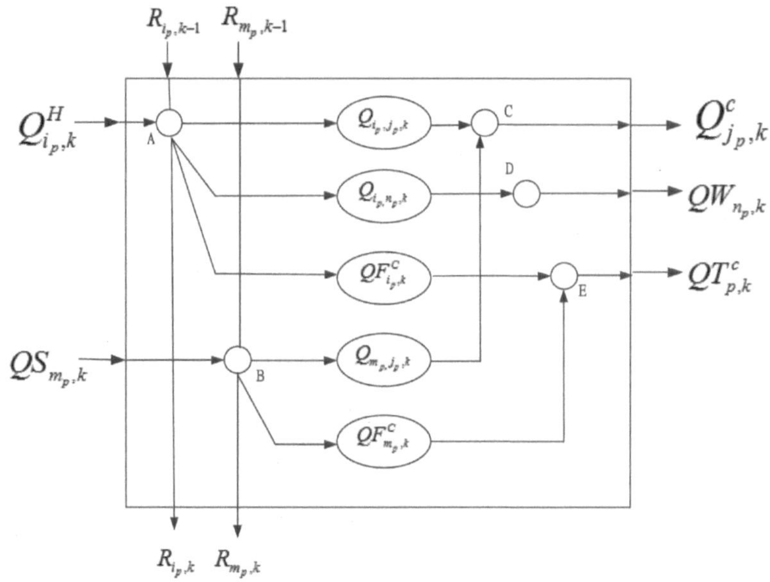

FIGURE 4.10 Interior heat-flow patterns of interval k in plant p with the intermediate fluid acting as a cold stream. (Reproduced with permission from *Energy* 2018, 148: 90–111. Copyright 2018 Elsevier.)

$$\text{B.}\quad R_{m_p,k} - R_{m_p,k-1} + \sum_{j_p \in C_k^p} Q_{m_p,j_p,k} + QF^C_{m_p,k} = QS_{m_p,k},\ \ m_p \in \tilde{S}_k^p; \tag{4.98}$$

$$\text{C.}\quad \sum_{i_p \in \tilde{H}_k^p} Q_{i_p,j_p,k} + \sum_{m_p \in \tilde{S}_k^p} Q_{m_p,j_p,k} = Q^C_{j_p,k},\ \ j_p \in C_k^p; \tag{4.99}$$

$$\text{E.}\quad \sum_{i_p \in \tilde{H}_k^p} Q_{i_p,n_p,k} = QW_{n_p,k},\ \ n_p \in W_k^p; \tag{4.100}$$

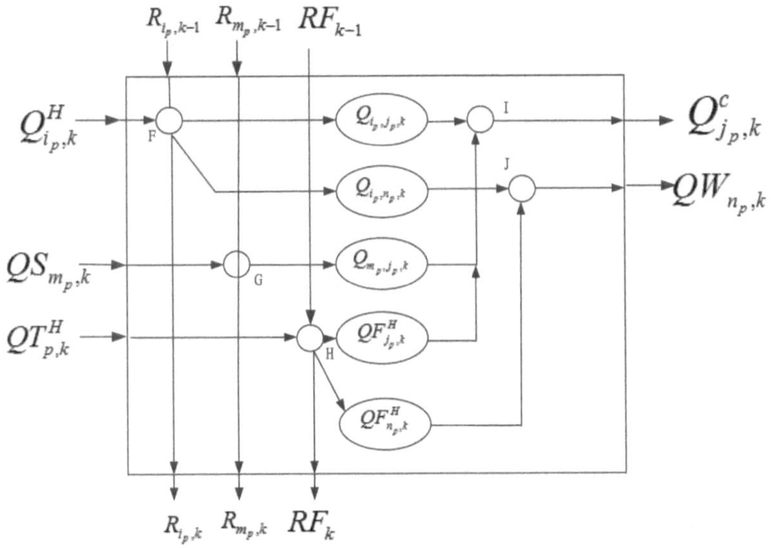

FIGURE 4.11 Interior heat-flow patterns of interval k in plant p with the intermediate fluid acting as a hot stream. (Reproduced with permission from *Energy* 2018, 148: 90–111. Copyright 2018 Elsevier.)

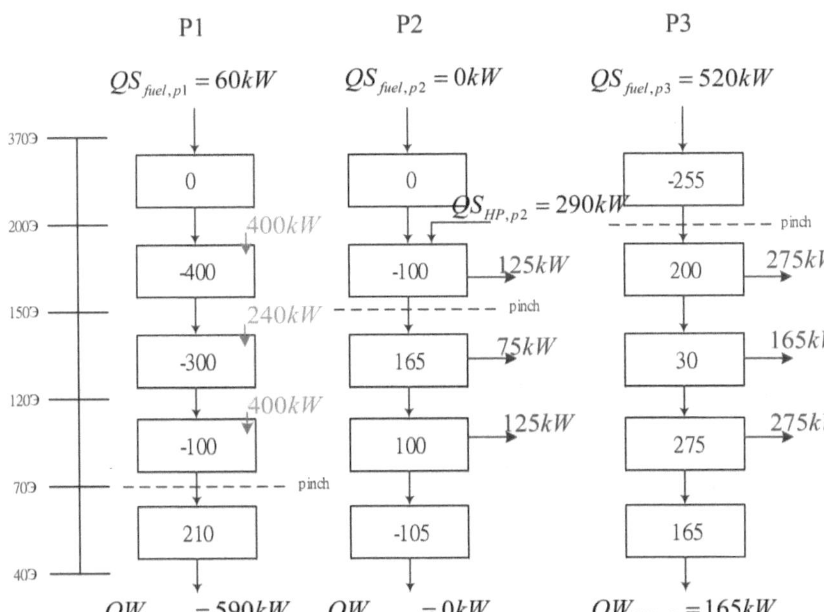

FIGURE 4.12 Integrated heat-flow cascades built at the first step of indirect strategy II in the illustrative example. (Reproduced with permission from *Energy* 2018, 148: 90–111. Copyright 2018 Elsevier.)

$$\text{F.}\quad \sum_{i_p \in \tilde{H}_k^p} QF^C_{i_p,k} + \sum_{m_p \in \tilde{S}_k^p} QF^C_{m_p,k} = FCP_p\left(T_k - T_{k-1}\right)w_k^{(1)} = QT^C_{p,k} \tag{4.101}$$

For $p \in PH$, the energy balances can be established according to Figure 4.13 as follows:

$$\text{A.}\quad R_{i_p,k} - R_{i_p,k-1} + \sum_{j_p \in C_k^p} Q_{i_p,j_p,k} + \sum_{n_p \in W_k^p} Q_{i_p,n_p,k} = Q^H_{i_p,k},\ i_p \in \tilde{H}_k^p; \tag{4.102}$$

$$\text{B.}\quad R_{m_p,k} - R_{m_p,k-1} + \sum_{j_p \in C_k^p} Q_{m_p,j_p,k} = QS_{m_p,k},\ m_p \in \tilde{S}_k^p; \tag{4.103}$$

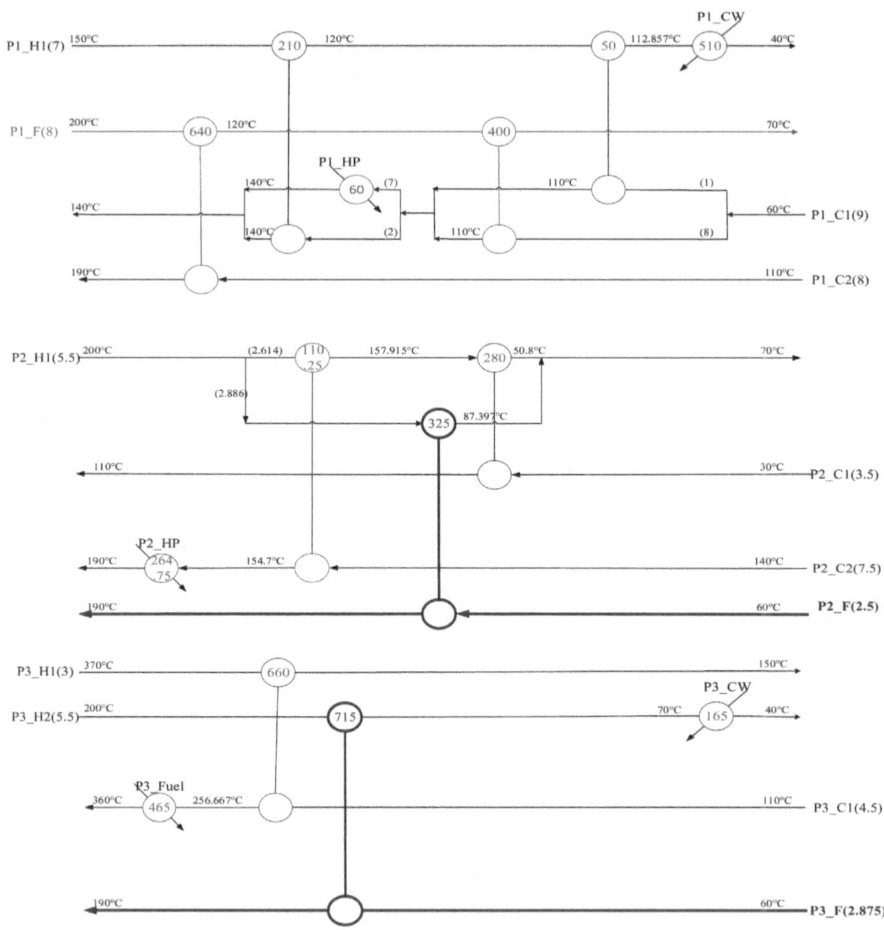

FIGURE 4.13 Interplant HEN design produced at the fourth step of indirect strategy II in the illustrative example. (Reproduced with permission from *Energy* 2018, 148: 90–111. Copyright 2018 Elsevier.)

$$\text{C.}\quad RF_k - RF_{k-1} + \sum_{j_p \in C_k^P} QF_{j_p,k}^H + \sum_{n_p \in W_k^P} QF_{n_p,k}^H = FCP_p\left(T_k - T_{k-1}\right)w_k^{(1)} = QT_{p,k}^H \tag{4.104}$$

$$\text{D.}\quad \sum_{i_p \in \tilde{H}_k^P} Q_{i_p,j_p,k} + \sum_{m_p \in \tilde{S}_k^P} Q_{m_p,j_p,k} + QF_{j_p,k}^H = Q_{j_p,k}^C,\ \ j_p \in C_k^P; \tag{4.105}$$

$$\text{E.}\quad \sum_{i_p \in \tilde{H}_k^P} Q_{i_p,n_p,k} + QF_{n_p,k}^H = QW_{n_p,k},\ \ n_p \in W_k^P; \tag{4.106}$$

In addition, the following equality constraint must also be imposed to maintain overall material balance on the intermediate fluid:

$$\sum_{p \in PH} FCP_p = \sum_{p \in PC} FCP_p \tag{4.107}$$

Since the actual temperature range of hot oil in each plant is not a priori given, a binary variable $w_k^{(1)} \in \{0,1\}$ is used in equations (4.101) and (4.104) to signify whether or not interval k exists. Furthermore, to allow this range to start and end at any two boundary temperatures of the partitioned intervals, two additional binary variables, i.e., $w_k^{(2)}$ and $w_k^{(3)}$, are introduced for use in the following constraint:

$$w_k^{(1)} = w_k^{(2)} - w_k^{(3)} \tag{4.108}$$

where k=1,2, …, K. In this equation, $w_k^{(2)} = 1$ means that interval k is higher than the low end of the temperature range, and $w_k^{(3)} = 1$ means that interval k is higher than the corresponding high end. To ensure convexity of the temperature range, two extra logic constraints should be imposed:

$$1 - w_k^{(2)} + w_{k+1}^{(2)} \le 1 \tag{4.109}$$

$$1 - w_k^{(3)} + w_{k+1}^{(3)} \le 1 \tag{4.110}$$

where k=1,2, …, K − 1 and $w_1^{(2)} = 1$. Equations (4.63) and (4.64) used in Step 1 of the Type-I procedure should also be included in the present model to compute the individual utility consumption rates, but the formulas for calculating the total utility costs should be slightly modified as follows:

$$\begin{aligned} Z_p' = &\sum_{m_p \in S^P} c_{m_p} QS_{m_p} + \sum_{n_p \in W^P} c_{n_p} QW_{n_p} \\ &+ c_M FCP_p \Delta T_{\min},\quad p \in PC, M \in S^P; \end{aligned} \tag{4.111}$$

$$Z'_p = \sum_{m_p \in S^p} c_{m_p} QS_{m_p} + \sum_{n_p \in W^p} c_{n_p} QW_{n_p}$$
$$+ c_N FCP_p \Delta T_{\min}, \quad p \in PC, N \in W^p. \tag{4.112}$$

If the intermediate fluid is adopted to play the role of a cold stream in plant p ($p \in PC$), its initial and target temperatures must coincide with the aforementioned range in cold-stream temperature. In order to reuse this spent cold stream of plant p as a fresh hot stream in another plant, it is necessary to further raise its temperature in plant p from the cold-stream target to a higher level to achieve an increase of ΔT_{min}. It is also assumed that this temperature change is accomplished by using utility $M \in S^p$ at the unit cost of c_M. Therefore, the third term on the right side of equation (4.111) is introduced to account for the corresponding hot utility cost. Similarly, if the intermediate fluid is used as a hot stream in plant p ($p \in PH$), it is necessary to further lower its temperature from the hot-stream target to achieve an additional decrease of ΔT_{min}, and the third term on the right side of equation (4.112) is adopted to account for the corresponding cold utility cost.

Finally, the MINLP model should include the equality constraints to stipulate zero residual heat flows entering the first and leaving the last temperature intervals, i.e., equation (4.66) and

$$RF_0 = RF_K = 0 \tag{4.113}$$

By solving the proposed model, one can obtain the optimal utility consumption rates of every plant (QS_{m_p} and QW_{n_p}), the heat-capacity flow rate in each plant (FCP_p), and its temperature range ($w_k^{(1)}$). Since all plants must be classified and grouped into two sets, i.e., PC and PH, in advance, it is necessary to construct and solve the proposed MINLP model repeatedly for $2^p - 2$ times, and the optimal solution corresponding to the smallest total utility cost should be chosen for use in the next step.

4.5.2 Step 2 of Indirect Strategy II

This step is also used to set the optimal trade prices. The general structure of the payoff matrix in equation (4.67) is still valid here, while its submatrices should be determined differently as follows:

If $p \in PC$, then the intermediate fluid is only used to facilitate energy export. Thus, $\mathbf{A}_{p,q_i}$ can be reduced to

$$\mathbf{A}_{p,q_i} = \begin{bmatrix} P_F^{pU,q_iU} & P_F^{pU,q_iL} \\ P_F^{pL,q_iU} & P_F^{pL,q_iL} \end{bmatrix} \tag{4.114}$$

If $p \in PH$, then the intermediate fluid is only used to facilitate energy import. Thus, $\mathbf{A}_{p,q_i}$ can be reduced to

$$\mathbf{A}_{p,q_i} = \begin{bmatrix} P_F^{q_iU,pU} & P_F^{q_iL,pU} \\ P_F^{q_iU,pL} & P_F^{q_iL,pL} \end{bmatrix} \tag{4.115}$$

As mentioned before, the ratio between the total amount of one particular type of heat exchange and that of all possible heat flows in and out of plant p is considered as a game strategy in this work. Therefore, equations (4.72)–(4.75) should be reformulated accordingly for the present case. For $p \in PC$, there are two strategies to export heat only, i.e.

$$\tilde{w}_p^{U,D} = \frac{1}{\tilde{Q}_p^{EC}} \sum_{k \in K_p^U} QT_{p,k}^C \tag{4.116}$$

$$\tilde{w}_p^{L,D} = \frac{1}{\tilde{Q}_p^{EC}} \sum_{k \in K_p^L} QT_{p,k}^C \tag{4.117}$$

where $\tilde{Q}_p^{EC} = \sum_{k=1}^{K} QT_{p,k}^C$ and $\tilde{w}_p^{U,D} + \tilde{w}_p^{L,D} = 1$. For $p \in PH$, there are two strategies to import heat only, i.e.,

$$\tilde{w}_p^{U,A} = \frac{1}{\tilde{Q}_p^{EH}} \sum_{k \in K_p^U} QT_{p,k}^H \tag{4.118}$$

$$\tilde{w}_p^{L,A} = \frac{1}{\tilde{Q}_p^{EH}} \sum_{k \in K_p^L} QT_{p,k}^H \tag{4.119}$$

where $\tilde{Q}_p^{EC} = \sum_{k=1}^{K} QT_{p,k}^H$ and $\tilde{w}_p^{U,A} + \tilde{w}_p^{L,A} = 1$.

For the purpose of setting the proper trade prices, the general framework of the NLP model used here is essentially the same as that described previously in Step 2 of the direct and indirect Type-I procedures, i.e., see equations (4.9)–(4.24), (4.30)–(4.32), (4.78)–(4.80), and (4.82)–(4.85). Other than equations (4.114)–(4.119) used in the present model, the only additional change is in computing the total trade revenue received by plant p, i.e., pf_p. Specifically, equation (4.81) should be replaced with the following two formulas:

$$pf_p = -\tilde{Q}_p^{EC} \sum_{\substack{q'=1 \\ q' \neq p}}^{P} \lambda_{p,q'}^{D} \left[\tilde{w}_{q'}^{U,A} \left(C_{trd}^{pU,q'U} \tilde{w}_p^{U,D} + C_{trd}^{pL,q'U} \tilde{w}_p^{L,D} \right) + \tilde{w}_{q'}^{L,A} \left(C_{trd}^{pU,q'L} \tilde{w}_p^{U,D} + C_{trd}^{pL,q'L} \tilde{w}_p^{L,D} \right) \right] \tag{4.120}$$

where $p \in PC$; $\lambda_{p,q'}^{D}$ denotes the fraction of total heat exported by plant p that is received by plant q', i.e., $\lambda_{p,q'}^{D} = f_{p,q'} / FCP_p$ and $FCP_p = \sum_{q' \in PH} f_{p,q'}$.

$$pf_p = \tilde{Q}_p^{EH} \sum_{\substack{q'=1 \\ q' \neq p}}^{P} \lambda_{q,p}^{A} \left[\tilde{w}_q^{U,D} \left(C_{trd}^{qU,pU} \tilde{w}_p^{U,A} + C_{trd}^{qU,pL} \tilde{w}_p^{L,A} \right) + \tilde{w}_q^{L,D} \left(C_{trd}^{qL,pU} \tilde{w}_p^{U,A} + C_{trd}^{qL,pL} \tilde{w}_p^{L,A} \right) \right] \tag{4.121}$$

where $p \in PH$; $\lambda_{q,p}^{A}$ denotes the fraction of total heat imported by plant p that is released by plant q, i.e., $\lambda_{q,p}^{A} = f_{q,p} / FCP_p$ and $FCP_p = \sum_{q \in PC} f_{q,p}$.

4.5.3 Step 3 of Indirect Strategy II

This step is to minimize the total number of matches. Specifically, the objective function should be expressed as

$$\min Z = \sum_{p=1}^{P} \left(\sum_{i_p \in H^P} \sum_{j_p \in C^P} z_{i_p,j_p} + \sum_{m_p \in S^P} \sum_{j_p \in C^P} z_{m_p,j_p} + \sum_{i_p \in H^P} \sum_{n_p \in W^P} z_{i_p,n_p} \right) + \sum_{p \in PC} \sum_{i_p \in H^P \cup S^P} zF_{i_p}^{C} + \sum_{p \in PH} \sum_{j_p \in C^P \cup W^P} zF_{j_p}^{H} \tag{4.122}$$

The first three terms on the right side of this equation are the same as those in equation (4.86) and they represent all possible inner-plant matches, while the remaining interplant matches are facilitated indirectly with the intermediate fluid and their heat duties are constrained as follows:

$$QF_{i_p}^{C} - zF_{i_p}^{C} UF_{i_p} \leq 0, p \in PC; \tag{4.123}$$

$$QF_{j_p}^{H} - zF_{j_p}^{H} UF_{j_p} \leq 0, p \in PH. \tag{4.124}$$

where $zF_{i_p}^{C} \in \{0,1\}$; $zF_{j_p}^{H} \in \{0,1\}$; $QF_{i_p}^{C} \left(= \sum_{k \in K} QF_{i_p,k}^{C} \right)$ denotes the heat duty of the match between hot stream i_p and the intermediate fluid acting as a cold stream; $QF_{j_p}^{H} \left(= \sum_{k \in K} QF_{j_p,k}^{H} \right)$ denotes the heat duty of the match between cold stream j_p and the intermediate fluid acting as a hot stream; UF_{i_p} and UF_{j_p} respectively denote the upper bounds of the corresponding heat duties.

On the other hand, the model constraints can be formulated with an approach which is very similar to that in Step 3 of the indirect Type-I procedure. The optimization results obtained previously in Steps 1 and 2 of the Type-II procedure can be used for building an MILP model to identify the optimal inner- and interplant matches and their heat duties. Specifically, the energy balances given in equations (4.97)–(4.107) should be included in this model as equality constraints. Notice that the following model parameters should become available after applying the aforementioned two steps:

- The heating utility consumption rates on the right sides of equations (4.98) and (4.103), i.e., $QS_{m_p,k}$
- The cooling utility consumption rates on the right sides of equations (4.100) and (4.106), i.e., $QW_{n_p,k}$
- The heat-capacity flow rate of intermediate medium used in each plant, i.e., FCP_p
- The temperature range of intermediate medium used in each plant, i.e., $w_k^{(1)}$

4.5.4 Step 4 of Indirect Strategy II

This step is to generate the optimal interplant HEN structure. The objective function of the NLP model used here is the same as that defined in equations (4.89)–(4.91), and the corresponding model constraints can be formulated according to the superstructure-based approach suggested by Floudas et al. (1986). As mentioned previously, the above-mentioned objective function is utilized to facilitate fair distribution of the TAC savings of all plants (S_p^T). If $p \in PC$, the TAC saving can be expressed as

$$S_p^T = S_p^U + Af\left(CL_p^{\min} - TC_p^{\text{inner}} - TPC_p^p - \sum_{q \in PH} TPH_q^p \right) \tag{4.125}$$

Notice that only two variables in this equation have not been defined before, i.e., TPC_p^p and TPH_q^p, and they are the capital costs of interplant heat exchangers (which are facilitated by the intermediate fluid) that must be paid by plant *p*. The former represents a fraction of the total capital cost of interplant matches in plant *p* and each can be denoted as (i_p, *F*), while the latter represents a fraction of the total capital cost of interplant matches in plant *q* and each is denoted as (*F*, j_q). On the other hand, the TAC saving of plant $p \in PH$ should be

$$S_p^T = S_p^U + Af\left(CL_p^{\min} - TC_p^{\text{inner}} - \sum_{q \in PC} TPC_q^p - TPH_p^p \right) \tag{4.126}$$

Again in this equation only the definitions of two variables have not been given before. In particular, TPC_q^p and TPH_p^p denote the capital costs of interplant heat exchangers (which are facilitated by the intermediate fluid) that must be paid by plant

p. The former is a fraction of the total capital cost of matches in plant q, i.e., (i_q, F), while the latter is a fraction of the total capital cost of matches in plant p, i.e., (F, j_p). More specifically, the aforementioned capital costs can be expressed in generalized forms as follows

$$TPH_q^p = \sum_{j_q \in C^q \cup W^q} zF_{j_q}^H \gamma F_{j_q}^p c_{F,j_q} \left\{ \frac{QF_{j_q}^H}{U_{F,j_q}\left[\theta_{F,j_q}^1 \theta_{F,j_q}^2 \left(\theta_{F,j_q}^1 + \theta_{F,j_q}^2\right)/2\right]^{\frac{1}{3}}} \right\}^{\beta} \quad (4.127)$$

$$TPC_q^p = \sum_{i_q \in H^q \cup S^q} zF_{i_q}^C \gamma F_{i_q}^p c_{i_q,F} \left\{ \frac{QF_{i_q}^C}{U_{i_q,F}\left[\theta_{i_q,F}^1 \theta_{i_q,F}^2 \left(\theta_{i_q,F}^1 + \theta_{i_q,F}^2\right)/2\right]^{\frac{1}{3}}} \right\}^{\beta} \quad (4.128)$$

where $p,q \in 1,2,\ldots,P$; $\gamma F_{j_q}^p$ denotes the fraction of the total capital cost of match (F, j_q) in plant $q\left(q \in PH\right)$ that is shared by plant p $(p \in PC)$; and $\gamma F_{i_q}^p$ denotes the fraction of the total capital cost of match (i_q, F) in plant $q\left(q \in PC\right)$ that is shared by plant p $(p \in PH)$. Clearly, additional equality constraints must be imposed upon these cost fractions as follows:

$$\gamma F_{i_p}^p + \sum_{q \in PH} \gamma F_{i_p}^q = 1, p \in PC \quad (4.129)$$

$$\gamma F_{j_p}^p + \sum_{q \in PH} \gamma F_{j_p}^q = 1, p \in PH \quad (4.130)$$

where $\gamma F_{i_p}^p$ in equation (4.129) denotes the fraction of the total capital cost of match (i_p, F) in plant p that is shared by plant p itself $(p \in PC)$, while $\gamma F_{i_p}^q$ represents the fraction of the same total capital cost that is shared by plant q; $\gamma F_{j_p}^p$ in equation (4.130) denotes the fraction of the total capital cost of match (F, j_p) in plant p that is shared by plant p itself $(p \in PH)$, while $\gamma F_{j_p}^q$ denotes the fraction of the same total capital cost that is shared by plant q. In other words,

- Equation (4.129) is applicable to match (i_p, F) in plant p when $p \in PC$.
- Equation (4.130) is applicable to match (F, j_p) in plant p when $p \in PH$.

4.5.5 Application of Indirect Strategy II for Multi-Plant HEN Synthesis

Let us consider the stream and utility data in Tables 4.1 and 4.2. Based on a minimum temperature approach of 10°C, the three individual heat-flow cascades in Figure 4.1 and the corresponding minimum utility costs for P1, P2, and P3, i.e., $66,100\left(\overline{Z}_{p1}\right)$, $6,600\left(\overline{Z}_{p2}\right)$, and $30,300\left(\overline{Z}_{p3}\right)$ USD/yr, can still be used for the present example.

- ***First step***
  By repeatedly solving the MINLP model six times in Step 1 of strategy II, one can produce the integrated heat-flow cascade in Figure 4.12. The temperature range of intermediate fluid is chosen to be [70,200]°C when it is viewed as a hot stream, and [60,190]°C if the intermediate fluid is a cold stream. Note that $PC=\{P2, P3\}$ and $PH=\{P1\}$, and it was found that $FCP_{P1}=8$, $FCP_{P2}=2.5$, and $FCP_{P3}=5.5$. Clearly, $FCP_{P1}=FCP_{P2}+FCP_{P3}$. The corresponding cost savings and the negotiation powers gained by implementing this step are summarized in Table 4.22.
- ***Second step***
  From Figure 4.14 and Table 4.22, the following fair trading prices of inter-plant heat exchanges can be obtained by solving the proposed NLP model:

$$C_{trd}^{2U,1U}=-90.0,\ C_{trd}^{2U,1L}=-30.0,$$

$$C_{trd}^{2L,1U}=-37.5,\ C_{trd}^{2L,1L}=+10.0,$$

$$C_{trd}^{3U,1U}=-60.0,\ C_{trd}^{3U,1L}=-18.5,$$

$$C_{trd}^{3L,1U}=-43.8,\ C_{trd}^{3L,1L}=+10.0.$$

The corresponding strategy vectors are:

$$\begin{bmatrix}\tilde{w}_1^{U,A}\\ \tilde{w}_1^{L,A}\end{bmatrix}=\begin{bmatrix}1\\0\end{bmatrix},\begin{bmatrix}\tilde{w}_2^{U,D}\\ \tilde{w}_2^{L,D}\end{bmatrix}=\begin{bmatrix}0.385\\0.615\end{bmatrix},\begin{bmatrix}\tilde{w}_3^{U,D}\\ \tilde{w}_3^{L,D}\end{bmatrix}=\begin{bmatrix}0\\1\end{bmatrix}.$$

Also, the ratios used in equations (4.120) and (4.121) were found to be: $\lambda_{2,1}^{A}=0.3125$, $\lambda_{3,1}^{A}=0.6875$, $\lambda_{2,1}^{D}=1.0$, and $\lambda_{3,1}^{D}=1.0$. The resulting net utility cost savings and the negotiation powers gained by implementing this step are summarized in Table 4.23.

TABLE 4.22
Utility Cost Savings and Negotiation Powers Established at the First Step of Indirect Strategy II in the Illustrative Example

Plant #	Utility Type	Utility Costs without Integration (USD/yr)	Utility Costs after Step 1 (USD/yr)	Total Utility Cost Savings after Step 1 (USD/yr)	Negotiation Powers Gained after Step 1 (ϖ_p)
P1	Fuel oil	64,000	4800	55,400	0.162
	Cooling water	2100	5900		
P2	High-pres. steam	3000	8700	−2100	1.318
	Cooling water	3600	0		
P3	Fuel oil	10,200	20,800	4550	0.850
	Cooling water	20,100	4950		

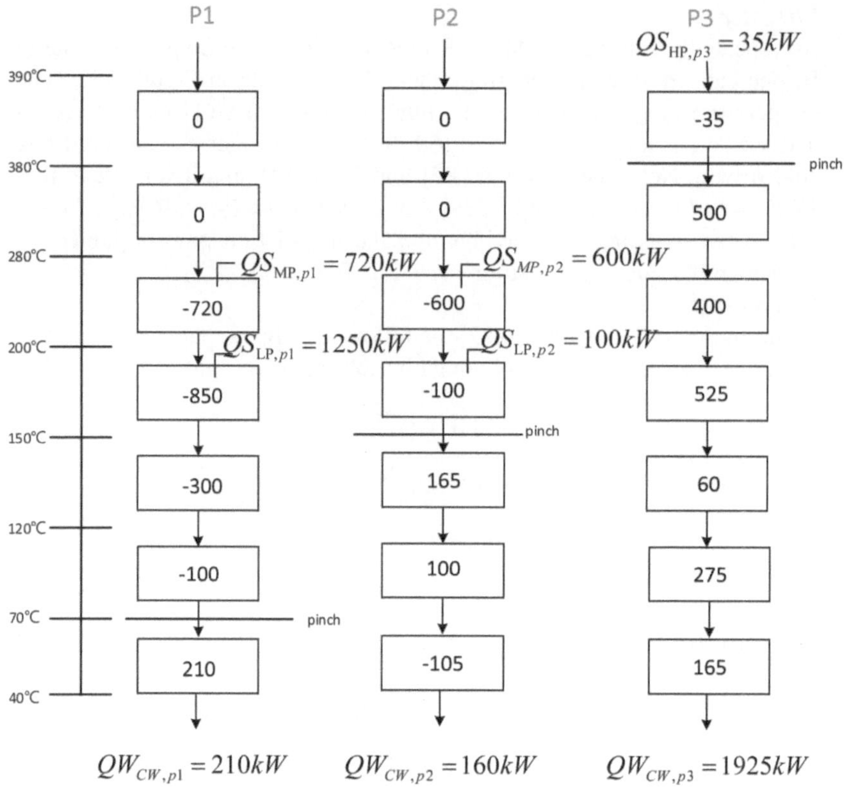

FIGURE 4.14 Standalone heat-flow cascades in the third example. (Reproduced with permission from *Energy* 2018, 148: 90–111. Copyright 2018 Elsevier.)

TABLE 4.23
Net Utility Cost Savings and Negotiation Powers Established at the Second Step of Indirect Strategy II in the Illustrative Example

Plant #	Total Utility Cost Savings after Step 1 (USD/yr)	Total Payoffs from Energy Trades in Step 2 (USD/yr)	Net Utility Cost Savings after Step 2 (USD/yr)	Negotiation Powers Gained after Step 2 (ω_p)
P1	55,400	–48,568	6832	0.61
P2	–2100	17,250	15,150	0.27
P3	4550	31,318	35,868	0.12

- ***Third step***
  The optimal matches obtained in this step can be found in Table 4.24. In each cell of this table a binary number is given to denote whether or not a corresponding match is present in the multi-plant HEN design. If a value of 1 is given, then the heat duty (kW) of the corresponding match is also specified

TABLE 4.24
Optimal Matches Obtained at the Third Step of Indirect Strategy II in the Illustrative Example

Cold \ Hot	P1_H1	P1_F	P1_Fuel	P2_H1	P2_HP	P3_H1	P3_H2	P3_Fuel
P1_C1	1 (260)	1 (400)	1 (60)	0	0	0	0	0
P1_C2	0	1 (640)	0	0	0	0	0	0
P1_CW	1 (510)	0	0	0	0	0	0	0
P2_C1	0	0	0	1 (280)	0	0	0	0
P2_C2	0	0	0	1 (110.25)	1 (264.75)	0	0	0
P2_F	0	0	0	1 (325)	0	0	0	0
P3_C1	0	0	0	0	0	1 (660)	0	1 (465)
P3_F	0	0	0	0	0	0	1 (715)	0
P3_CW	0	0	0	0	0	0	1 (165)	0

in a parenthesis in the same cell. It can be observed from Table 4.24 that the inner-plant heat exchanges are facilitated with four heat exchangers, two furnaces, one heater (using high-pressure steam), and two coolers (using cooling water). On the other hand, the interplant heat flows are realized with four heat-transfer units. The intermediate fluid is used as the cold stream in one such unit in plant P2 (P2_F) and also in another in plant P3 (P3_F), while it is the hot stream in two separate units in plant P1 (P1_F).

- ***Fourth step***

  On the basis of the optimal matches listed in Table 4.24, a unique superstructure can be built for every process stream in the three plants under consideration. The objective function defined by equations (4.89)–(4.90), (4.92), and (4.125)–(4.130) can then be established according to the negotiation powers given in Table 4.23, while the other model constraints can be obtained by applying the basic principles of material and energy balances to the superstructures (Floudas et al., 1986). The annualization factor in equations (4.125) and (4.126), i.e., Af, was also set at 0.1349. All cost coefficients in equations (4.92), (4.127), and (4.128) were chosen to be 670 USD/$m^{1.66}$ and $\beta = 0.83$, while all overall heat-transfer coefficients were taken to be 1 W/m^2·K. Finally, the temperature rise of cooling water in every cooler was set to be 5°C.

TABLE 4.25
Total Annual Cost Savings Achieved with Indirect Strategy II in the Illustrative Example

Plant #	Net Utility Cost Savings after Step 2 (USD/yr)	Total Capital Costs without Integration (USD/yr)	Total Capital Costs after Step 4 (USD/yr)	Total Capital Cost Savings after Step 4 (USD/yr)	TAC Savings after Step 4 (USD/yr)
P1	6832	5891	4273	1618	8450
P2	15,150	5861	12,027	−6166	8984
P3	35,868	7865	8999	−1134	34,734

By solving the corresponding NLP program, one could generate the interplant HEN design presented in Figure 4.13 and the final economic assessment in Table 4.25. It can be observed that, although the capital investments of plants P2 and P3 must be increased to realize the required inner- and interplant heat exchanges, this interplant heat integration project should still be quite feasible since the TAC savings of all plants are positive and reasonably distributed.

4.6 DISTINCT FEATURES IN APPLICATION RESULTS OF ILLUSTRATIVE EXAMPLE

In the aforementioned sections, three alternative direct and indirect multi-plant HEN synthesis strategies have been developed based on non-cooperative games. From the application results reported in Section 4.2 and Sections 4.4.5 and 4.5.5, several obvious and unique features can be identified

- By comparing the net savings in utility costs achieved with the direct and indirect synthesis strategies in the illustrative example (see Tables 4.5, 4.19, and 4.23), one could see that in general the overall energy cost of a direct heat integration scheme should be lower than any of its indirect counterparts and, also, the sum of net utility cost savings achieved with an intermediate fluid should be larger than that with the heating and cooling utilities.
- From Figure 2.2 in Chapter 2 and Figure 4.8, it can be observed that the minimum consumption rates of steams and cooling water of the standalone plants are basically the same as those of the utility-facilitated indirect heat integration scheme, i.e., indirect strategy I. Since every available hot/cold utility in the latter case is essentially a common heat source/sink shared by all plants, the total utility cost saving reported in Table 4.18, i.e., 55,400 (= 55,700 − 20,400 + 20,100) USD/yr, is achieved in Step 1 of the indirect strategy I simply by replacing the utilities used in standalone cascades (Figure 4.1) with their cheapest alternatives (Figure 4.18).

- One can also see from Figures 4.8 and 4.12 that, instead of sharing utilities across the plant boundaries, the indirect interplant heat integration scheme can be made thermodynamically more efficient by using an intermediate fluid to produce better matches.
- Furthermore, one would expect that a direct interplant heat integration scheme can be adopted to reap the largest possible energy saving as long as the required interplant heat exchanges are realizable in practice.
- Finally, in the illustrative example discussed above, the total capital cost of the HEN synthesized with indirect strategy I is lower than that with indirect strategy II and, also, the former is structurally simpler and should be considered as a more operable and controllable design.

4.7 EXTRA CASE STUDIES ON INDIRECT STRATEGIES

Since the indirect strategies are more practical, a third example is presented below to further confirm the findings given in the previous section. Tables 4.26 and 4.27 respectively show the stream and utility data used in this example. Based on a minimum temperature approach of 10 K, the corresponding standalone heat-flow cascades can be built and they are given in Figure 4.14. The minimum utility costs of P1, P2, and P3 can be determined respectively to be $88,100\left(\overline{Z}_{p1}\right)$, $30,100\left(\overline{Z}_{p2}\right)$, and $60,500\left(\overline{Z}_{p3}\right)$ USD/yr.

4.7.1 Indirect Strategy I

The integrated heat-flow cascade given in Figure 4.15 was produced at the first step of strategy I. Although the required utility consumption rates are the same as those in the standalone heat-flow cascades (see Figure 4.14), the cheapest utility is adopted at every occasion in this integrated scheme, i.e., the cooling water is always taken

TABLE 4.26
Stream Data Used in the Third Example

Plant #	Stream #	T_{in} (°C)	T_{in} (°C)	F_{cp} (kW/°C)
P1	H1	150	40	7
	C1	60	270	9
	C2	110	190	8
P2	H1	200	70	5.5
	C1	30	110	3.5
	C2	140	190	7.5
P3	H1	380	150	8.5
	H2	200	40	5.5
	C1	110	380	3.5

TABLE 4.27
Utility Data Used in the Third Example

Plant #	Utility Type	Temperature (°C)	Unit Cost (USD/kW·yr)	Upper Limit (kW)
P1	Cooling water	25	10	5000
	High p. steam	400	70	5000
	Medium p. steam	280	50	5000
	Low p. steam	200	40	5000
P2	Cooling water	25	22.5	5000
	High p. steam	400	60	5000
	Medium p. steam	280	40	5000
	Low p. steam	200	25	5000
P3	Cooling water	25	30	5000
	High p. steam	400	80	5000
	Medium p. steam	280	35	5000
	Low p. steam	200	30	5000

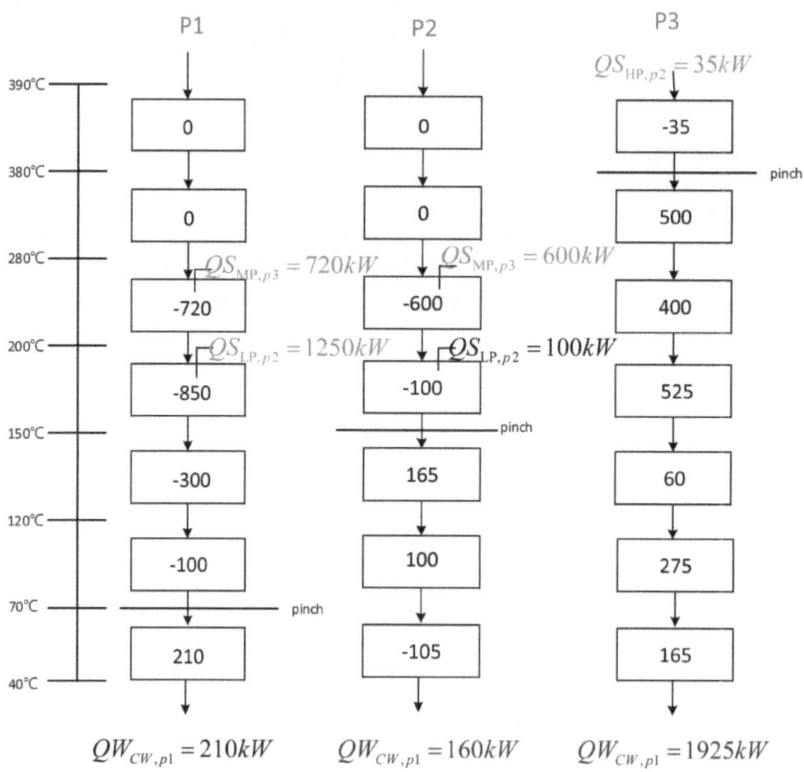

FIGURE 4.15 Integrated heat-flow cascades built at the first step of procedure I in the third example. (Reproduced with permission from *Energy* 2018, 148: 90–111. Copyright 2018 Elsevier.)

from P1, the high- and low-pressure steams are from P2, and the medium-pressure steam is from P3. The resulting interplant HEN design is presented in Figure 4.16, while a complete economic summary is provided in Table 4.28.

4.7.2 Indirect Strategy II

The integrated heat-flow cascade in Figure 4.17 was produced by applying the first step of strategy II. The temperature range of hot oil in this case can be chosen to be [70,280]°C if it is used as a hot stream, and [60,270]°C if otherwise. Note that $PC = \{P2, P3\}$ and $PH = \{P1\}$, and it was determined that $FCP_{P1} = 9$, $FCP_{P2} = 0{,}619$, and $FCP_{P3} = 8.381$. Note that, by using the intermediate fluid, a significant amount, 1459.5 kW, can be reduced respectively from the hot and cold utility consumption rates of the integrated scheme obtained with procedure I (see Figures 4.15 and 4.17). By adopting the same parameter values presented previously in the illustrative example (see Section 4.4.5 or 4.5.5), one can produce the interplant HEN design in Figure 4.18 and the corresponding cost summary can be found in Table 4.29.

4.7.3 Cost Analysis

From Tables 4.28 and 4.29, it can be observed that these optimization results clearly reaffirm the conclusions drawn from the illustrative example (see the beginning of Section 4.6). Notice that the total utility cost saving achieved with the heating and cooling utilities (73,750 USD/yr) is much smaller than that with the intermediate fluid (116,410 USD/yr). This difference is probably due to their roles in facilitating heat flows between plants. Although various hot and cold utilities may be available in different plants, they are all supposed to be treated equally without distinguishing their origins. Thus, if the interplant heat exchanges are facilitated with utilities, then the total hot and cold utility consumption rates in every plant should respectively remain unchanged. A considerable utility cost saving may still be achievable in this scenario due to the use of cheaper alternatives. On the other hand, notice that the intermediate fluid is treated as either an *extra* hot or cold process stream in every individual plant. These additional streams are selected in Step 1 of indirect strategy II for optimally adjusting the grand composite curve of each plant so as to make the corresponding total utility cost lower than that of the standalone counterpart. As a result, the overall consumption rates of hot and cold utilities of the entire site can both be reduced to lower levels as well.

However, since the intermediate fluid is present in a much wider temperature range than that of any hot or cold utility, the interplant HEN design synthesized with the intermediate fluid requires a slightly higher total capital cost than the utility-facilitated design. This tendency is also consistent with that found in the illustrative example. Finally, note that the eventual feasibility of each interplant heat integration program can also be ensured by the reasonably distributed TAC savings among all participating members.

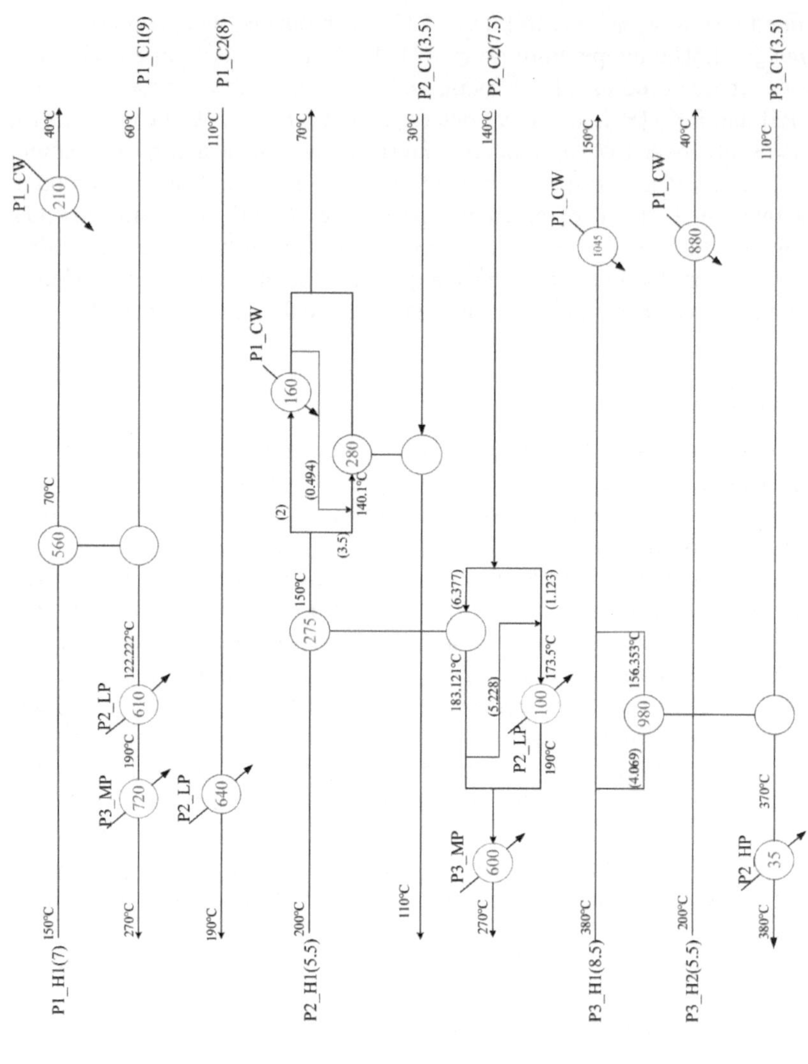

FIGURE 4.16 Interplant HEN design produced with procedure I in the third example. (Reproduced with permission from *Energy* 2018, 148: 90–111. Copyright 2018 Elsevier.)

TABLE 4.28
Total Annual Cost Savings Achieved with Indirect Strategy I in the Third Example

Plant #	Net Utility Cost Savings after Step 2 (USD/yr)	Total Capital Costs without Integration (USD/yr)	Total Capital Costs after Step 4 (USD/yr)	Total Capital Cost Savings after Step 4 (USD/yr)	TAC Savings after Step 4 (USD/yr)
P1	9748	6066	8367	−2500	7248
P2	38,783	6321	8152	−1882	36,901
P3	25,219	4533	6674	−1363	23,856

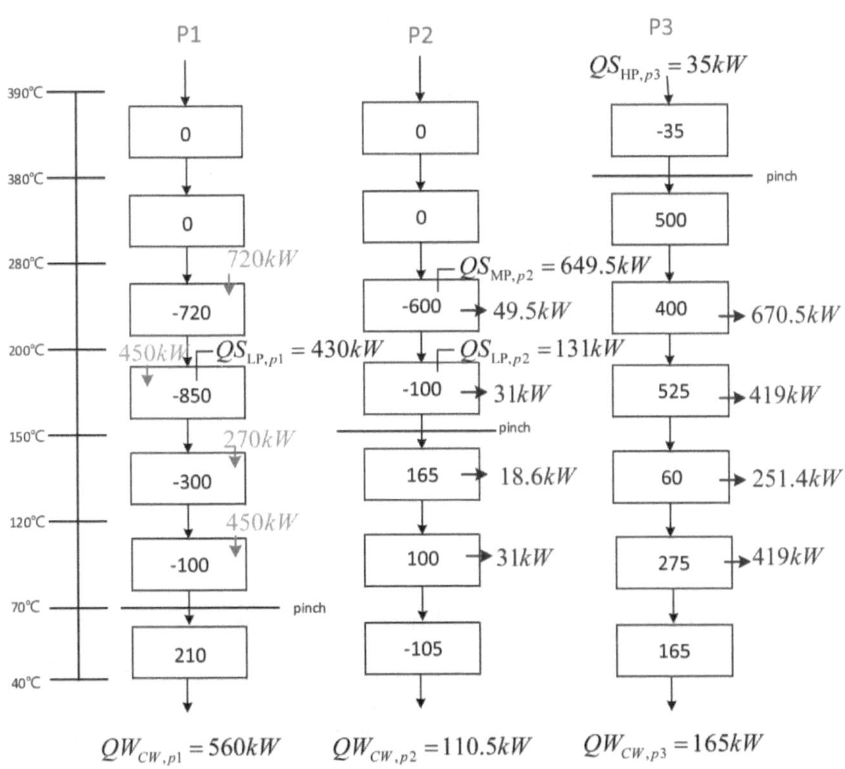

FIGURE 4.17 Integrated heat-flow cascades built at the first step of indirect strategy II in the third example. (Reproduced with permission from *Energy* 2018, 148: 90–111. Copyright 2018 Elsevier.)

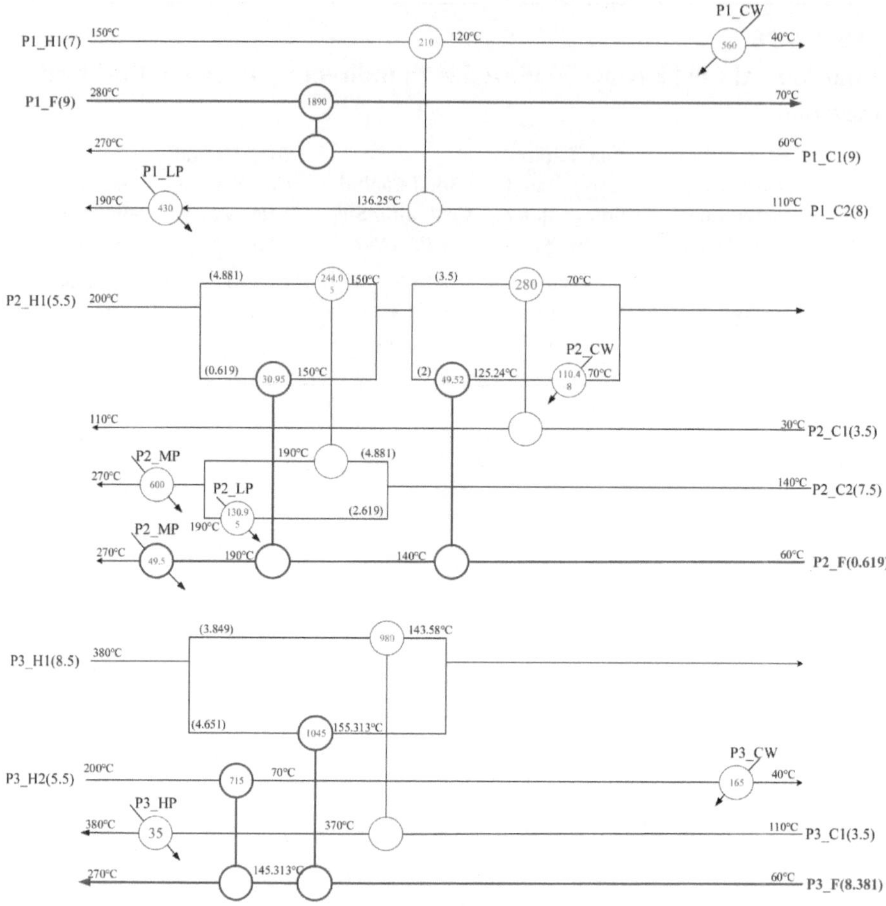

FIGURE 4.18 Interplant HEN design produced with indirect strategy II in the third example. (Reproduced with permission from *Energy* 2018, 148: 90–111. Copyright 2018 Elsevier.)

TABLE 4.29
Total Annual Cost Savings Achieved with Indirect Strategy II in the Third Example

Plant #	Net Utility Cost Savings after Step II_{ii} (USD/yr)	Total Capital Costs without Integration (USD/yr)	Total Capital Costs after Step II_{iv} (USD/yr)	Total Capital Cost Savings after Step I_{iv} (USD/yr)	TAC Savings after Step II_{iv} (USD/yr)
P1	73,208	6066	11,841	−5775	67,433
P2	6965	6321	7096	−775	6190
P3	36,237	4533	4691	−158	36,079

4.8 CONCLUDING REMARKS

This chapter aims to improve the practical applicability of interplant heat integration schemes. An existing non-cooperative-game-theory based sequential optimization strategy has been improved on the basis of the individual negotiation power of every participating plant to stipulate the "fair" price for each energy trade, to determine the "reasonable" proportions of capital cost to be shouldered by the involved parties of every interplant heat exchanger, and to produce an acceptable distribution of TAC savings. Also, to address various safety and operational concerns about direct heat transfers across plant boundaries, the interplant heat flows have been facilitated in the proposed integration schemes with either the available utilities or an extra intermediate fluid. Because of these additional practical features in interplant heat integration, realization of the resulting financial and environmental benefits should become more likely. Furthermore, from the optimization results obtained in case studies, one can find that feasible schemes can indeed be synthesized with the above two procedures. The interplant HEN design generated with procedure I should be simpler and more operable, while its counterpart created by procedure II is usually more energy efficient.

REFERENCES

Bagajewicz, M.J., Rodera, H. 2002. Multiple plant heat integration in a total site. *AIChE Journal* 48: 2255–2270.

Chang, C., Chen, X., Wang, Y., Feng, X. 2017. Simultaneous optimization of multi-plant heat integration using intermediate fluid circles. *Energy* 121: 306–317.

Chang, H.H., Chang, C.T., Li, B.H. 2018. Game-theory based optimization strategies for stepwise development of indirect interplant heat integration plans. *Energy* 148: 90–111.

Cheng, S.L., Chang, C.T., Jiang, D. 2014. A game-theory based optimization strategy to configure inter-plant heat integration schemes. *Chemical Engineering Science* 118: 60–73.

Floudas, C.A., Ciric, A.R. 1989. Strategies for overcoming uncertainties in heat exchanger network synthesis. *Computers and Chemical Engineering* 13: 1133–1152.

Floudas, C.A., Ciric, A.R., Grossmann, I.E. 1986. Automatic synthesis of optimum heat exchanger network configurations. *AIChE Journal* 32: 276–290.

Floudas, C.A., Grossmann, I.E. 1987. Automatic generation of multiperiod heat exchanger network configurations. *Computers and Chemical Engineering* 11: 123–142.

Hackl, R., Harvey, S. 2015. From heat integration targets toward implementation—a TSA(total site analysis)-based design approach for heat recovery systems in industrial clusters. *Energy* 90: 163–172.

Hiete, M., Ludwig, J., Schultmann, F. 2012. Intercompany energy integration adaptation of thermal pinch analysis and allocation of savings. *Journal of Industrial Ecology* 16: 689–698.

Kralj, A.K. 2008. Heat integration between two biodiesel processes using a simple method. *Energy & Fuels* 22: 1972–1979.

Lakshmanan, A., Rooney, W.C., Biegler, L.T. 1999. A case study for reactor network synthesis: the vinyl chloride process. *Computers and Chemical Engineering* 23: 479–495.

Liew, P.Y., Alwi, S. R, Klemes, J.J., Varbanov, P.S., Manan, Z.A. 2014a. Algorithmic targeting for total site heat integration with variable energy supply/demand. *Applied Thermal Engineering* 70: 1073–1083.

Liew, P.Y., Lim, J.S., Alwi, S.R., Manan, Z.A., Varbanov, P.S., Klemes, J.J. 2014b. A retrofit framework for total site heat recovery systems. *Applied Energy* 135: 778–790.
Liew, P.Y., Theo, W.L., Alwi, S.R.W., Lim, J.S., Manan, Z.A., Klemes, J.J., Varbanov, P.S. 2017. Total site heat integration planning and design for industrial, urban and renewable systems. *Renewable and Sustainable Energy Reviews* 68: 964–985.
Matsuda, K., Hirochi, Y., Tatsumi, H., Shire, T. 2009. Applying heat integration total site based pinch technology to a large industrial area in Japan to further improve performance of highly efficient process plants. *Energy* 34: 1687–1692.
Papoulias, S.A., Grossmann, I.E. 1983. A structural optimization approach in process synthesis—II. Heat recovery networks. *Computers and Chemical Engineering* 7: 707–721.
Quintas, L.G. 1989. A note on poly-matrix games. *International Journal of Game Theory* 18: 261–272.
Tarighaleslami, A.H., Walmsley, T.G., Atkins, M.J., Walmsley, M.R.W., Liew, P.Y., Neale, J.R. 2017. A unified total site heat integration targeting method for isothermal and non-isothermal utilities. *Energy* 119: 10–25.
Wang, Y., Chang, C., Feng, X. 2015. A systematic framework for multi-plants heat integration combining direct and indirect heat integration methods. *Energy* 90: 56–67.
Yee, T.F., Grossmann, I.E. 1990. Simultaneous optimization models for heat integration—II. Heat exchanger network synthesis. *Computers and Chemical Engineering* 14: 1165–1184.
Yee, T.F., Grossmann, I.E., Kravanja, Z. 1990. Simultaneous optimization models for heat integration—I. Area and energy targeting and modeling of multi-stream exchangers. *Computers and Chemical Engineering* 14: 1151–1164.
Zhang, B.J., Li, J., Zhang, Z.L., Wang, K., Chen, Q.L. 2016. Simultaneous design of heat exchanger network for heat integration using hot direct discharges/feeds between process plants. *Energy* 109: 400–411.

5 Fair Benefit Allocation to Facilitate Interplant Heat Integration Based on Cooperative Games

Since the assumption of a non-cooperative game may not always be valid and, also, the HEN design produced with the sequential optimization method cannot always reach a true optimum, there are incentives for further development of an alternative TSHI strategy to circumvent these drawbacks. Hiete et al. (2012) treated TSHI problem as a *cooperative* game, but their design procedure still required heuristic judgments. Jin et al. (2018) proposed a rigorous synthesis procedure for producing a fair cost-sharing plan in the spirit of a cooperative game so as to facilitate realization of the multi-plant HEN design obtained with the simultaneous optimization strategy. The proposed approach is implemented basically in two stages. The minimum total annual cost (TAC) of each and every potential coalition was first determined with a modified version of the conventional MINLP model. An important implication of this model is that, in the resulting multi-plant HEN, the interplant heat flows are primarily facilitated with direct matches between hot and cold process streams. On the other hand, the benefit allocation issue is addressed in the second stage on the basis of the risk-based Shapley values. An effective cost-sharing scheme is constructed in this stage according to the core solution of a cooperative game and the risk-based Shapley values of all players. The former ensures solution stability, while the latter yields a reasonable cost-distribution plan. The justifications for adopting the aforementioned two-stage implementation strategy are twofold, i.e., (1) the simultaneous HEN synthesis method usually yields a better trade-off than the sequential counterpart and (2) the Shapley value of any given coalition can be computed in a straightforward fashion in the second stage with only a single aggregated index, i.e., the minimum TACs obtained in the first stage.

An obvious weakness of the above approach can be attributed to the assumption of grass-root designs. In many cases in the real world, the existing plants on an industrial park were built primarily to meet the targeted market demands which arose during different periods in the past, and each should have already been equipped with a HEN by the time of its completion. Therefore, the above benefit allocation problem occurs usually when a multi-plant HEN retrofit project is initiated for the purpose of gaining extra energy savings. Since an existing HEN is present in each plant, a modified objective function, i.e., the extra TAC saving, is utilized in the model formulation of the multi-plant counterpart. The subsequent allocation approach taken in the present study is basically the same as that adopted in Jin et al. (2018). In particular, the task of devising the benefit distribution scheme is considered to be analogous to that of engaging in a cooperative game. More specifically, an effective cost-sharing plan is stipulated according to the core of a cooperative game and the Shapley values of all players.

The discussions presented in this chapter are divided into three parts. Firstly, the computation procedure of risk-based Shapley values is described in detail. This cooperative-game-based approach is next applied to grass-root designs of the multi-plant heat exchanger networks, while the same benefit allocation strategy is adopted to facilitate implementation of the retrofit projects in the third part.

5.1 RISK-BASED SHAPLEY VALUES

5.1.1 Core

"Core" is the solution set of a cooperative game. Each solution in the set depicts a realizable plan for every member of the joint venture to receive a reasonable portion of the total profit of a capital investment project. To facilitate illustration, let us use $\Theta = \{1,2,\ldots,n\}$ to represent the set of all players in a game and $\Sigma \subseteq \Theta$ may form a so-called "coalition." Then the collection of all possible coalitions should be the power set of Θ (denoted as 2^Θ) and a characteristic function $\nu(\cdot)$ can be defined accordingly as the mapping $\nu : 2^\Theta \rightarrow R$. The function value $\nu(\Sigma)$, where $\Sigma \in 2^\Theta$, is the total profit realized by the coalition as a whole. To ensure function consistency, it is required that $\nu(\varnothing) = 0$. Note also that, since $\Theta \in 2^\Theta$, Θ is referred to as the *grand coalition* in this chapter. Let us further denote the financial benefit allocated to plant $i \in \Theta$ in coalition $\Sigma \subseteq \Theta$ as $x_{\Sigma,i}$. Thus, $\nu(\Sigma) = \sum_{i \in \Sigma} x_{\Sigma,i}$ and $\mathbf{x}_\Sigma = [x_{\Sigma,1}, x_{\Sigma,2}, \ldots]$ is referred to as the *pre-imputation* vector of coalition Σ. The pre-imputation vector of the grand coalition, i.e., $\mathbf{x}_\Theta$, in the core $C(\nu)$ should possess the following properties:

1. *Individual rationality*: The financial benefit allocated to player i in the grand coalition Θ should be larger than or equal to that received by a single player working alone, i.e.,

$$x_{\Theta,i} \geq \nu(i), \forall i \in \Theta \tag{5.1}$$

2. *Group rationality*: The total financial benefit realized by the grand coalition should be entirely allocated to all its members, i.e.,

$$\sum_{i \in \Theta} x_{\Theta,i} = \nu(\Theta) \tag{5.2}$$

3. *Coalition rationality*: The total financial benefit realized by a sub-coalition Σ of grand coalition Θ ($\Sigma \subseteq \Theta$) should not be greater than the sum of

financial benefit allocated to the members of this sub-coalition by the grand coalition, i.e.,

$$\sum_{i\in\Sigma} x_{\Theta,i} \geq \nu(\Sigma), \forall\Sigma \subseteq \Theta \tag{5.3}$$

4. *No cross subsidization*: The financial benefit allocated to player i by grand coalition Θ should be smaller than or equal to the marginal contribution of player i to the total financial benefit of Θ, i.e.,

$$x_{\Theta,i} \leq \nu(\Theta) - \nu(\Theta \setminus i), \forall i \in \Theta. \tag{5.4}$$

Notice that $\nu(\Theta \setminus i)$ above denotes the total financial benefit achieved by the sub-coalition $\Theta\setminus i$ of the grand coalition Θ and this sub-coalition is formed by excluding player i from Θ. Equation (5.4) is necessary because, if otherwise, the members in $\Theta\setminus i$ do not have the incentive to accept player i. Finally, it should be noted that the core defined by equations (5.1)–(5.4) only represents a solution set for the *benefit allocation* problems. If the opposite *cost-sharing* problems are under consideration, then the inequalities in equations (5.1), (5.3), and (5.4) should be modified by changing the embedded operators from "$\geq$" to "$\leq$" and vice versa. This practice allows the function value $\nu(\cdot)$ to remain nonnegative so as to facilitate easy calculation and analysis.

5.1.2 Conventional Shapley Values

The core $C(\nu)$ clearly represents a feasible region, while a one-point solution in this region can be obtained by computing the Shapley values. This allocation approach essentially calls for dividing and distributing the total benefit realized (or cost incurred) by a coalition according to the contribution level of each participating member. Before evaluating the Shapley values, it is necessary to calculate the profits generated (or cost needed) by all possible coalitions $\Sigma \subseteq \Theta$. To enumerate all scenarios exhaustively, let us first consider the $n!$ permutations of the n players in Θ and collect the corresponding sequences in a set denoted as $\Pi(\Theta)$. Let us further express an element in $\Pi(\Theta)$ as $\boldsymbol{\pi}_\sigma$ (where $\sigma = 1, 2, \ldots, n!$), i.e., $\Pi(\Theta) = \{\pi_1, \pi_2, \ldots, \pi_{n!}\}$, while a particular sequence σ' in $\Pi(\Theta)$ may be written explicitly as $\pi_{\sigma'} = (\pi_{\sigma'}(1), \pi_{\sigma'}(2), \ldots, \pi_{\sigma'}(n))$.

Furthermore, a sequence of marginal contributions to the total benefit or cost (denoted as $\mathbf{m}^\sigma$) can be computed next according to each sequence $\boldsymbol{\pi}_\sigma$ in $\Pi(\Theta)$, i.e.,

$$\mathbf{m}^\sigma = \left(m^\sigma_{\pi_\sigma(1)}, m^\sigma_{\pi_\sigma(2)}, \ldots, m^\sigma_{\pi_\sigma(k)}, \ldots, m^\sigma_{\pi_\sigma(n)}\right) \tag{5.5}$$

where

$$m^\sigma_{\pi_\sigma(1)} = \nu(\pi_\sigma(1)) - \nu(\varnothing) \tag{5.6}$$

$$m_{\pi_\sigma(k)}^{\sigma} = \nu\left(\pi_\sigma(1),\ldots,\pi_\sigma(k)\right) - \nu\left(\pi_\sigma(1),\ldots,\pi_\sigma(k-1)\right) \quad (5.7)$$

and $k = 2,3,\ldots,n$.

It should be noted that the precedence order of the elements in sequence $\mathbf{m}^\sigma$ corresponds to that in sequence $\boldsymbol{\pi}_\sigma$. These n elements can be rearranged according their original order in Θ and then placed in another column vector $\mathbf{o}^\sigma$. After obtaining $\mathbf{o}^\sigma$ that stores the rearranged marginal contributions for every sequence $\boldsymbol{\pi}_\sigma$ in Π(Θ), one can finally compute their arithmetic averages and store them in the Shapley-value vector φ_Θ as follows

$$\varphi_\Theta = \frac{1}{n!} \sum_{\pi_\sigma \in \Pi(\Theta)} \mathbf{o}^\sigma \quad (5.8)$$

where $\varphi_\Theta = \left[\varphi_{\Theta,1} \quad \varphi_{\Theta,2} \quad \cdots \quad \varphi_{\Theta,n}\right]^T$ and $\varphi_{\Theta,i}$ $(i = 1,2,\ldots,n)$ denotes the average benefit/cost allocated to player i by grand coalition Θ. It should also be noted that the above notation on the Shapley values for Θ can be extended to any subset of the grand coalition, i.e., $\Sigma \subseteq \Theta$. If the players in Σ form a coalition, then the Shapley value of player i ($\forall i \in \Sigma$) can be written as $\varphi_{\Sigma,i}$ and computed with the same procedure.

5.1.3 Potential Fallout of Coalition

Notice that the above computation procedure ignores the potential risk of coalition collapse. As a general rule, the reliability of an interconnected engineering system should be regarded as an important design issue. If the interplant heat integration scheme is viewed as a coalition in a cooperative game, then the potential risk of its collapse must also be considered seriously. An unplanned plant shutdown, which may be due to a wide variety of equipment failures and/or human errors, obviously disables some of the hot and cold streams in the multi-plant HEN and forces the other functional plants in coalition to take remedial actions. Thus, it is necessary to adjust the aforementioned conventional Shapley values by assessing penalties for the potential fallouts of interplant heat integration.

The characteristic function $\nu(\Sigma)$ should yield the minimum total annual cost (TAC) of a *functional* multi-plant HEN for coalition $\Sigma \subseteq \Theta$ if this coalition is a grass-root venture, while $\nu(\Sigma)$ should be the total saving in TAC if a retrofit project is under consideration. In either case, the shutdown risk of plant $i \in \Sigma$ can be expressed with its unreliability over a year (denoted as α_i). Note that, since the failure rate of any hardware item may be extracted from the historical maintenance data, e.g., see Hoyland and Rausand (1994), the unreliability should be viewed as an intrinsic parameter of the given plant. On the other hand, the financial penalties of other plants in coalition can only be estimated on a case-by-case basis.

5.1.4 The Defective Coalition

As mentioned before, the characteristic function $\nu(\Sigma)$ determines the maximum benefit or minimum cost of coalition $\Sigma \subseteq \Theta$. If for some reason player $j \in \Sigma$ drops out of the coalition, the financial benefit of defective coalition $\Sigma\backslash j$ clearly cannot be higher (or the cost burden of $\Sigma\backslash j$ cannot be lower) than $v(\Sigma \setminus j)$ because of the potential extra penalty incurred due to coalition breakup. To facilitate precise explanation, let us use Σ_{+i} and Ξ_{+i} to represent two subsets of the grand coalition that both contain player i. The former subset $\Sigma_{+i} \subseteq \Theta$ denotes the original coalition, while the latter $\Xi_{+i} \subseteq \Sigma_{+i}$ represents the defective coalition formed by excluding all members in $\Sigma_{+i} \setminus \Xi_{+i}$.

Let us next write the total benefit (or cost) of defective coalition Ξ_{+i} in a more informative notation as $w\left(\Xi_{+i} : \Xi_{+i}^{\Sigma_{+i}}\right)$ and the corresponding individual benefit (or cost) allocated to player $i \in \Xi_{+i}$ as $w\left(i : \Xi_{+i}^{\Sigma_{+i}}\right)$. It must be noted that the latter is not the same as player i working alone, i.e., $w\left(i : \Xi_{+i}^{\Sigma_{+i}}\right) \neq w\left(i : i^{\Sigma_{+i}}\right)$.

5.1.5 Benefits/Costs Allocated to Members of a Defective Coalition

The computation procedure of $w\left(i : \Xi_{+i}^{\Sigma_{+i}}\right)$ is in principle similar to that for the conventional Shapley values. First of all, the group rationality must be followed, i.e.,

$$\sum_{j \in \Xi_{+i}} w\left(j : \Xi_{+i}^{\Sigma_{+i}}\right) = w\left(\Xi_{+i} : \Xi_{+i}^{\Sigma_{+i}}\right) \tag{5.9}$$

If the defective coalition Ξ_{+i} further collapses into $\Upsilon \subseteq \Xi_{+i}$, the corresponding total profit (or cost) of sub-coalition Υ can be denoted as $w\left(\Upsilon : \Xi_{+i}^{\Sigma_{+i}}\right)$. Notice that Υ may or may not include player i. Notice also that $w\left(\Upsilon : \Xi_{+i}^{\Sigma_{+i}}\right)$ can be used to replace $\nu(\Upsilon)$ in computing the conventional Shapley values if coalition breakup is not an issue. Notice finally that in most cases $w\left(\Upsilon : \Upsilon^{\Sigma_{+i}}\right)$ may be determined independently with other means in practical applications.

As a result, a fair benefit/cost allocation scheme can be determined for the defective coalition Ξ_{+i} by calculating $w\left(i : \Xi_{+i}^{\Sigma_{+i}}\right)$ as follows

$$w\left(i : \Xi_{+i}^{\Sigma_{+i}}\right) = \sum_{\Upsilon \subseteq \Xi_{+i}} \varpi(\Upsilon) M(\Upsilon) \tag{5.10}$$

$$\varpi(\Upsilon) = \frac{\left(|\Upsilon| - 1\right)!\left(|\Xi_{+i}| - |\Upsilon|\right)!}{|\Xi_{+i}|!} \tag{5.11}$$

$$M(\Upsilon) = w\left(\Upsilon : \Upsilon^{\Sigma_{+i}}\right) - w\left(\{\Upsilon \setminus i\} : \{\Upsilon \setminus i\}^{\Sigma_{+i}}\right) \tag{5.12}$$

where $\varpi(\Upsilon)$ is the weighting factor of $\Upsilon \subseteq \Xi_{+i}$; $M(\Upsilon)$ is the marginal contribution of player i in the defective sub-coalition Υ; $|\Upsilon|$ and $|\Xi_{+i}|$ denote the cardinalities of sets Υ and Ξ_{+i} respectively. The weighted average $w\left(i : \Xi_{+i}^{\Sigma_{+i}}\right)$ represents the contribution level of player i in the defective coalition Ξ_{+i}. It should be noted that the above calculations should be carried out without sorting the marginal contributions since all possible scenarios are inherently included. Finally, the individual benefit/cost of player i in the original non-defective coalition Σ_{+i} should be identical to its conventional Shapley value, i.e.,

$$w\left(i:\Sigma_{+i}^{\Sigma_{+i}}\right)=\varphi_{\Sigma_{+i},i} \tag{5.13}$$

As a result, it is clear that

$$w\left(\Sigma_{+i}:\Sigma_{+i}^{\Sigma_{+i}}\right)=v\left(\Sigma_{+i}\right) \tag{5.14}$$

5.1.6 Expected Loss Due to Unscheduled Plant Shutdown(s)

Let us consider plant $i\in\Theta$ in a coalition $\Sigma_{+i}\subseteq\Theta$. If all other plants can be assumed to always stay within the coalition, the marginal contribution of plant i joining Σ_{+i} should be $v\left(\Sigma_{+i}\right)-v\left(\Sigma_{+i}\setminus i\right)$. However, as mentioned previously, since the unreliability of plant j, i.e., α_j ($\forall j\in\Sigma_{+i}$), is not negligible, it is necessary to evaluate the expected loss of plant i. To facilitate analysis, let us again divide Σ_{+i} into two subsets as before, that is, the defective coalition Ξ_{+i} and also the subset formed by the dropouts ($\Sigma_{+i}\setminus\Xi_{+i}$). By assuming that the plants in $\Sigma_{+i}\setminus\Xi_{+i}$ break down independently, the conditional probability of this scenario can be expressed as

$$p\left(\Xi_{+i}\mid\Sigma_{+i}\right)=\left(\prod_{j\in\left\{\Sigma_{+i}\setminus\Xi_{+i}\right\}}\alpha_j\right)\left(\prod_{k\in\Xi_{+i}}\beta_k\right) \tag{5.15}$$

where $\beta_k=1-\alpha_k$.

As mentioned before, the total profit (or cost) of defective coalition Ξ_{+i} can be written as $w\left(\Xi_{+i}:\Xi_{+i}^{\Sigma_{+i}}\right)$ and the benefit (or cost) allocated to player $i\in\Xi_{+i}$ is expressed as $w\left(i:\Xi_{+i}^{\Sigma_{+i}}\right)$. This allocated value should be smaller than $\varphi_{\Xi_{+i},i}$ (see equation (5.8)) in a benefit allocation problem, while larger in the cost-sharing problem. Consequently, the "risk" of loss in the former case can be expressed as

$$\begin{aligned}r_b\left(i:\Xi_{+i}^{\Sigma_{+i}}\right)&=p\left(\Xi_{+i}\mid\Sigma_{+i}\right)\left(\varphi_{\Xi_{+i},i}-w\left(i:\Xi_{+i}^{\Sigma_{+i}}\right)\right)\\&=\left(\prod_{j\in\left\{\Sigma_{+i}\setminus\Xi_{+i}\right\}}\alpha_j\right)\left(\prod_{k\in\Xi_{+i}}\beta_k\right)\left(\varphi_{\Xi_{+i},i}-w\left(i:\Xi_{+i}^{\Sigma_{+i}}\right)\right)\end{aligned} \tag{5.16}$$

By enumerating all possible defective coalitions in Σ_{+i}, one can then construct a formula for computing the expected loss due to unplanned plant shutdowns:

$$\begin{aligned}E\left[r_b\left(i:\Sigma_{+i}\right)\right]&=\sum_{\Xi_{+i}\subseteq\Sigma_{+i}}r_b\left(i:\Xi_{+i}^{\Sigma_{+i}}\right)\\&=\sum_{\Xi_{+i}\subseteq\Sigma_{+i}}\left\{\left(\prod_{j\in\left\{\Sigma_{+i}\setminus\Xi_{+i}\right\}}\alpha_j\right)\left(\prod_{k\in\Xi_{+i}}\beta_k\right)\left(\varphi_{\Xi_{+i},i}-w\left(i:\Xi_{+i}^{\Sigma_{+i}}\right)\right)\right\}\end{aligned} \tag{5.17}$$

In computing the conventional Shapley value with equation (5.7), one can see that the marginal contribution $\nu(\Sigma_{+i})-\nu(\Sigma_{+i}\setminus i)$ is supposed to be allocated to player *i*. Since the risk of coalition collapse is not negligible in the present case, player *i* is entitled to ask for more to compensate the loss resulting from unplanned plant shutdowns. A reasonable adjustment in this scenario is to add the expected loss to the original benefit allocated to player *i*, i.e., $\nu(\Sigma_{+i})-\nu(\Sigma_{+i}\setminus i)+E\left[r_b\left(i:\Sigma_{+i}\right)\right]$.

On the other hand, the risk and the corresponding expected loss in the cost-sharing case can be respectively written as

$$\begin{aligned} r_c\left(i:\Xi_{+i}^{\Sigma_{+i}}\right) &= p\left(\Xi_{+i}\mid\Sigma_{+i}\right)\left(w\left(i:\Xi_{+i}^{\Sigma_{+i}}\right)-\varphi_{\Xi_{+i},i}\right)\\ &= \left(\prod_{j\in\{\Sigma_{+i}\setminus\Xi_{+i}\}}\alpha_j\right)\left(\prod_{k\in\Xi_{+i}}\beta_k\right)\left(w\left(i:\Xi_{+i}^{\Sigma_{+i}}\right)-\varphi_{\Xi_{+i},i}\right) \end{aligned} \tag{5.18}$$

$$\begin{aligned} E\left[r_c\left(i:\Sigma_{+i}\right)\right] &= \sum_{\Xi_{+i}\subseteq\Sigma_{+i}} r_c\left(i:\Xi_{+i}^{\Sigma_{+i}}\right)\\ &= \sum_{\Xi_{+i}\subseteq\Sigma_{+i}}\left\{\left(\prod_{j\in\{\Sigma_{+i}\setminus\Xi_{+i}\}}\alpha_j\right)\left(\prod_{k\in\Xi_{+i}}\beta_k\right)\left(w\left(i:\Xi_{+i}^{\Sigma_{+i}}\right)-\varphi_{\Xi_{+i},i}\right)\right\} \end{aligned} \tag{5.19}$$

Therefore, the adjusted cost to be shared by player *i* in this case should be less and can be expressed as $\nu(\Sigma_{+i})-\nu(\Sigma_{+i}\setminus i)-E\left[r_c\left(i:\Sigma_{+i}\right)\right]$.

5.1.7 Computation of Risk-Based Shapley Values

Let us again consider the grand coalition $\Theta=\{1,2,\ldots,n\}$ mentioned before in Section 5.1.1 and also consider sequence σ in the set $\Pi(\Theta)$ defined in Section 5.1.2, i.e., $\pi_\sigma=\left(\pi_\sigma(1),\pi_\sigma(2),\ldots,\pi_\sigma(n)\right)$. A sequence of risk-adjusted marginal contributions can be computed according to each sequence $\boldsymbol{\pi}_\sigma$ in $\Pi(\Theta)$, i.e.,

$$\mathbf{h}^\sigma=\left(h^\sigma_{\pi_\sigma(1)},h^\sigma_{\pi_\sigma(2)},\ldots,h^\sigma_{\pi_\sigma(k)},\ldots,h^\sigma_{\pi_\sigma(n)}\right) \tag{5.20}$$

One can then try to determine the corresponding profit allocated to *every* player in a coalition according to equation (5.17) in the benefit allocation case. All elements in the sequence defined in equation (5.20) can be determined as follows:

$$h^\sigma_{\pi_\sigma(1)}=\nu\left(\pi_\sigma(1)\right)-\nu(\varnothing)+E\left[r_b\left(\pi_\sigma(1):\Xi_{+\pi_\sigma(1)}^{\{\pi_\sigma(1)\}}\right)\right] \tag{5.21}$$

$$\begin{aligned} h^\sigma_{\pi_\sigma(k)} &= \nu\left(\pi_\sigma(1),\ldots,\pi_\sigma(k)\right)-\nu\left(\pi_\sigma(1),\ldots,\pi_\sigma(k-1)\right)\\ &\quad+E\left[r_b\left(\pi_\sigma(k):\Xi_{+\pi_\sigma(k)}^{\{\pi_\sigma(1),\ldots,\pi_\sigma(k)\}}\right)\right] \end{aligned} \tag{5.22}$$

where $k = 2,3,\cdots,n$; $\nu(\cdot)$ is the characteristic function defined in Section 5.1.1.

On the other hand, one can also try to determine the cost shared by *every* player in a coalition according to equation (5.19) in the cost-sharing scenario. All elements in equation (5.20) can be determined as follows:

$$h^{\sigma}_{\pi_\sigma(1)} = \nu\left(\pi_\sigma(1)\right) - \nu(\varnothing) - E\left[r_c\left(\pi_\sigma(1) : \Xi^{\{\pi_\sigma(1)\}}_{+\pi_\sigma(1)}\right)\right] \tag{5.23}$$

$$\begin{aligned} h^{\sigma}_{\pi_\sigma(k)} &= \nu\left(\pi_\sigma(1),\ldots,\pi_\sigma(k)\right) - \nu\left(\pi_\sigma(1),\ldots,\pi_\sigma(k-1)\right) \\ &\quad - E\left[r_c\left(\pi_\sigma(k) : \Xi^{\{\pi_\sigma(1),\cdots,\pi_\sigma(k)\}}_{+\pi_\sigma(k)}\right)\right] \end{aligned} \tag{5.24}$$

where $k = 2,3,\ldots,n$ and $\nu(\cdot)$ is the characteristic function defined in Section 5.1.1.

The precedence order of the elements in sequence $\mathbf{h}^\sigma$ must correspond to that in $\boldsymbol{\pi}_\sigma$, and these elements should be rearranged next according to the original precedence order in Θ and then placed in a column vector $\boldsymbol{\theta}^\sigma$. A column vector of the risk-based distribution coefficients can be determined according to the following formula

$$\Psi_\Theta = \frac{1}{n!}\sum_{\pi_\sigma \in \Pi(\Theta)} \theta^\sigma \tag{5.25}$$

Notice that this vector is not yet the finalized allocation plan since the requirement of group rationality, i.e., equation (5.2), is not satisfied. To address this need, Ψ_Θ should be normalized so as to produce the risk-based Shapley values as follows

$$\hat{\varphi}_\Theta = \frac{\Psi_\Theta}{\Psi_\Theta{}^T \mathbf{1}_{n\times 1}} \nu(\Theta) \tag{5.26}$$

where $\hat{\varphi}_\Theta = \left[\hat{\varphi}_{\Theta,1}, \hat{\varphi}_{\Theta,2}, \ldots, \hat{\varphi}_{\Theta,n}\right]^T$ and $\hat{\varphi}_{\Theta,i}$ $(i = 1,2,\ldots,n)$ denotes the risk-based Shapley value of plant $i \in \Theta$; $\mathbf{1} = \left[1\cdots1\right]^T$; $\nu(\Theta)$ is the total profit or cost of grand coalition Θ.

5.2 GRASS-ROOT DESIGNS

As mentioned before, the heat exchanger network in a single plant can be designed with either a sequential synthesis procedure or a simultaneous optimization strategy. Although the former approach is simpler, the total annual cost of the final solution may not be truly minimized. Since the interplant heat integration scheme is viewed in this chapter as a coalition in a cooperative game, the latter strategy is adopted for synthesizing the HEN structure across the plant boundaries so as to achieve an optimal trade-off between utility and capital costs. To this end, the conventional MINLP model has to be reformulated and this modified version is presented below.

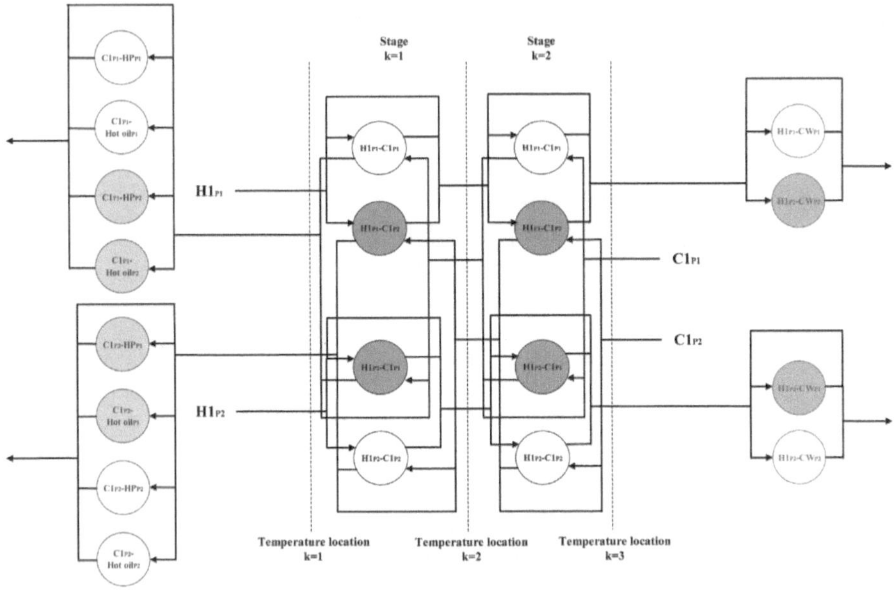

FIGURE 5.1 An example of superstructure for two-plant HEN synthesis.

5.2.1 Superstructure of Multi-Plant Heat Exchanger Networks

An example of the multi-plant counterpart of traditional single-plant superstructure (Yee and Grossmann, 1990) is shown in Figure 5.1. This superstructure has been adopted here to address the need to incorporate design options for placing interplant matches and for consuming external utilities across plant boundaries. For illustration convenience, Figure 5.1 only shows the structure used for two fictitious plants (say P1 and P2). Plant P1 is equipped with two hot utilities (HP_{p1} and $Fuel_{p1}$), one cold utility (CW_{p1}), one hot stream ($H1_{p1}$), and one cold stream ($C1_{p1}$), while P2 has two hot utilities (HP_{p2} and $Fuel_{p2}$), one cold utility (CW_{p2}), one hot stream ($H1_{p2}$), and one cold stream ($C1_{p2}$). Notice that the interplant and inner-plant matches in this superstructure are represented with colored and uncolored circles, respectively. Since there are two hot streams and two cold streams in the multi-plant HEN, the number of stages (NOK) should set to be two. Notice also that, in order to introduce design flexibility, an optional bypass is placed on each process stream in every stage and also at the end of the stream. If this option is not desired, the corresponding terms in the following model constraints can certainly be removed.

5.2.2 Model Constraints

The material and energy balances for stages, matches, and mixing nodes in superstructure can be formulated in a straightforward fashion. Let us assume that a total of *P* plants are interested in taking part in the interplant heat integration project. Together with all M&E balance equations, the model constraints corresponding to plant p $(p = 1,2,\ldots,P)$ are presented in detail here. To facilitate clear illustration, all sets, parameters, and variables utilized in these constraints are first defined below:

Sets:

C^p	Set of cold process streams in plant p $(p = 1,2,\ldots,P)$
CU^p	Set of cold utilities in plant p $(p = 1,2,\ldots,P)$
H^p	Set of hot process streams in plant p $(p = 1,2,\ldots,P)$
HU^p	Set of hot utilities in plant p $(p = 1,2,\ldots,P)$
ST	Set of all stages in superstructure

Parameters:

F_{i_p}	Heat-capacity flowrate of hot stream $i_p \in H^p$
F_{j_p}	Heat-capacity flowrate of cold stream $j_p \in C^p$
P	Total number of plants
TIN_{i_p}	Starting temperature of hot stream $i_p \in H^p$
TIN_{j_p}	Starting temperature of cold stream $j_p \in C^p$
$TOUT_{i_p}$	Target temperature of hot stream $i_p \in H^p$
$TOUT_{j_p}$	Target temperature of cold stream $j_p \in C^p$
U_{i_p,j_p}	Overall heat transfer coefficient of an inner-plant heat exchanger between hot stream $i_p \in H^p$ and cold stream $j_p \in C^p$
$U_{i_p,j_{q'}}$	Overall heat transfer coefficient of an interplant heat exchanger between hot stream $i_p \in H^p$ and cold stream $j_{q'} \in C^{q'}$
U_{i_q,j_p}	Overall heat transfer coefficient of an interplant heat exchanger between hot stream $i_q \in H^q$ and cold stream $j_p \in C^p$
U_{i_p,n_p}	Overall heat transfer coefficient of an inner-plant cooler between hot stream $i_p \in H^p$ and cold utility $n_p \in CU^p$
$U_{i_p,n_{q'}}$	Overall heat transfer coefficient of an interplant cooler between hot stream $i_p \in H^p$ and cold utility $n_{q'} \in CU^{q'}$
U_{m_p,j_p}	Overall heat transfer coefficient of an inner-plant heater between hot utility $m_p \in HU^p$ and cold stream $j_p \in C^p$
U_{m_q,j_p}	Overall heat transfer coefficient of an interplant heater between hot utility $m_q \in HU^q$ and cold stream $j_p \in C^p$
Γ_{i_p,j_p}	A large enough constant which is not smaller than the largest temperature approach corresponding to the inner-plant match (i_p, j_p), where $i_p \in H^p$ and $j_p \in C^p$
$\Gamma_{i_p,j_{q'}}$	A large enough constant which is not smaller than the largest temperature approach corresponding to the interplant match $(i_p, j_{q'})$, where i_p $i_p \in H^p$ and $j_{q'} \in C^{q'}$
Γ_{i_q,j_p}	A large enough constant which is not smaller than the largest temperature approach corresponding to the interplant match (i_q, j_p), where $i_q \in H^q$ and $j_p \in C^p$

Γ_{i_p,n_p} A large enough constant which is not smaller than the largest temperature approach corresponding to the inner-plant match (i_p, n_p), where $i_p \in H^p$ and $n_p \in CU^p$

$\Gamma_{i_p,n_{q'}}$ A large enough constant which is not smaller than the largest temperature approach corresponding to the interplant match $(i_p, n_{q'})$, where $i_p \in H^p$ and $n_{q'} \in CU^{q'}$

Γ_{m_p,j_p} A large enough constant which is not smaller than the largest temperature approach corresponding to the inner-plant match (m_p, j_p), where $m_p \in HU^p$ and $j_p \in C^p$

Γ_{m_q,j_p} A large enough constant which is not smaller than the largest temperature approach corresponding to the interplant match (m_q, j_p), where $m_q \in HU^q$ and $j_p \in C^p$

Ω_{i_p,j_p} A large enough constant which is not smaller than the largest heat duty of heat exchanger corresponding to inner-plant match (i_p, j_p), where $i_p \in H^p$ and $j_p \in C^p$

$\Omega_{i_p,j_{q'}}$ A large enough constant which is not smaller than the largest heat duty of heat exchanger corresponding to interplant match $(i_p, j_{q'})$, where $i_p \in H^p$ and $j_{q'} \in C^{q'}$

Ω_{i_q,j_p} A large enough constant which is not smaller than the largest heat duty of heat exchanger corresponding to interplant match (i_q, j_p), where $i_q \in H^q$ and $j_p \in C^p$

$\Omega_{i_p,n_{q'}}$ A large enough constant which is not smaller than the largest heat duty of cooler corresponding to interplant match $(i_p, n_{q'})$, where $i_p \in H^p$ and $n_{q'} \in CU^{q'}$

Ω_{m_q,j_p} A large enough constant which is not smaller than the largest heat duty of heater corresponding to interplant match (m_q, j_p), where $m_q \in HU^q$ and $j_p \in C^p$

Variables:

$A_{i_p,j_p,k}$ Heat-transfer area needed to facilitate the inner-plant match (i_p, j_p) at stage k, where $i_p \in H^p$, $j_p \in C^p$, and $k \in ST$

$A_{i_p,j_{q'},k}$ Heat-transfer area needed to facilitate the interplant match $(i_p, j_{q'})$ at stage k, where $i_p \in H^p$, $j_{q'} \in C^{q'}$, and $k \in ST$

$A_{i_q,j_p,k}$ Heat-transfer area needed to facilitate the interplant match (i_q, j_p) at stage k, where $i_q \in H^q$, $j_p \in C^p$, and $k \in ST$

A_{i_p,n_p} Heat-transfer area needed to facilitate the inner-plant cooler match (i_p, n_p), where $i_p \in H^p$ and $n_p \in CU^p$

$A_{i_p,n_{q'}}$ Heat-transfer area needed to facilitate the interplant cooler match $(i_p, n_{q'})$, where $i_p \in H^p$ and $n_{q'} \in CU^{q'}$

A_{m_p,j_p} Heat-transfer area needed to facilitate the inner-plant heater match (m_p, j_p), where $m_p \in HU^p$ and $j_p \in C^p$

A_{m_q,j_p} Heat-transfer area needed to facilitate the interplant heater match (m_q, j_p), where $m_q \in HU^q$ and $j_p \in C^p$

$dt_{i_p,j_p,k}$ Hot and cold temperature approach at the high-temperature end of the heat exchanger corresponding to match (i_p, j_p) at stage k, where $i_p \in H^p$, $j_p \in C^p$, and $k \in ST$

$dt_{i_p,j_p,k+1}$ Hot and cold temperature approach at the low-temperature end of the heat exchanger corresponding to match (i_p, j_p) at stage k where $i_p \in H^p$, $j_p \in C^p$, and $k \in ST$

$dt_{i_p,j_{q'},k}$ Hot and cold temperature approach at the high-temperature end of the heat exchanger corresponding to match $\left(i_p, j_{q'}\right)$ at stage k, where $i_p \in H^p$, $j_{q'} \in C^{q'}$, and $k \in ST$

$dt_{i_p,j_{q'},k+1}$ Hot and cold temperature approach at the low-temperature end of the heat exchanger corresponding to match $\left(i_p, j_{q'}\right)$ at stage k, where $i_p \in H^p$, $j_{q'} \in C^{q'}$, and $k \in ST$

$dt_{i_q,j_p,k}$ Hot and cold temperature approach at the high-temperature end of the heat exchanger corresponding to match (i_q, j_p) at stage k where $i_q \in H^q$, $j_p \in C^p$, and $k \in ST$

$dt_{i_q,j_p,k+1}$ Hot and cold temperature approach at the low-temperature end of the heat exchanger corresponding to match $\left(i_q, j_p\right)$ at stage k, where $i_q \in H^q$, $j_p \in C^p$, and $k \in ST$

$dtin_{i_p,n_p}$ Hot and cold temperature approach at hot stream inlet of cooler match $\left(i_p, n_p\right)$, where $i_p \in H^p$ and $n_p \in CU^p$

$dtin_{i_p,n_{q'}}$ Hot and cold temperature approach at hot steam inlet of cooler match $\left(i_p, n_{q'}\right)$, where $i_p \in H^p$ and $n_{q'} \in CU^{q'}$

$dtin_{m_p,j_p}$ Hot and cold temperature approach at cold stream inlet of heater match $\left(m_p, j_p\right)$, where $m_p \in HU^p$ and $j_p \in C^p$

$dtin_{m_q,j_p}$ Hot and cold temperature approach at cold stream inlet of heater match $\left(m_q, j_p\right)$, where $m_q \in HU^q$ and $j_p \in C^p$

$dtout_{i_p,n_p}$ Hot and cold temperature approach at hot stream outlet of cooler match $\left(i_p, n_p\right)$, where $i_p \in H^p$ and $n_p \in CU^p$

$dtout_{i_p,n_{q'}}$ Hot and cold temperature approach at hot stream outlet of cooler match $\left(i_p, n_{q'}\right)$, where $i_p \in H^p$ and $n_{q'} \in CU^{q'}$

$dtout_{m_p,j_p}$ Hot and cold temperature approach at cold stream outlet of heater match $\left(m_p, j_p\right)$, where $m_p \in HU^p$ and $j_p \in C^p$

$dtout_{m_q,j_p}$ Hot and cold temperature approach at cold stream outlet of heater match $\left(m_q, j_p\right)$, where $m_q \in HU^q$ and $j_p \in C^p$

$q_{i_p,j_p,k}$ Heat duty of heat exchanger corresponding to inner-plant match $\left(i_p, j_p\right)$ at stage k in plant p, where $i_p \in H^p$, $j_p \in C^p$, and $k \in ST$

$q_{i_p,j_{q'},k}$ Heat duty of heat exchanger corresponding to interplant match $\left(i_p, j_{q'}\right)$ at stage k, where $i_p \in H^p$, $j_{q'} \in C^{q'}$, and $k \in ST$

$q_{i_q,j_p,k}$	Heat duty of heat exchanger corresponding to interplant match (i_q, j_p) at stage k, where $i_q \in H^q$, $j_p \in C^p$, and $k \in ST$
q_{i_p,n_p}	Heat duty of inner-plant cooler corresponding to match (i_p, n_p), where $i_p \in H^p$ and $n_p \in CU^p$
$q_{i_p,n_{q'}}$	Heat duty of interplant cooler corresponding to match $(i_p, n_{q'})$, where $i_p \in H^p$ and $n_{q'} \in CU^{q'}$
q_{m_p,j_p}	Heat duty of inner-plant heater corresponding to match (m_p, j_p), where $m_p \in HU^p$ and $j_p \in C^p$
q_{m_q,j_p}	Heat duty of interplant heater corresponding to match (m_q, j_p), where $m_q \in HU^q$ and $j_p \in C^p$
$rc_{i_p,j_p,k}$	Flow fraction of cold stream j_p in match (i_p, j_p) at stage k, where $i_p \in H^p$, $j_p \in C^p$, and $k \in ST$
$rc_{i_q,j_p,k}$	Flow fraction of cold stream j_p in match (i_q, j_p) at stage k, where $i_q \in H^q$, $j_p \in C^p$, and $k \in ST$
rcu_{m_p,j_p}	Flow fraction of cold stream j_p in heater match (m_p, j_p), where $m_p \in HU^p$ and $j_p \in C^p$
rcu_{m_q,j_p}	Flow fraction of cold stream j_p in heater match (m_q, j_p), where $m_q \in HU^q$ and $j_p \in C^p$
$rh_{i_p,j_p,k}$	Flow fraction of hot stream i_p in match (i_p, j_p) at stage k, where $i_p \in H^p$, $j_p \in C^p$, and $k \in ST$
$rh_{i_p,j_{q'},k}$	Flow fraction of hot stream i_p in match $(i_p, j_{q'})$ at stage k, where $i_p \in H^p$, $j_{q'} \in C^{q'}$, and $k \in ST$
rhu_{i_p,n_p}	Flow fraction of hot stream i_p in cooler match (i_p, n_p), where $i_p \in H^p$ and $n_p \in CU^p$
$rhu_{i_p,n_{q'}}$	Flow fraction of hot stream i_p in cooler match $(i_p, n_{q'})$, where $i_p \in H^p$ and $n_{q'} \in CU^{q'}$
$rrc_{j_p,k}$	Bypass flow fraction of cold stream $j_p \in C^p$ at stage $k \in ST$
$rrcu_{j_p}$	Bypass flow fraction of cold stream $j_p \in C^p$ at the end of stream
$rrh_{i_p,k}$	Bypass flow fraction of hot stream $i_p \in H^p$ at stage $k \in ST$
$rrhu_{i_p}$	Bypass flow fraction of hot stream $i_p \in H^p$ at the end of stream
$t_{i_p,k}$	Temperature of hot stream i_p at location k in superstructure
$t_{j_p,k}$	Temperature of cold stream j_p at location k in superstructure
$tcs_{i_p,j_p,k}$	Exit temperature of cold stream j_p from match (i_p, j_p) at stage k, where $i_p \in H^p$, $j_p \in C^p$, and $k \in ST$
$tcs_{i_q,j_p,k}$	Exit temperature of cold stream j_p from match (i_q, j_p) at stage k, $i_q \in H^q$, $j_p \in C^p$, and $k \in ST$
$tcus_{m_p,j_p}$	Exit temperature of cold stream j_p from heater match (m_p, j_p), where $m_p \in HU^p$ and $j_p \in C^p$

$tcus_{m_q,j_p}$ Exit temperature of cold stream j_p from heater match (m_q, j_p), where $m_q \in HU^q$ and $j_p \in C^p$

$ths_{i_p,j_p,k}$ Exit temperature of hot stream i_p from match (i_p, j_p) at stage k, where $i_p \in H^p$, $j_p \in C^p$, and $k \in ST$

$ths_{i_p,j_{q'},k}$ Exit temperature of hot stream i_p from match $(i_p, j_{q'})$ at stage k, where $i_p \in H^p$, $j_{q'} \in C^{q'}$, and $k \in ST$

$thus_{i_p,n_p}$ Exit temperature of hot stream i_p from cooler match (i_p, n_p), where $i_p \in H^p$ and $n_p \in CU^p$

$thus_{i_p,n_{q'}}$ Exit temperature of hot stream i_p from cooler match $(i_p, n_{q'})$, where $i_p \in H^p$ and $n_{q'} \in CU^{q'}$

$\xi_{i_p,j_p,k}$ Binary variable used for determining whether or not the inner-plant exchanger match (i_p, j_p) is present at stage k, where $i_p \in H^p$, $j_p \in C^p$, and $k \in ST$

$\xi_{i_p,j_{q'},k}$ Binary variable used for determining whether or not the interplant exchanger match $(i_p, j_{q'})$ is present at stage k, where $i_p \in H^p$, $j_{q'} \in C^{q'}$, and $k \in ST$

$\xi_{i_q,j_p,k}$ Binary variable used for determining whether or not the interplant exchanger match (i_q, j_p) is present at stage k, where $i_q \in H^q$, $j_p \in C^p$, and $k \in ST$

$\xi^{CU}_{i_p,n_{q'}}$ Binary variable used for determining whether or not the cooler match $(i_p, n_{q'})$ is present, where $i_p \in H^p$ and $n_{q'} \in CU^{q'}$

$\xi^{HU}_{m_q,j_p}$ Binary variable used for determining whether or not the heater match (m_q, j_p) is present, where $m_q \in HU^q$ and $j_p \in C^p$

On the basis of the aforementioned definitions, the model constraints can be written as follows:

- Overall heat balance of every process stream

$$\left(TIN_{i_p} - TOUT_{i_p}\right)F_{i_p} = \sum_{k \in ST}\sum_{j_p \in C^p} q_{i_p,j_p,k} + \sum_{\substack{q'=1 \\ q' \neq p}}^{P}\sum_{k \in ST}\sum_{j_{q'} \in C^{q'}} q_{i_p,j_{q'},k} + \sum_{n_p \in CU^p} q_{i_p,n_p} + \sum_{\substack{q'=1 \\ q' \neq p}}^{P}\sum_{n_{q'} \in CU^{q'}} q_{i_p,n_{q'}};\ i_p \in H^p. \tag{5.27}$$

$$\left(TOUT_{j_p} - TIN_{j_p}\right)F_{j_p} = \sum_{k \in ST} \sum_{i_p \in H^p} q_{i_p,j_p,k} + \sum_{\substack{q=1 \\ q \neq p}}^{P} \sum_{k \in ST} \sum_{i_q \in H^q} q_{i_q,j_p,k} + \sum_{m_p \in HU^p} q_{m_p,j_p} + \sum_{\substack{q=1 \\ q \neq p}}^{P} \sum_{m_q \in HU^q} q_{m_q,j_p}; j_p \in C^p. \tag{5.28}$$

- Heat balances around every stage in superstructure

$$\left(t_{i_p,k} - t_{i_p,k+1}\right)F_{i_p} = \sum_{j_p \in C^p} q_{i_p,j_p,k} + \sum_{\substack{q'=1 \\ q' \neq p}}^{P} \sum_{j_{q'} \in C^{q'}} q_{i_p,j_{q'},k}; k \in ST; i_p \in H^p. \tag{5.29}$$

$$\left(t_{j_p,k} - t_{j_p,k+1}\right)F_{j_p} = \sum_{i_p \in H^p} q_{i_p,j_p,k} + \sum_{\substack{q=1 \\ q \neq p}}^{P} \sum_{i_q \in H^q} q_{i_q,j_p,k}; k \in ST; j_p \in C^p. \tag{5.30}$$

- Heat balances around heat exchangers

$$\left(t_{i_p,k} - ths_{i_p,j_p,k}\right) rh_{i_p,j_p,k} F_{i_p} = q_{i_p,j_p,k}; k \in ST; i_p \in H^p; j_p \in C^p. \tag{5.31}$$

$$\left(t_{i_p,k} - ths_{i_p,j_{q'},k}\right) rh_{i_p,j_{q'},k} F_{i_p} = q_{i_p,j_{q'},k}; q' \neq p; k \in ST; i_p \in H^p; j_{q'} \in C^{q'}. \tag{5.32}$$

$$\left(tcs_{i_p,j_p,k} - t_{j_p,k+1}\right) rc_{i_p,j_p,k} F_{j_p} = q_{i_p,j_p,k}; k \in ST; i_p \in H^p; j_p \in C^p. \tag{5.33}$$

$$\left(tcs_{i_q,j_p,k} - t_{j_p,k+1}\right) rc_{i_q,j_p,k} F_{j_p} = q_{i_q,j_p,k}; q \neq p; k \in ST; i_q \in H^q; j_p \in C^p. \tag{5.34}$$

- Heat balances around coolers

$$\left(t_{i_p,NOK+1} - thus_{i_p,n_p}\right) rhu_{i_p,n_p} F_{i_p} = q_{i_p,n_p}; i_p \in H^p; n_p \in CU^p. \tag{5.35}$$

$$\left(t_{i_p,NOK+1} - thus_{i_p,n_{q'}}\right) rhu_{i_p,n_{q'}} F_{i_p} = q_{i_p,n_{q'}}; q' \neq p; i_p \in H^p; n_{q'} \in CU^{q'}. \tag{5.36}$$

- Heat balances around heaters

$$\left(tcus_{m_p,j_p} - t_{j_p,1}\right) rcu_{m_p,j_p} F_{j_p} = q_{m_p,j_p}; j_p \in C^p; m_p \in HU^p \quad (5.37)$$

$$\left(tcus_{m_q,j_p} - t_{j_p,1}\right) rcu_{m_q,j_p} F_{j_p} = q_{m_q,j_p}; q \neq p; j_p \in C^p; m_q \in HU^q \quad (5.38)$$

- Sums of flow fractions

$$rrh_{i_p,k} + \sum_{q'=1}^{P} \sum_{j_{q'} \in C^{q'}} rh_{i_p,j_{q'},k} = 1; i_p \in H^p; k \in ST \quad (5.39)$$

$$rrhu_{i_p} + \sum_{q'=1}^{P} \sum_{n_{q'} \in CU^{q'}} rhu_{i_p,n_{q'}} = 1; i_p \in H^p \quad (5.40)$$

$$rrc_{j_p,k} + \sum_{q=1}^{P} \sum_{i_q \in H^q} rc_{i_q,j_p,k} = 1; j_p \in C^p; k \in ST. \quad (5.41)$$

$$rrcu_{j_p} + \sum_{q=1}^{P} \sum_{m_q \in HU^q} rcu_{m_q,j_p} = 1; j_p \in C^p. \quad (5.42)$$

Notice that, due to the presence of bypasses, the above constraints on the flow fractions are slightly different from the traditional versions. If bypasses are not needed, then the first term on the left-hand side of each of the above equations should be dropped.

- Hot and cold stream temperatures at the boundary locations of each stage

$$rrh_{i_p,k} t_{i_p,k} + \sum_{q'=1}^{P} \sum_{j_{q'} \in C^{q'}} rh_{i_p,j_{q'},k} ths_{i_p,j_{q'},k} = t_{i_p,k+1}; i_p \in H^p; k \in ST. \quad (5.43)$$

$$rrc_{j_p,k} t_{j_p,k+1} + \sum_{q=1}^{P} \sum_{i_q \in H^q} rc_{i_q,j_p,k} tcs_{i_q,j_p,k} = t_{j_p,k}; j_p \in C^p; k \in ST. \quad (5.44)$$

- Constraints on the exit temperatures of the coolers and heaters

$$rrhu_{i_p} t_{i_p,NOK+1} + \sum_{q'=1}^{P} \sum_{n_{q'} \in CU^{q'}} rhu_{i_p,n_{q'}} thus_{i_p,n_{q'}} = TOUT_{i_p}; i_p \in H^p \quad (5.45)$$

$$rrcu_{j_p} t_{j_p,1} + \sum_{q=1}^{P} \sum_{m_q \in HU^q} rcu_{m_q,j_p} tcus_{m_q,j_p} = TOUT_{j_p}; j_p \in C^p \quad (5.46)$$

- Total amount of utility consumed by a process stream

$$\left(t_{i_p,NOK+1} - TOUT_{i_p}\right)F_{i_p} = \sum_{q'=1}^{P} \sum_{n_{q'} \in CU^{q'}} q_{i_p,n_{q'}}; i_p \in H^p. \tag{5.47}$$

$$\left(TOUT_{j_p} - t_{j_p,1}\right)F_{j_p} = \sum_{q=1}^{P} \sum_{m_q \in HU^q} q_{m_q,j_p}; j_p \in C^p. \tag{5.48}$$

- Logic constraint imposed upon heat duty of each match
  - Heat exchangers

$$q_{i_p,j_p,k} - \Omega_{i_p,j_p}\xi_{i_p,j_p,k} \le 0; i_p \in H^p; j_p \in C^p; k \in ST \tag{5.49}$$

$$q_{i_p,j_{q'},k} - \Omega_{i_p,j_{q'}}\xi_{i_p,j_{q'},k} \le 0; i_p \in H^p; j_{q'} \in \bigcup_{\substack{q'=1 \\ q' \neq p}}^{P} C^{q'}; k \in ST \tag{5.50}$$

$$q_{i_q,j_p,k} - \Omega_{i_q,j_p}\xi_{i_q,j_p,k} \le 0; i_q \in \bigcup_{\substack{q=1 \\ q \neq p}}^{P} H^q; j_p \in C^p; k \in ST \tag{5.51}$$

  - Coolers and heaters

$$q_{i_p,n_{q'}} - \Omega_{i_p,n_{q'}}\xi^{CU}_{i_p,n_{q'}} \le 0; \quad i_p \in H^p; n_{q'} \in \bigcup_{q'=1}^{P} CU^{q'} \tag{5.52}$$

$$q_{m_q,j_p} - \Omega_{m_q,j_p}\xi^{HU}_{m_q,j_p} \le 0; \quad m_q \in \bigcup_{q=1}^{P} HU^q; j_p \in C^p. \tag{5.53}$$

- Every process stream is allowed to be matched with at most one utility stream

$$\sum_{q'=1}^{P} \sum_{n_{q'} \in CU^{q'}} \xi^{CU}_{i_p,n_{q'}} \le 1; i_p \in H^p \tag{5.54}$$

$$\sum_{q=1}^{P} \sum_{m_q \in HU^q} \xi^{HU}_{m_q,j_p} \le 1; j_p \in C^p \tag{5.55}$$

- Monotonic temperature decreasing constraints on hot streams and monotonic temperature increasing constraints on cold streams
  - Hot streams

$$t_{i_p,1} = TIN_{i_p}; i_p \in H^p. \tag{5.56}$$

$$t_{i_p,k} \geq t_{i_p,k+1}; i_p \in H^p; k \in ST. \tag{5.57}$$

$$t_{i_p,NOK+1} \geq TOUT_{i_p}; i_p \in H^p. \tag{5.58}$$

- Cold streams

$$t_{j_p,NOK+1} = TIN_{j_p}; j_p \in C^p. \tag{5.59}$$

$$t_{j_p,k+1} \leq t_{j_p,k}; j_p \in C^p; k \in ST. \tag{5.60}$$

$$t_{j_p,1} \leq TOUT_{j_p}; j_p \in C^p \tag{5.61}$$

- Heat transfer areas of heat exchangers, coolers, and heaters
  - Heat exchangers

$$A_{i_p,j_p,k} = \frac{q_{i_p,j_p,k}}{U_{i_p,j_p}\left[dt_{i_p,j_p,k}dt_{i_p,j_p,k+1}\left(dt_{i_p,j_p,k} + dt_{i_p,j_p,k+1}\right)/2\right]^{1/3}}; \tag{5.62}$$

$$i_p \in H^p; j_p \in C^p; k \in ST.$$

$$A_{i_p,j_{q'},k} = \frac{q_{i_p,j_{q'},k}}{U_{i_p,j_{q'}}\left[dt_{i_p,j_{q'},k}dt_{i_p,j_{q'},k+1}\left(dt_{i_p,j_{q'},k} + dt_{i_p,j_{q'},k+1}\right)/2\right]^{1/3}};. \tag{5.63}$$

$$i_p \in H^p; j_{q'} \in C^{q'}; k \in ST.$$

$$A_{i_q,j_p,k} = \frac{q_{i_q,j_p,k}}{U_{i_q,j_p}\left[dt_{i_q,j_p,k}dt_{i_q,j_p,k+1}\left(dt_{i_q,j_p,k} + dt_{i_q,j_p,k+1}\right)/2\right]^{1/3}};. \tag{5.64}$$

$$i_q \in H^q; j_p \in C^p; k \in ST.$$

- Coolers

$$A_{i_p,n_p} = \frac{q_{i_p,n_p}}{U_{i_p,n_p}\left[dtin_{i_p,n_p}dtout_{i_p,n_p}\left(dtin_{i_p,n_p} + dtout_{i_p,n_p}\right)/2\right]^{1/3}}; \tag{5.65}$$

$$i_p \in H^p; n_p \in CU^p.$$

$$A_{i_p,n_{q'}} = \frac{q_{i_p,n_{q'}}}{U_{i_p,n_{q'}}\left[dtin_{i_p,n_{q'}}dtout_{i_p,n_{q'}}\left(dtin_{i_p,n_{q'}} + dtout_{i_p,n_{q'}}\right)/2\right]^{1/3}}; \tag{5.66}$$

$$i_p \in H^p; n_{q'} \in CU^{q'}$$

- Heaters

$$A_{m_p,j_p} = \frac{q_{m_p,j_p}}{U_{m_p,j_p}\left[dtin_{m_p,j_p}dtout_{m_p,j_p}\left(dtin_{m_p,j_p}+dtout_{m_p,j_p}\right)/2\right]^{1/3}}; \quad (5.67)$$

$$m_p \in HU^p; j_p \in C^p.$$

$$A_{m_q,j_p} = \frac{q_{m_q,j_p}}{U_{m_q,j_p}\left[dtin_{m_q,j_p}dtout_{m_q,j_p}\left(dtin_{m_q,j_p}+dtout_{m_q,j_p}\right)/2\right]^{1/3}}; \quad (5.68)$$

$$m_q \in HU^q; j_p \in C^p.$$

Notice that, to enhance the computation efficiency, the LMTDs in equations (5.62)–(5.68) are approximated according to Chen (1987).

- Temperature approaches in heat exchangers, coolers, and heaters
  - Heat exchangers in which the minimum temperature approach (ΔT_{min}) is fixed
    - Inner-plant match between $i_p \in H^p$ and $j_p \in C^p$ at stage $k \in ST$

$$dt_{i_p,j_p,k} \geq \Delta T_{min} \quad (5.69)$$

$$dt_{i_p,j_p,k} \leq t_{i_p,k} - tcs_{i_p,j_p,k} + \Gamma_{i_p,j_p}\left(1-\xi_{i_p,j_p,k}\right) \quad (5.70)$$

$$dt_{i_p,j_p,k+1} \leq ths_{i_p,j_p,k} - t_{j_p,k+1} + \Gamma_{i_p,j_p}\left(1-\xi_{i_p,j_p,k}\right) \quad (5.71)$$

- Interplant match between $i_p \in H^p$ and $j_{q'} \in C^{q'}$ $(q' \neq p)$ at stage $k \in ST$

$$dt_{i_p,j_{q'},k} \geq \Delta T_{min} \quad (5.72)$$

$$dt_{i_p,j_{q'},k} \leq t_{i_p,k} - tcs_{i_p,j_{q'},k} + \Gamma_{i_p,j_{q'}}\left(1-\xi_{i_p,j_{q'},k}\right) \quad (5.73)$$

$$dt_{i_p,j_{q'},k+1} \leq ths_{i_p,j_{q'},k} - t_{j_{q'},k+1} + \Gamma_{i_p,j_{q'}}\left(1-\xi_{i_p,j_{q'},k}\right) \quad (5.74)$$

- Interplant match between $i_q \in H^q$ $(q \neq p)$ and $j_p \in C^p$ at stage $k \in ST$

$$dt_{i_q,j_p,k} \geq \Delta T_{min} \quad (5.75)$$

$$dt_{i_q,j_p,k} \leq t_{i_q,k} - tcs_{i_q,j_p,k} + \Gamma_{i_q,j_p}\left(1-\xi_{i_q,j_p,k}\right) \quad (5.76)$$

$$dt_{i_q,j_p,k+1} \leq ths_{i_q,j_p,k} - t_{j_p,k+1} + \Gamma_{i_q,j_p}\left(1-\xi_{i_q,j_p,k}\right) \quad (5.77)$$

- Coolers in which the minimum temperature approach (ΔT_{min}) is fixed
  - Inner-plant match between $i_p \in H^p$ and $n_p \in CU^p$

$$\Delta T_{\min} \le dtin_{i_p,n_p} \le t_{i_p,NOK+1} - TOUT_{n_p} + \Gamma_{i_p,n_p}\left(1-\xi^{CU}_{i_p,n_p}\right) \tag{5.78}$$

$$\Delta T_{\min} \le dtout_{i_p,n_p} \le thus_{i_p,n_p} - TIN_{n_p} + \Gamma_{i_p,n_p}\left(1-\xi^{CU}_{i_p,n_p}\right) \tag{5.79}$$

  - Interplant match between $i_p \in H^p$ and $n_{q'} \in CU^{q'}$ $(q' \neq p)$

$$\Delta T_{\min} \le dtin_{i_p,n_{q'}} \le t_{i_p,NOK+1} - TOUT_{n_{q'}} + \Gamma_{i_p,n_{q'}}\left(1-\xi^{CU}_{i_p,n_{q'}}\right) \tag{5.80}$$

$$\Delta T_{\min} \le dtout_{i_p,n_{q'}} \le thus_{i_p,n_{q'}} - TIN_{n_{q'}} + \Gamma_{i_p,n_{q'}}\left(1-\xi^{CU}_{i_p,n_{q'}}\right) \tag{5.81}$$

- Heaters in which the minimum temperature approach (ΔT_{min}) is fixed
  - Inner-plant match between $m_p \in HU^p$ and $j_p \in C^p$

$$\Delta T_{\min} \le dtin_{m_p,j_p} \le TOUT_{m_p} - t_{j_p,1} + \Gamma_{m_p,j_p}\left(1-\xi^{HU}_{m_p,j_p}\right) \tag{5.82}$$

$$\Delta T_{\min} \le dtout_{m_p,j_p} \le TIN_{m_p} - tcus_{m_p,j_p} + \Gamma_{m_p,j_p}\left(1-\xi^{HU}_{m_p,j_p}\right) \tag{5.83}$$

  - Interplant match between $m_q \in HU^q$ $(q \neq p)$ and $j_p \in C^p$

$$\Delta T_{\min} \le dtin_{m_q,j_p} \le TOUT_{m_q} - t_{j_p,1} + \Gamma_{m_q,j_p}\left(1-\xi^{HU}_{m_q,j_p}\right) \tag{5.84}$$

$$\Delta T_{\min} \le dtout_{m_q,j_p} \le TIN_{m_q} - tcus_{m_q,j_p} + \Gamma_{m_q,j_p}\left(1-\xi^{HU}_{m_q,j_p}\right) \tag{5.85}$$

5.2.3 Computation Procedure

The required inputs of the suggested two-stage computation procedure are: (1) the labels of all *n* plants in the grand coalition; (2) the stream data (i.e., the heat capacity flow rates of all process streams and their initial and target temperatures) and utility specifications (i.e., the temperature ranges and unit costs of all hot and cold utilities) of each plant; (3) the drop-out probabilities of all plants.

In the first stage, the modified MINLP model described in Section 5.2.2 is first solved repeatedly to generate optimal HEN designs for $2^n - 1$ combinations of potential coalitions. The objective function is the total annual cost (TAC) of the corresponding coalition. This *TAC* is approximated as the sum of the total annual utility cost (*TAUC*) and the total annualized capital cost (*TACC*), i.e., $TAC = TAUC + TACC$.

As an example, the latter two costs for the grand coalition can be expressed explicitly as follows:

$$TAUC = \sum_{i\in H}\sum_{n\in CU} CQ_{i,n}q_{i,n} + \sum_{j\in C}\sum_{m\in HU} CQ_{j,m}q_{j,m} \tag{5.86}$$

$$\begin{aligned}
TACC = &\sum_{i\in H}\sum_{j\in C}\sum_{k\in ST} CF_{i,j,k}z_{i,j,k} + \sum_{i\in H}\sum_{n\in CU} CF_{i,n}z_{i,n} + \sum_{j\in C}\sum_{m\in HU} CF_{j,m}z_{j,m} \\
&+\sum_{i\in H}\sum_{j\in C}\sum_{K\in ST} CA_{i,j}\left(\frac{q_{i,j,k}}{U_{i,j}\left(dt_{i,j,k}dt_{i,j,k+1}\left(dt_{i,j,k}+dt_{i,j,k+1}\right)/2\right)^{1/3}}\right)^{\beta} \\
&+\sum_{i\in H}\sum_{n\in CU} CA_{i,n}\left(\frac{q_{i,n}}{U_{i,n}\left(dtin_{i,n}dtout_{i,n}\left(dtin_{i,n}+dtout_{i,n}\right)/2\right)^{1/3}}\right)^{\beta} \\
&+\sum_{j\in C}\sum_{m\in HU} CA_{j,m}\left(\frac{q_{j,m}}{U_{j,m}\left(dtin_{j,m}dtout_{j,m}\left(dtin_{j,m}+dtout_{j,m}\right)/2\right)^{1/3}}\right)^{\beta}
\end{aligned} \tag{5.87}$$

The resulting minimum TAC of each coalition is treated as its characteristic value in a cooperative game. The actual cost burdens of sub-coalitions in all possible defective coalitions, i.e., $w\left(P : P^{S_{+i}}\right)$s, and the corresponding costs of players remaining in the defective coalitions, i.e., $w\left(i : L_{+i}^{S_{+i}}\right)$, can then be computed according to equations (5.9)–(5.14).

The computations in stage 2 address the allocation issues. Equation (5.19) determines the expected loss due to random plant shutdowns in the present case, while equations (5.20), (5.23)–(5.26) yield the corresponding risk-based Shapley values. Finally, it is necessary to check whether or not the solution at hand is inside the core. If so, then the implementation procedure can be terminated. If not, then the current grand coalition should be partitioned into two non-overlapping sub-coalitions. These two sub-coalitions should be treated as the grand coalition in turn and the aforementioned steps repeated respectively. This iteration process should be continued until the feasibility check is satisfied.

5.2.4 Example 5.1

A simple example is presented below to illustrate the aforementioned computation procedure. Let us consider the stream and utility data given in Table 5.1 and Table 5.2 respectively.

A minimum temperature approach (ΔT_{min}) of 10°C was assumed in designing heat exchangers. The annualized capital investment of every unit was calculated

TABLE 5.1
Stream Data Used in Example 5.1

Plant	Stream	*TIN* (°C)	*TOUT* (°C)	*F* (kW/°C)
P1	H1	150	40	7.0
P1	C1	60	140	9.0
P1	C2	110	190	8.0
P2	H1	200	70	5.5
P2	C1	30	110	3.5
P2	C2	140	190	7.5
P3	H1	370	150	3.0
P3	H2	200	40	5.5
P3	C1	110	360	4.5

TABLE 5.2
Utility Data Used in Example 5.2

Plant	Utility	*TI* (°C)	*CQ* ($/kW·yr)
P1	Cooling water	25	100
P1	HP steam	200	800
P1	Hot oil	500	1200
P2	Cooling water	25	150
P2	HP steam	200	900
P2	Hot oil	500	1300
P3	Cooling water	25	80
P3	HP steam	200	850
P3	Hot oil	500	1100

according to the following formula: $10000 + 670 \times area^{0.83}$ ($/yr). All multi-plant HEN designs in this example were generated according to the MINLP model formulation presented in Sections 5.2.2 and 5.2.3, while all optimization runs were performed with solver COUENNE in GAMS 24.0.

- **Conventional Shapley values**
  By performing repeated optimization runs to generate the interplant HENs in all possible coalitions, one can determine the corresponding TACs in Table 5.3. Their network structures were presented previously in Chapter 2 (see Figures 2.10–2.13) and also in Figure 5.2–Figure 5.4.

  From the above results, the conventional Shapley values can be computed according to equations (5.5)–(5.8), and these values are presented in Table 5.4. Notice that the individual cost of each plant in coalition {1, 2, 3} is much lower than that of the standalone counterpart. However, as

TABLE 5.3
Minimum TACs of HEN Designs in All Possible Coalitions ($/yr)

$v(1) = 725{,}433.4$	$v(2) = 168{,}593.8$	$v(3) = 404{,}900.8$	$v(1, 2) = 696{,}886.1$
$v(1, 3) = 880{,}416.7$	$v(2, 3) = 463{,}990.1$	$v(1, 2, 3) = 887{,}932.4$	$v(\varnothing) = 0$

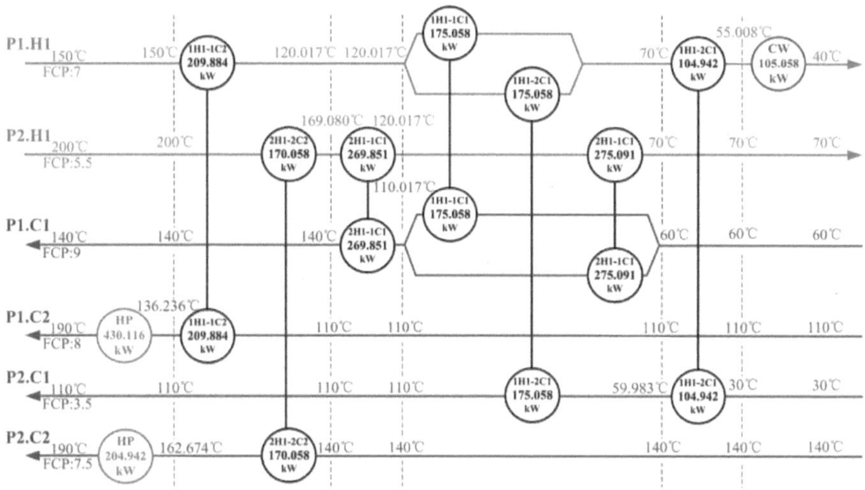

FIGURE 5.2 Optimal HEN design of coalition {P1, P2} in Example 5.1. (Reproduced with permission from *Applied Energy* 2018, 211: 904–920. Copyright 2018 Elsevier.)

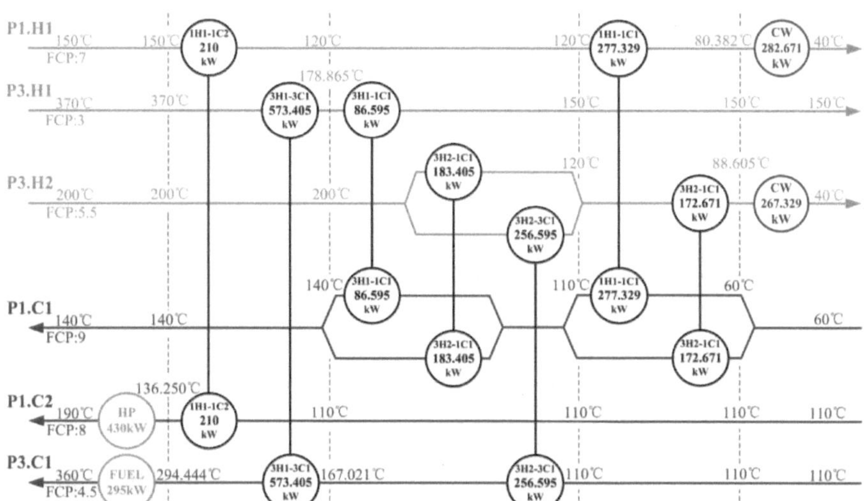

FIGURE 5.3 Optimal HEN design of coalition {P1, P3} in Example 5.1. (Reproduced with permission from *Applied Energy* 2018, 211: 904–920. Copyright 2018 Elsevier.)

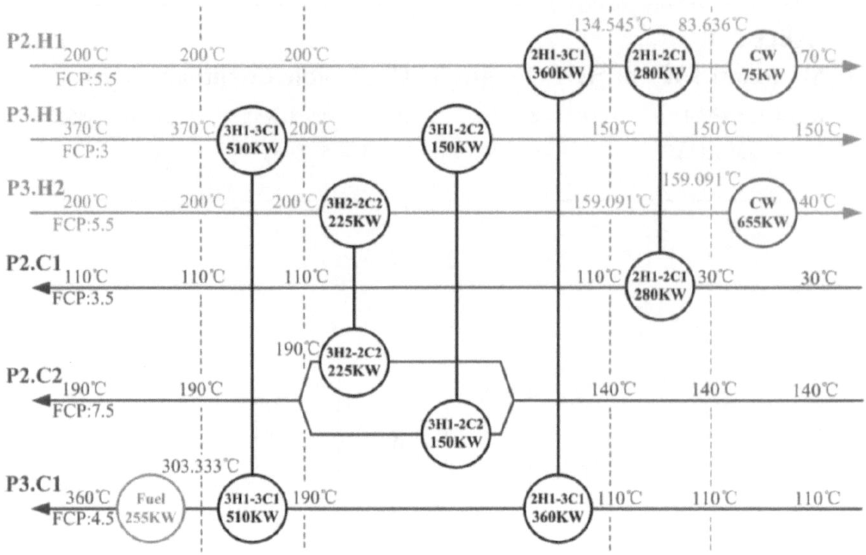

FIGURE 5.4 Optimal HEN design of coalition {P2, P3} in Example 5.1. (Reproduced with permission from *Applied Energy* 2018, 211: 904–920. Copyright 2018 Elsevier.)

TABLE 5.4
Cost Allocations: Shapley Values in Different Coalitions

Coalition	Cost Shared by P1 (\$/yr)	Cost Shared by P2 (\$/yr)	Cost Shared by P3 (\$/yr)
{1}	$\varphi_{\{1\},1} = 725,433.4$	0	0
{2}	0	$\varphi_{\{2\},2} = 168,593.8$	0
{3}	0	0	$\varphi_{\{3\},3} = 404,900.8$
{1,2}	$\varphi_{\{1,2\},1} = 626,862.8$	$\varphi_{\{1,2\},2} = 70,023.2$	0
{1,3}	$\varphi_{\{1,3\},1} = 600,474.7$	0	$\varphi_{\{1,3\},3} = 279,942.1$
{2,3}	0	$\varphi_{\{2,3\},2} = 113,841.5$	$\varphi_{\{2,3\},3} = 350,148.5$
{1,2,3}	$\varphi_{\{1,2,3\},1} = 550,426.6$	$\varphi_{\{1,2,3\},2} = 63,793.5$	$\varphi_{\{1,2,3\},3} = 273,712.3$

indicated previously in Section 5.1.3, it may not be reasonable to ask all three participating members to share the costs accordingly since there is a chance of unexpected plant shutdown.

- **Risk-based Shapley values**
  As mentioned in Section 5.1.3, the unreliability of each plant could be assumed to be available in advance and, specifically, $\alpha_1 = 0.1$, $\alpha_2 = 0.05$, and $\alpha_3 = 0.15$ were adopted for the present example. To facilitate clear illustration of the computation procedure, let us consider a simple scenario in which only P1 and P2 form a two-plant coalition. It is assumed that every

interplant heat exchange can be facilitated with the utility supplied by P2 if P1 breaks down, and vice versa if otherwise. More specifically, let us consider the defective coalitions evolved from the original coalition in Figure 5.2 (see Figure 5.5 and Figure 5.6). These two scenarios can be described respectively as follows:

- P1 is forced to operate the interplant heat exchangers with high-pressure steam (544.9 kW) and cooling water (280.0 kW) due to P2 outage. The resulting financial burden of P1 should be increased from a Shapley value of $\varphi_{\{1,2\},1} = 626{,}862.8$ to $w\left(1:1^{\{1,2\}}\right) = 949{,}223.0$ \$/yr.
- P2 is forced to operate the interplant heat exchangers with high-pressure steam (280.0 kW) and cooling water (544.9 kW) due to P1 outage. The resulting financial burden of P2 should be increased from a Shapley value of $\varphi_{\{1,2\},2} = 70{,}023.2$ to $w\left(2:2^{\{1,2\}}\right) = 635{,}617.7$ \$/yr.

To facilitate clearer understanding, let us analyze the second scenario in further detail. It can be observed from Table 5.3 that the total cost burden of coalition {P1, P2} is $v(1,2) = 696{,}886.1$. According to Table 5.4, P1 and

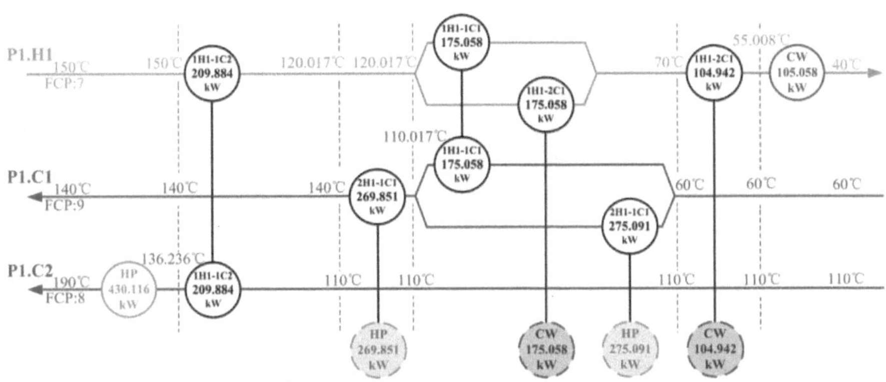

FIGURE 5.5 HEN of defective coalition {P1} evolved from coalition {P1, P2}. (Reproduced with permission from *Applied Energy* 2018, 211: 904–920. Copyright 2018 Elsevier.)

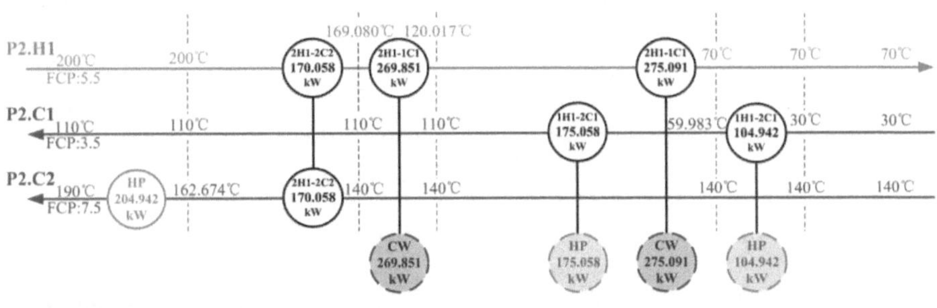

FIGURE 5.6 HEN of defective coalition {P2} evolved from coalition {P1, P2}. (Reproduced with permission from *Applied Energy* 2018, 211: 904–920. Copyright 2018 Elsevier.)

P2 are required to shoulder 626,862.8 (i.e., $\varphi_{\{1,2\},1}$) and 70,023.2 (i.e., $\varphi_{\{1,2\},2}$) respectively. However, if the risks of plant breakdowns are taken into consideration, the corresponding expected expenditures of both plants should obviously be higher. Specifically, the expected costs of P1 and P2 can be determined respectively as follows

$$\left(1-\alpha_2\right)\varphi_{\{1,2\},1}+\alpha_2 w\left(1:1^{\{1,2\}}\right)=642{,}980.81\leq v\left(1\right)$$

$$\left(1-\alpha_1\right)\varphi_{\{1,2\},2}+\alpha_1 w\left(2:2^{\{1,2\}}\right)=126{,}582.65\leq v\left(2\right)$$

Although in this case the expected expenditures of both plants are smaller than those of their standalone counterparts, there are still unsettled subtle allocation issues. Further calculations should reveal that the aforementioned cost increases differ significantly. P1 is financially penalized by only 3%, but the expected expenditure of P2 is raised 81% higher. Therefore, instead of computing the expected cost burdens mentioned above, the risk-based Shapley values must be determined to produce a fairer cost allocation scheme.

The cost burdens of every sub-coalition in all possible scenarios can be computed according to equations (5.10)–(5.12) and all results are given in Table 5.5, while the corresponding risk-based Shapley values can be found in Table 5.6. Notice that the conventional Shapley values are also presented in Table 5.6 to facilitate convenient comparison. It can be observed that, due to considerations of potential shutdowns, only plant P2 is allowed to reduce its share of the financial burden and the other two plants, P1 and P3, are both required to shoulder larger portions of TAC. This is reasonable since the cost hikes of P2 due to various plant shutdowns in coalition are all relatively large and other member(s) may be willing to shoulder a heavier burden within the core.

5.3 RETROFIT DESIGNS

Four revamp strategies are presented in the sequel for the facilitation of interplant heat integration. Basically, each differs from the others mainly in the reclaimed energy and in the capital investments and re-piping costs of the resulting multi-plant HEN structure. The design guidelines adopted in these strategies are described in the sequel.

- **Strategy 0**: New interplant exchanger matches are not allowed in the revamp design. The existing exchanger matches located within each plant must be kept unchanged, while the existing cooler and heater matches may adopt utilities from any plant in the multi-plant HEN. Every inner-plant exchanger match should be housed in its original unit. The inner- or interplant cooler and heater matches should be facilitated with the existing units.
- **Strategy 1**: New interplant exchanger, cooler, and heater matches can be introduced into the revamp design. The existing exchanger matches located within each plant must be kept unchanged, while the existing cooler and

TABLE 5.5
Cost Burdens of Individual Plants in Defective Coalitions in All Scenarios

$w\left(1:1^{1}\right)=725{,}433.4$	$w\left(3:3^{\{1,3\}}\right)=477{,}863.2$
$w\left(2:2^{2}\right)=168{,}593.8$	$w\left(2:2^{\{2,3\}}\right)=479{,}079.9$
$w\left(3:3^{3}\right)=404{,}900.8$	$w\left(3:3^{\{2,3\}}\right)=764{,}239.8$
$w\left(1:\{1,2\}^{\{1,2\}}\right)=626{,}862.8$	$w\left(1:1^{\{1,2,3\}}\right)=1{,}103{,}470.5$
$w\left(2:\{1,2\}^{\{1,2\}}\right)=70{,}023.2$	$w\left(2:2^{\{1,2,3\}}\right)=797{,}944.4$
$w\left(1:\{1,3\}^{\{1,3\}}\right)=600{,}474.7$	$w\left(3:3^{\{1,2,3\}}\right)=658{,}271.2$
$w\left(3:\{1,3\}^{\{1,3\}}\right)=279{,}942.1$	$w\left(\{1,2\}:\{1,2\}^{\{1,2,3\}}\right)=1{,}085{,}074.7$
$w\left(2:\{2,3\}^{\{2,3\}}\right)=113{,}841.5$	$w\left(\{1,3\}:\{1,3\}^{\{1,2,3\}}\right)=1{,}528{,}182.4$
$w\left(3:\{2,3\}^{\{2,3\}}\right)=350{,}148.5$	$w\left(\{2,3\}:\{2,3\}^{\{1,2,3\}}\right)=877{,}561.5$
$w\left(1:\{1,2,3\}^{\{1,2,3\}}\right)=550{,}426.6$	$w\left(1:\{1,2\}^{\{1,2,3\}}\right)=695{,}300.3$
$w\left(2:\{1,2,3\}^{\{1,2,3\}}\right)=63{,}793.5$	$w\left(2:\{1,2\}^{\{1,2,3\}}\right)=389{,}774.3$
$w\left(3:\{1,2,3\}^{\{1,2,3\}}\right)=273{,}712.3$	$w\left(1:\{1,3\}^{\{1,2,3\}}\right)=986{,}690.8$
$w\left(1:1^{\{1,2\}}\right)=949{,}223.0$	$w\left(3:\{1,3\}^{\{1,2,3\}}\right)=541{,}491.6$
$w\left(2:2^{\{1,2\}}\right)=635{,}617.7$	$w\left(2:\{2,3\}^{\{1,2,3\}}\right)=508{,}617.3$
$w\left(1:1^{\{1,3\}}\right)=831{,}061.5$	$w\left(3:\{2,3\}^{\{1,2,3\}}\right)=368{,}944.1$

TABLE 5.6
Conventional and Risk-Based Shapley Values (\$/yr)

i	1	2	3
$\varphi_{\{1,2,3\},i}$	550,426.6	63,793.5	273,712.3
$\hat{\varphi}_{\{1,2,3\},i}$	578,443.7	23,388.7	286,100.0

heater matches may adopt utilities from any plant in the multi-plant HEN. The interplant exchanger matches should be housed in new units purchased externally. Every inner-plant exchanger match should be housed in its original unit and the inner- or interplant cooler and heater matches can be facilitated with either existing or purchased units of the same type. If a larger heat-transfer area is called for when an existing unit is used for any match, this unit should be connected with an extra one in series according to Figure 5.7 (Ciric and Floudas, 1990).

- **Strategy 2**: New interplant exchanger, cooler, and heater matches are allowed in the revamp design. The existing exchanger matches located within each plant must be kept unchanged, while the existing cooler and heater matches may adopt utilities from any plant in the multi-plant HEN. The interplant exchanger matches should be housed in new units purchased externally. Every inner-plant exchanger match can be housed in either its original heat exchanger *or another existing one of the same type within the same plant.* The inner- or interplant cooler and heater matches can be facilitated with either existing or purchased units of the same type. If a larger heat-transfer area is called for when an existing unit is used for any match, this unit should be connected with an extra one in series according to Figure 5.7 (Ciric and Floudas, 1990).
- **Strategy 3**: New inner-plant and interplant exchangers, coolers, and heaters matches can be introduced into the revamp design, while some of the existing ones may not be utilized. Every existing heat exchanger should be kept within the plant where it is originally located. Any match in revamp design can be housed either in a purchased unit or an existing one and, if a larger heat-transfer area is called for in the latter case, this unit can be connected with another heat exchanger in series according to Figure 5.7 (Ciric and Floudas, 1990).

It should be noted from the outset that, since Strategy 0 yields only trivial designs, i.e., the cheapest utilities are adopted by all plants while the original single-plant HEN structures are kept intact, this strategy will not be discussed in the present chapter. Other than the constraints described in Section 5.2.2, it is necessary to incorporate additional ones in the programming models for the realization of the other three revamp strategies. Since a greater financial gain can usually be obtained by solving an MINLP with fewer constraints, it can be expected that Strategy 3 reaps the most benefit, Strategy 1 the least, and Strategy 2 yields a cost saving that lies between the above two. However, it should also be noted that selecting an appropriate strategy depends more on other practical issues, e.g., safety concerns, spatial limits and operability, etc. The aforementioned additional constraints are outlined below.

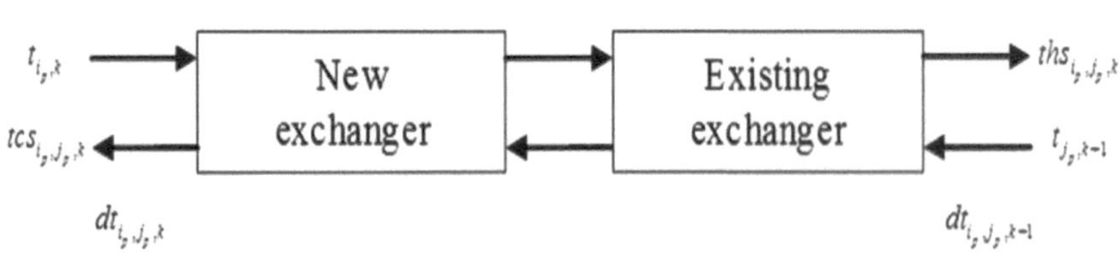

FIGURE 5.7 Connections between an existing heat exchanger and a new one to enlarge the heat-transfer area. (Reproduced with permission from *Ind. Eng. Chem. Res.* 2020, 59: 18088–18105. Copyright 2020 American Chemical Society.)

5.3.1 Extra Constraints Needed for Implementing Strategy 1

In addition to the sets, parameters, and variables defined in Section 5.2.2, more are introduced below to facilitate illustration of the proposed model formulation.

Sets:

CL_p Set of inner-plant matches between cold utilities and hot streams in the original HEN of plant p

$\overline{CL}_p$ Set of inner-plant matches between cold utilities and hot streams that do not consume cold utilities in the original HEN of plant p, i.e., $\{(i_p, n_p) \mid i_p \in H^p \setminus H^p_{CL}; n_p \in CU^p\}$

C^p_{HT} Set of cold streams that consume hot utilities in the original HEN of plant p

G_p Set of all interplant matches between process streams in plant p and process streams in other plants in the multi-plant HEN

GC_p Set of all interplant matches between hot streams in plant p and cold utilities in other plants in the multi-plant HEN

GH_p Set of all interplant matches between cold streams in plant p and hot utilities in other plants in the multi-plant HEN

HT_p Set of inner-plant matches between hot utilities and cold streams in the original HEN of plant p

$\overline{HT}_p$ Set of inner-plant matches between hot utilities and cold streams that do not consume hot utilities in the original HEN of plant p, i.e., $\{(m_p, j_p) \mid m_p \in HU^p; j_p \in C^p \setminus C^p_{HT}\}$

H^p_{CL} Set of hot streams that consume cold utilities in the original HEN of plant p

Y^p Set of existing inner-plant matches in the original HEN of plant p

Parameters:

$A^{EX}_{y_{i_p,j_p}}$ Heat transfer area of existing heat exchanger y_{i_p,j_p} ($y_{i_p,j_p} = 1, 2, \ldots, N_{i_p,j_p}$) used for housing match $(i_p, j_p) \in Y^p$ in the original HEN of plant p

$A^{EX}_{i_p,n_p}$ Heat transfer area of existing cooler used for housing match $(i_p, n) \in CL^p$ in the original HEN of plant p

$A^{EX}_{m_p,j_p}$ Heat transfer area of existing heater used for housing match $(m_p, j_p) \in HT^p$ in the original HEN of plant p

N_{i_p,j_p} Total number of existing heat exchangers used for housing match $(i_p, j_p) \in Y^p$ in the original HEN of plant p

Λ_{i_p,j_p} A large enough constant which is not smaller than the largest heat-transfer area of the augmented unit for housing match $(i_p, j_p) \in Y^p$ in the multi-plant HEN

Λ_{i_p,n_p} A large enough constant which is not smaller than the largest heat-transfer area of the augmented unit for housing cooler match $(i_p, n_p) \in CL^p$ in the multi-plant HEN

Λ_{m_p,j_p} A large enough constant which is not smaller than the largest heat-transfer area of the augmented unit for housing heater match $(m_p, j_p) \in HT^p$ in the multi-plant HEN

Variables:

$A_{i_p,j_p,k}$ Heat-transfer area of match (i_p, j_p) at stage k in the multi-plant HEN

A_{i_p,n_p} Heat-transfer area of cooler match (i_p, n_p) in the multi-plant HEN

A_{m_p,j_p} Heat-transfer area of heater match (m_p, j_p) in the multi-plant HEN

$e_{i_p,j_p,k}^{y_{ip,jp}}$ Binary variable used for determining whether or not the existing heat exchanger y_{i_p,j_p} $(y_{i_p,j_p} = 1,2,\ldots,N_{i_p,j_p})$ adopted for housing match $(i_p, j_p) \in Y^p$ in the original HEN of plant p can be used to house the same match at stage k of the multi-plant HEN

$rz_{i_p,k}$ Binary variable used for determining whether or not hot stream $i_p \in H^p$ requires a bypass stream at stage k of the multi-plant HEN

$rz_{j_p,k}$ Binary variable used for determining whether or not cold stream $j_p \in C^p$ requires a bypass stream at stage k of the multi-plant HEN

$rzhu_{i_p}$ Binary variable used for determining whether or not hot stream $i_p \in H^p$ requires a bypass stream for its cooler in the multi-plant HEN

$rzcu_{j_p}$ Binary variable used for determining whether or not cold stream $j_p \in C^p$ requires a bypass stream for its heater in the multi-plant HEN

$X_{i_p,j_p,k}^{y_{ip,jp}}$ Heat-transfer area of the augmented unit of the existing heat exchanger y_{i_p,j_p} $(y_{i_p,j_p} = 1,2,\ldots,N_{i_p,j_p})$ adopted for housing match $(i_p, j_p) \in Y^p$ at stage k of the multi-plant HEN

X_{i_p,n_p} Heat-transfer area of the augmented unit of the existing cooler for housing match $(i_p, n_p) \in CL^p$ in the multi-plant HEN

X_{m_p,j_p} Heat-transfer area of the augmented unit of the existing cooler for housing match $(m_p, j_p) \in HT^p$ in the multi-plant HEN

$\sigma_{i_p,j_p,k}^{y_{ip,jp}}$ Binary variable used for determining whether or not the existing heat exchanger y_{i_p,j_p} $(y_{i_p,j_p} = 1,2,\ldots,N_{i_p,j_p})$ adopted for housing match $(i_p, j_p) \in Y^p$ in the original HEN of plant p can be used to house the same match at stage k of the multi-plant HEN by enlarging its heat transfer area according to Figure 5.7

σ_{i_p,n_p} Binary variable used for determining whether or not the existing cooler for housing match $(i_p, n_p) \in CL^p$ in the original HEN of plant p can be used to house the same match in the multi-plant HEN by enlarging its heat transfer area according to Figure 5.7

σ_{m_p,j_p} Binary variable used for determining whether or not the existing heater for housing match $(m_p, j_p) \in HT^p$ in the original HEN of plant p can be used to house the same match in the multi-plant HEN by enlarging its heat transfer area according to Figure 5.7

Other than all the constraints mentioned in Section 5.2.2, extra ones should be included for the implementation of Strategy 1. Firstly, to house the existing matches in their original heat exchangers, the following constraints can be imposed:

$$\sum_{k \in ST} \xi_{i_p,j_p,k} \le N_{i_p,j_p} \; ; i_p \in H^p; j_p \in C^p; (i_p, j_p) \in Y_p. \tag{5.88}$$

$$\xi^{CU}_{i_p,n_p} \le 1; \quad i_p \in H^p; \quad n_p \in CU^p; \quad (i_p, n_p) \in CL_p. \tag{5.89}$$

$$\xi^{HU}_{m_p,j_p} \le 1; \quad m_p \in HU^p; \quad j_p \in C^p; \quad (m_p, j_p) \in HT_p. \tag{5.90}$$

where $p = 1, 2, \cdots, P$. Note also that the binary variables $\xi_{i_p,j_p,k}$, $\xi^{CU}_{i_p,n_p}$, and $\xi^{HU}_{m_p,j_p}$ should be set to zero if they are not associated with the existing matches, i.e.,

$$\xi_{i_p,j_p,k} = 0; \quad i_p \in H^p; j_p \in C^p; (i_p, j_p) \notin Y_p. \tag{5.91}$$

$$\xi^{CU}_{i_p,n_p} = 0; \quad i_p \in H^p; \quad n_p \in CU^p; \quad (i_p, n_p) \notin CL_p. \tag{5.92}$$

$$\xi^{HU}_{m_p,j_p} = 0; \quad m_p \in HU^p; \quad j_p \in C^p; \quad (m_p, j_p) \notin HT_p. \tag{5.93}$$

Exactly one existing heat exchanger (which is originally used to house an existing match in the single-plant HEN) can be used to house the same match in a distinct stage in the multi-plant HEN, i.e.,

$$\sum_{y_{i_p,j_p}=1}^{N_{i_p,j_p}} e^{y_{i_p,j_p}}_{i_p,j_p,k} - \xi_{i_p,j_p,k} = 0; i_p \in H^p; j_p \in C^p; (i_p, j_p) \in Y_p; k \in ST. \tag{5.94}$$

$$\sum_{k \in ST} e^{y_{i_p,j_p}}_{i_p,j_p,k} = 1; i_p \in H^p; j_p \in C^p; (i_p, j_p) \in Y_p; k \in ST; y_{i_p,j_p} = 1, 2, \ldots, N_{i_p,j_p} \tag{5.95}$$

Also, the heat-transfer areas of the units in the multi-plant HEN should be constrained as follows:

$$\sum_{k \in ST} A_{i_p,j_p,k} \ge \sum_{y_{i_p,j_p}=1}^{N_{i_p,j_p}} A^{EX}_{y_{i_p,j_p}} ; i_p \in H^p; j_p \in C^p; (i_p, j_p) \in Y_p. \tag{5.96}$$

$$A_{i_p,n_p} \ge A^{EX}_{i_p,n_p}; \; i_p \in H^p; \; n_p \in CU^p; \; (i_p, n_p) \in CL_p. \tag{5.97}$$

$$A_{m_p,j_p} \ge A^{EX}_{m_p,j_p}; \; m_p \in HU^p; \; j_p \in C^p; \; (m_p, j_p) \in HT_p. \tag{5.98}$$

Since an inner-plant match in the multi-plant HEN is housed in an existing unit which is used to house the same match in the single-plant HEN and a new heat exchanger may or may not be added according to Figure 5.7, the heat-transfer area of this augmented unit ($X_{i_p,j_p,k}^{y_{i_p,j_p}}$) can be determined as follows

$$X_{i_p,j_p,k}^{y_{i_p,j_p}} = \max\left\{\left(A_{i_p,j_p,k} - A_{y_{i_p,j_p}}^{EX}\right)e_{i_p,j_p,k}^{y_{i_p,j_p}}, 0\right\};$$

$$i_p \in H^p;\ j_p \in C^p;\ \left(i_p, j_p\right) \in Y_p;\ k \in ST;\ y_{i_p,j_p} = 1,2,\ldots,N_{i_p,j_p}. \tag{5.99}$$

On the other hand, the inner-plant coolers and heaters in the multi-plant HEN are always *not* smaller than the existing ones due to equations (5.97) and (5.98). Their augmented heat-transfer areas (X_{i_p,n_p} and X_{m_p,j_p}) can be expressed as

$$X_{i_p,n_p} = A_{i_p,n_p} - A_{i_p,n_p}^{EX};\ i_p \in H^p;\ n_p \in CU^p;\ \left(i_p, n_p\right) \in CL_p. \tag{5.100}$$

$$X_{m_p,j_p} = A_{m_p,j_p} - A_{m_p,j_p}^{EX};\ m_p \in HU^p;\ j_p \in C^p;\ \left(m_p, j_p\right) \in HT_p. \tag{5.101}$$

To facilitate calculation of the capital cost of augmented unit for each existing match in the multi-plant HEN, the following logic constraints must be imposed:

$$X_{i_p,j_p,k}^{y_{i_p,j_p}} - \Lambda_{i_p,j_p}\sigma_{i_p,j_p,k}^{y_{i_p,j_p}} \le 0;$$

$$i_p \in H^p;\ j_p \in C^p;\ \left(i_p, j_p\right) \in Y_p;\ k \in ST;\ y_{i_p,j_p} = 1,2,\ldots,N_{i_p,j_p} \tag{5.102}$$

$$X_{i_p,n_p} - \Lambda_{i_p,n_p}\sigma_{i_p,n_p} \le 0;\ i_p \in H^p;\ n_p \in CU^p;\ \left(i_p, n_p\right) \in CL_p. \tag{5.103}$$

$$X_{m_p,j_p} - \Lambda_{m_p,j_p}\sigma_{m_p,j_p} \le 0;\ m_p \in HU^p;\ j_p \in C^p;\ \left(m_p, j_p\right) \in HT_p. \tag{5.104}$$

Similarly, to facilitate calculation of the capital cost of bypasses in the multi-plant HEN, the following logic constraints must also be imposed:

$$rrh_{i_p,k} - rz_{i_p,k} \le 0;\ i_p \in H^p;\ k \in ST. \tag{5.105}$$

$$rrc_{j_p,k} - rz_{j_p,k} \le 0;\ j_p \in C^p;\ k \in ST. \tag{5.106}$$

$$rrhu_{i_p} - rzhu_{i_p} \le 0;\ i_p \in H^p. \tag{5.107}$$

$$rrcu_{j_p} - rzcu_{j_p} \le 0;\ j_p \in C^p. \tag{5.108}$$

5.3.2 Extra Constraints Needed for Implementing Strategy 2

In addition to the sets, parameters, and variables defined in Section 5.2.2 and in Section 5.3.1, more should be introduced below to facilitate illustration of the proposed model formulation for implementing Strategy 2.

Sets:

$CL_p^{w_p}$ Set of existing cooler matches (i_p, n_p) of type w_p, i.e., $CL_p^{w_p} \subset CL_p$ and $\bigcup_{w_p=1}^{W_p} CL_p^{w_p} = CL_p$, in plant p

$HT_p^{l_p}$ Set of existing heater matches (m_p, j_p) of type l_p, i.e., $HT_p^{l_p} \subset HT_p$ and $\bigcup_{l_p=1}^{L_p} HT_p^{l_p} = HT_p$, in plant p

Parameters:

L_p Total number of heater types in plant p

N_p Total number of existing heat exchangers in the original HEN of plant p, i.e.,

$$N_p = \sum_{(i_p, j_p) \in Y^p} N_{i_p, j_p}$$

$N_{CL}^{w_p}$ Total number of existing coolers of type w_p in the original HEN of plant p

$N_{HT}^{l_p}$ Total number of existing heaters of type l_p in the original HEN of plant p

W_p Total number of cooler types in plant p

Variables:

$e_{\tilde{i}_p, \tilde{j}_p, k}^{y_{i_p, j_p}}$ Binary variable used for determining whether or not the existing heat exchanger y_{i_p, j_p} $(y_{i_p, j_p} = 1, 2, \cdots, N_{i_p, j_p})$ adopted for housing match $(i_p, j_p) \in Y^p$ in the original HEN of plant p can be used to house another match $(\tilde{i}_p, \tilde{j}_p) \in Y^p$ at stage k of the multi-plant HEN

$e_{\tilde{i}_p, \tilde{n}_p}^{i_p, n_p, w_p}$ Binary variable used for determining whether or not an existing cooler of type w_p, i.e., $(i_p, n_p) \in CL_p^{w_p}$, in the original HEN of plant p can be used to house another cooler match $(\tilde{i}_p, \tilde{n}_p) \in CL_p^{w_p}$ in the multi-plant HEN

$e_{\tilde{m}_p, \tilde{j}_p}^{m_p, j_p, l_p}$ Binary variable used for determining whether or not an existing heater of type l_p, i.e., $(m_p, j_p) \in HT_p^{l_p}$, in the original HEN of plant p can be used to house another heater match $(\tilde{m}_p, \tilde{j}_p) \in HT_p^{l_p}$ in the multi-plant HEN

$X_{\tilde{i}_p, \tilde{j}_p, k}^{y_{i_p, j_p}}$ Heat-transfer area of the augmented unit of the existing heat exchanger y_{i_p, j_p} $(y_{i_p, j_p} = 1, 2, \ldots, N_{i_p, j_p})$ of match$(i_p, j_p) \in Y^p$ in the original single-plant HEN for housing another match $(\tilde{i}_p, \tilde{j}_p) \in Y^p$ at stage k of the multi-plant HEN

$X_{\tilde{i}_p, \tilde{n}_p}^{i_p, n_p, w_p}$ Heat-transfer area of the augmented unit of the existing cooler of type w_p, i.e., $(i_p, n_p) \in CL_p^{w_p}$, in the original HEN of plant p for housing another cooler match $(\tilde{i}_p, \tilde{n}_p) \in CL_p^{w_p}$ in the multi-plant HEN

$X_{\tilde{m}_p,\tilde{j}_p}^{m_p,j_p,l_p}$ Heat-transfer area of the augmented unit of the existing heater of type l_p, i.e., $(m_p, j_p) \in HT_p^{l_p}$, in the original HEN of plant p for housing another heater match $(\tilde{m}_p, \tilde{j}_p) \in HT_p^{l_p}$ in the multi-plant HEN

$\sigma_{\tilde{i}_p,\tilde{j}_p,k}^{y_{i_p,j_p}}$ Binary variable used for determining whether or not the existing heat exchanger y_{i_p,j_p} $(y_{i_p,j_p} = 1,2,\ldots,N_{i_p,j_p})$ adopted for housing match $(i_p, j_p) \in Y^p$ in the original HEN of plant p can be used to house another match $(\tilde{i}_p, \tilde{j}_p) \in Y^p$ at stage k of the multi-plant HEN by enlarging its heat transfer area according to Figure 5.7

$\sigma_{\tilde{i}_p,\tilde{n}_p}^{i_p,n_p,w_p}$ Binary variable used for determining whether or not the existing cooler of type w_p, i.e., $(i_p, n_p) \in CL_p^{w_p}$, in the original HEN of plant p can be used to house another cooler match $(\tilde{i}_p, \tilde{n}_p) \in CL_p^{w_p}$ in the multi-plant HEN by enlarging its heat transfer area according to Figure 5.7

$\sigma_{\tilde{m}_p,\tilde{j}_p}^{m_p,j_p,l_p}$ Binary variable used for determining whether or not the existing heater of type l_p, i.e., $(m_p, j_p) \in HT_p^{l_p}$, in the original HEN of plant p can be used to house another heater match $(\tilde{m}_p, \tilde{j}_p) \in HT_p^{l_p}$ in the multi-plant HEN by enlarging its heat transfer area according to Figure 5.7

Together with all the constraints mentioned in Section 5.2.2, additional ones should be included for implementing Strategy 2. The needed constraints are described below:

First of all, for housing the existing exchanger matches of the multi-plant HEN with the available heat exchangers, the following inequality should be imposed:

$$\sum_{(i_p,j_p)\in Y_p} \sum_{k\in ST} \xi_{i_p,j_p,k} \le N_p \tag{5.109}$$

where $p = 1,2,\ldots,P$. On the other hand, for housing the existing cooler and heater matches of multi-plant HEN with the available coolers and heaters of the same types, the following constraints should be used:

$$\sum_{(i_p,n_p)\in CL_p^{w_p}} \xi_{i_p,n_p}^{CU} \le N_{CL}^{w_p} \tag{5.110}$$

$$\sum_{(m_p,j_p)\in HT_p^{l_p}} \xi_{m_p,j_p}^{HU} \le N_{HT}^{l_p} \tag{5.111}$$

In addition, equations (5.91)–(5.93) are still valid in the present case. Since every existing match can be housed in either its original heat exchanger or another existing one of the same type within the same plant, the corresponding constraints should be written as

$$\sum_{(i_p,j_p)\in Y^P} \sum_{y_{i_p,j_p}=1}^{N_{i_p,j_p}} e_{\tilde{i}_p,\tilde{j}_p,k}^{y_{i_p,j_p}} - \xi_{\tilde{i}_p,\tilde{j}_p,k} = 0;$$

$$\tilde{i}_p \in H^p; \tilde{j}_p \in C^p; \left(\tilde{i}_p, \tilde{j}_p\right) \in Y^p; k \in ST. \tag{5.112}$$

$$\sum_{(i_p,n_p)\in CL_p^{w_p}} e_{\tilde{i}_p,\tilde{n}_p}^{i_p,n_p,w_p} - \xi_{\tilde{i}_p,\tilde{n}_p}^{CU} = 0;$$

$$\tilde{i}_p \in H^p; \tilde{n}_p \in CU^p; \left(\tilde{i}_p, \tilde{n}_p\right) \in CL_p^{w_p}; w_p = 1,2,\ldots,W_p \tag{5.113}$$

$$\sum_{(m_p,j_p)\in HT_p^{l_p}} e_{\tilde{m}_p,\tilde{j}_p}^{m_p,j_p,l_p} - \xi_{\tilde{m}_p,\tilde{j}_p}^{HU} = 0;$$

$$\tilde{m}_p \in HU^p; \tilde{j}_p \in C^p; \left(\tilde{m}_p, \tilde{j}_p\right) \in HT_p^{l_p}; l_p = 1,2,\ldots,L_p \tag{5.114}$$

Furthermore, each existing heat exchanger should be used to house exactly one match in the multi-plant HEN, i.e.,

$$\sum_{(\tilde{i}_p,\tilde{j}_p)\in Y^P} \sum_{k\in ST} e_{\tilde{i}_p,\tilde{j}_p,k}^{y_{i_p,j_p}} = 1;$$

$$i_p \in H^p; j_p \in C^p; \left(i_p, j_p\right) \in Y^p; y_{i_p,j_p} = 1,2,\ldots,N_{i_p,j_p} \tag{5.115}$$

Since an inner-plant match in the multi-plant HEN may be housed in any existing unit and a new heat exchanger may or may not be added according to Figure 5.11, the heat-transfer area of this augmented unit ($X_{\tilde{i}_p,\tilde{j}_p,k}^{y_{i_p,j_p}}$) can be determined with the following equations:

$$X_{\tilde{i}_p,\tilde{j}_p,k}^{y_{i_p,j_p}} = \max\left\{\left(A_{\tilde{i}_p,\tilde{j}_p,k} - A_{y_{i_p,j_p}}^{EX}\right) e_{\tilde{i}_p,\tilde{j}_p,k}^{y_{i_p,j_p}}, 0\right\};$$

$$i_p, \tilde{i}_p \in H^p; j_p, \tilde{j}_p \in C^p; \left(i_p, j_p\right), \left(\tilde{i}_p, \tilde{j}_p\right) \in Y^p; k \in ST;$$

$$y_{i_p,j_p} = 1,2,\ldots,N_{i_p,j_p}. \tag{5.116}$$

$$X_{\tilde{i}_p,\tilde{n}_p}^{i_p,n_p,w_p} = \max\left\{\left(A_{\tilde{i}_p,\tilde{n}_p} - A_{i_p,n_p}^{EX}\right) e_{\tilde{i}_p,\tilde{n}_p}^{i_p,n_p,w_p}, 0\right\};$$

$$i_p, \tilde{i}_p \in H^p; n_p, \tilde{n}_p \in CU^p; \left(i_p, n_p\right), \left(\tilde{i}_p, \tilde{n}_p\right) \in CL_p^{w_p}; w_p = 1,2,\ldots,W_p. \tag{5.117}$$

$$X_{\tilde{m}_p,\tilde{j}_p}^{m_p,j_p,l_p} = \max\left\{\left(A_{\tilde{m}_p,\tilde{j}_p} - A_{m_p,j_p}^{EX}\right)e_{\tilde{m}_p,\tilde{j}_p}^{m_p,j_p,l_p}, 0\right\};$$

$$m_p,\tilde{m}_p \in HU^p;\ j_p,\tilde{j}_p \in C^p;\ (m_p,j_p),(\tilde{m}_p,\tilde{j}_p) \in HT_p^{l_p};\ l_p = 1,2,\ldots,L_p. \quad (5.118)$$

To facilitate calculation of the capital cost of an augmented unit for each existing match in the multi-plant HEN, the following logic constraints must be imposed:

$$X_{\tilde{i}_p,\tilde{j}_p,k}^{y_{i_p,j_p}} - \Lambda_{\tilde{i}_p,\tilde{j}_p}\sigma_{\tilde{i}_p,\tilde{j}_p,k}^{y_{i_p,j_p}} \leq 0;$$

$$i_p,\tilde{i}_p \in H^p;\ j_p,\tilde{j}_p \in C^p;(i_p,j_p),(\tilde{i}_p,\tilde{j}_p) \in Y^p;k \in ST;$$

$$y_{i_p,j_p} = 1,2,\ldots,N_{i_p,j_p}. \quad (5.119)$$

$$X_{\tilde{i}_p,\tilde{n}_p}^{i_p,n_p,w_p} - \Lambda_{\tilde{i}_p,\tilde{n}_p}\sigma_{\tilde{i}_p,\tilde{n}_p}^{i_p,n_p,w_p} \leq 0;$$

$$i_p,\tilde{i}_p \in H^p;\ n_p,\tilde{n}_p \in CU^p;\ (i_p,n_p),(\tilde{i}_p,\tilde{n}_p) \in CL_p^{w_p};\ w_p = 1,2,\ldots,W_p. \quad (5.120)$$

$$X_{\tilde{m}_p,\tilde{j}_p}^{m_p,j_p,l_p} - \Lambda_{\tilde{m}_p,\tilde{j}_p}\sigma_{\tilde{m}_p,\tilde{j}_p}^{m_p,j_p,l_p} \leq 0;$$

$$m_p,\tilde{m}_p \in HU^p;\ j_p,\tilde{j}_p \in C^p;\ (m_p,j_p),(\tilde{m}_p,\tilde{j}_p) \in HT_p^{l_p};\ l_p = 1,2,\ldots,L_p. \quad (5.121)$$

Finally, notice that equations (5.105)–(5.108) should also be included in the present application.

5.3.3 Extra Constraints Needed for Implementing Strategy 3

In addition to the sets, parameters, and variables defined in Section 5.2.2 and in Sections 5.3.1 and 5.3.2, extra definitions are introduced below to facilitate model formulation for implementing Strategy 3.

Sets:

- Z_p — Set of all possible inner-plant matches between hot stream $\tilde{i}_p \in H^p$ and cold stream $\tilde{j}_p \in C^p$ in plant p
- $ZCL_p^{w_p}$ — Set of all possible cooler matches between hot stream $\tilde{i}_p \in H^p$ and cold utility $\tilde{n}_p \in CU^p$ which can be housed in coolers of type w_p
- $ZHT_p^{l_p}$ — Set of all possible heater matches between hot utility $\tilde{m}_p \in H^p$ and cold stream $\tilde{j}_p \in C^p$ which can be housed in heaters of type l_p

Variables:

$e_{\tilde{i}_p,\tilde{j}_p,k}^{y_{ip,jp}}$ Binary variable used for determining whether or not the existing heat exchanger y_{i_p,j_p} ($y_{i_p,j_p} = 1,2,\ldots,N_{i_p,j_p}$) adopted for housing match $(i_p, j_p) \in Y^p$ in the original HEN of plant p can be used to house another match $(\tilde{i}_p, \tilde{j}_p) \in Z^p$ at stage k of the multi-plant HEN

$e_{\tilde{i}_p,\tilde{n}_p}^{i_p,n_p,w_p}$ Binary variable used for determining whether or not an existing cooler of type w_p, i.e., $(i_p, n_p) \in CL_p^{w_p}$, in the original HEN of plant p can be used to house another cooler match $(\tilde{i}_p, \tilde{n}_p) \in ZCL_p^{w_p}$ in the multi-plant HEN

$e_{\tilde{m}_p,\tilde{j}_p}^{m_p,j_p,l_p}$ Binary variable used for determining whether or not an existing heater of type l_p, i.e., $(m_p, j_p) \in HT_p^{l_p}$, in the original HEN of plant p can be used to house another heater match $(\tilde{m}_p, \tilde{j}_p) \in ZHT_p^{l_p}$ in the multi-plant HEN

$u_{\tilde{i}_p,\tilde{j}_p,k}$ Binary variable used for determining whether or not a new heat exchanger should be purchased to house match $(\tilde{i}_p, \tilde{j}_p) \in Z^p$ at stage k of the multi-plant HEN

$u_{\tilde{i}_p,\tilde{n}_p}^{CU,w_p}$ Binary variable used for determining whether or not a new cooler of type w_p should be purchased to house cooler match $(\tilde{i}_p, \tilde{n}_p) \in ZCL_p^{w_p}$ in the multi-plant HEN

$u_{\tilde{m}_p,\tilde{j}_p}^{HU,l_p}$ Binary variable used for determining whether or not a new heater of type l_p should be purchased to house heater match $(\tilde{m}_p, \tilde{j}_p) \in ZHT_p^{l_p}$ in the multi-plant HEN

$X_{\tilde{i}_p,\tilde{j}_p,k}^{y_{ip,jp}}$ Heat-transfer area of the augmented unit for the existing heat exchanger y_{i_p,j_p} ($y_{i_p,j_p} = 1,2,\ldots,N_{i_p,j_p}$) of match$(i_p, j_p) \in Y^p$ in the original single-plant HEN used for housing another match $(\tilde{i}_p, \tilde{j}_p) \in Z^p$ at stage k in the multi-plant HEN

$X_{\tilde{i}_p,\tilde{n}_p}^{i_p,n_p,w_p}$ Heat-transfer area of the augmented unit of the existing cooler of type w_p, i.e., $(i_p, n_p) \in CL_p^{w_p}$, in the original HEN of plant p for housing another cooler match $(\tilde{i}_p, \tilde{n}_p) \in ZCL_p^{w_p}$ in the multi-plant HEN

$X_{\tilde{m}_p,\tilde{j}_p}^{m_p,j_p,l_p}$ Heat-transfer area of the augmented unit of the existing heater of type l_p, i.e., $(m_p, j_p) \in HT_p^{l_p}$, in the original HEN of plant p for housing another heater match $(\tilde{m}_p, \tilde{j}_p) \in ZHT_p^{l_p}$ in the multi-plant HEN

$\sigma_{\tilde{i}_p,\tilde{j}_p,k}^{y_{ip,jp}}$ Binary variable used for determining whether or not the existing heat exchanger y_{i_p,j_p} ($y_{i_p,j_p} = 1,2,\cdots,N_{i_p,j_p}$) originally adopted for housing match $(i_p, j_p) \in Y^p$ in the single-plant HEN of plant p can be used to house another match $(\tilde{i}_p, \tilde{j}_p) \in Z^p$ at stage k of the multi-plant HEN by enlarging its heat transfer area according to Figure 5.7

$\sigma_{\tilde{i}_p,\tilde{n}_p}^{i_p,n_p,w_p}$ Binary variable used for determining whether or not the existing cooler of type w_p, i.e., $(i_p, n_p) \in CL_p^{w_p}$, in the original HEN of plant p can be used to house another cooler match $(\tilde{i}_p, \tilde{n}_p) \in ZCL_p^{w_p}$ in the multi-plant HEN by enlarging its heat transfer area according to Figure 5.7

$\sigma_{\tilde{m}_p,\tilde{j}_p}^{m_p,j_p,l_p}$ Binary variable used for determining whether or not the existing heater of type l_p, i.e., $(m_p,j_p)\in HT_p^{l_p}$, in the original HEN of plant p can be used to house another heater match $(\tilde{m}_p,\tilde{j}_p)\in ZHT_p^{l_p}$ in the multi-plant HEN by enlarging its heat transfer area according to Figure 5.7

Other than all the constraints mentioned in Section 5.2.2, additional ones should be included for the implementation of Strategy 3. Such constraints are given below:

For housing the exchanger matches of the multi-plant HEN with either available or new heat exchangers, the following inequality should be imposed:

$$\sum_{(i_p,j_p)\in Y^p}\sum_{y_{i_p,j_p}=1}^{N_{i_p,j_p}} e_{\tilde{i}_p,\tilde{j}_p,k}^{y_{i_p,j_p}} + u_{\tilde{i}_p,\tilde{j}_p,k} - \xi_{\tilde{i}_p,\tilde{j}_p,k} = 0;$$

$$\tilde{i}_p \in H^p;\tilde{j}_p \in C^p;(\tilde{i}_p,\tilde{j}_p)\in Z^p;k\in ST. \tag{5.122}$$

On the other hand, for housing the cooler and heater matches in multi-plant HEN with either available or new utility users of the same types, the following constraints should be used:

$$\sum_{(i_p,n_p)\in CL_p^{w_p}} e_{\tilde{i}_p,\tilde{n}_p}^{i_p,n_p,w_p} + u_{\tilde{i}_p,\tilde{n}_p}^{CU,w_p} - \xi_{\tilde{i}_p,\tilde{n}_p}^{CU} = 0;$$

$$\tilde{i}_p \in H^p;\tilde{n}_p \in CU^p;(\tilde{i}_p,\tilde{n}_p)\in ZCL_p^{w_p};w_p = 1,2,\ldots,W_p \tag{5.123}$$

$$\sum_{(m_p,j_p)\in HT_p^{l_p}} e_{\tilde{m}_p,\tilde{j}_p}^{m_p,j_p,l_p} + u_{\tilde{m}_p,\tilde{j}_p}^{HU,l_p} - \xi_{\tilde{m}_p,\tilde{j}_p}^{HU} = 0;$$

$$\tilde{m}_p \in HU^p;\tilde{j}_p \in C^p;(\tilde{m}_p,\tilde{j}_p)\in ZHT_p^{l_p};l_p = 1,2,\ldots,L_p \tag{5.124}$$

In the multi-plant HEN, the total numbers of heat exchangers, coolers, and heaters that are housed in existing units should be bounded, i.e.,

$$\sum_{(i_p,j_p)\in Y^p}\sum_{y_{i_p,j_p}=1}^{N_{i_p,j_p}}\sum_{(\tilde{i}_p,\tilde{j}_p)\in Z^p}\sum_{k\in ST} e_{\tilde{i}_p,\tilde{j}_p,k}^{y_{i_p,j_p}} \le N_p \tag{5.125}$$

$$\sum_{(i_p,n_p)\in CL_p^{w_p}}\sum_{(\tilde{i}_p,\tilde{n}_p)\in ZCL_p^{w_p}} e_{\tilde{i}_p,\tilde{n}_p}^{i_p,n_p,w_p} \le N_{CL}^{w_p};w_p = 1,2,\ldots,W_p. \tag{5.126}$$

$$\sum_{(m_p,j_p)\in HT_p^{l_p}}\sum_{(\tilde{m}_p,\tilde{j}_p)\in ZHT_p^{l_p}} e_{\tilde{m}_p,\tilde{j}_p}^{m_p,j_p,l_p} \le N_{HT}^{l_p};l_p = 1,2,\ldots,L_p. \tag{5.127}$$

Every existing heat exchanger should of course be used to house exactly one exchanger match in the multi-plant HEN, i.e.,

$$\sum_{(\tilde{i}_p,\tilde{j}_p)\in Z^p}\sum_{k\in ST} e_{\tilde{i}_p,\tilde{j}_p,k}^{y_{i_p,j_p}} = 1; (i_p, j_p)\in Y^p; y_{i_p,j_p} = 1,2,\ldots,N_{i_p,j_p}. \tag{5.128}$$

The heat-transfer areas of the augmented units can be expressed as

$$X_{\tilde{i}_p,\tilde{j}_p,k}^{y_{i_p,j_p}} = \max\left\{\left(A_{\tilde{i}_p,\tilde{j}_p,k} - A_{y_{i_p,j_p}}^{EX}\right)e_{\tilde{i}_p,\tilde{j}_p,k}^{y_{i_p,j_p}}, 0\right\};$$

$$i_p,\tilde{i}_p \in H^p; j_p,\tilde{j}_p \in C^p; (i_p,j_p)\in Y^p; (\tilde{i}_p,\tilde{j}_p)\in Z_p; k\in ST;$$

$$y_{i_p,j_p} = 1,2,\ldots,N_{i_p,j_p}. \tag{5.129}$$

$$X_{\tilde{i}_p,\tilde{n}_p}^{i_p,n_p,w_p} = \max\left\{\left(A_{\tilde{i}_p,\tilde{n}_p} - A_{i_p,n_p}^{EX}\right)e_{\tilde{i}_p,\tilde{n}_p}^{i_p,n_p,w_p}, 0\right\};$$

$$i_p,\tilde{i}_p \in H^p; n_p,\tilde{n}_p \in CU^p; (i_p,n_p)\in CL_p^{w_p}; (\tilde{i}_p,\tilde{n}_p)\in ZCL_p^{w_p}; w_p = 1,2,\ldots,W_p. \tag{5.130}$$

$$X_{\tilde{m}_p,\tilde{j}_p}^{m_p,j_p,l_p} = \max\left\{\left(A_{\tilde{m}_p,\tilde{j}_p} - A_{m_p,j_p}^{EX}\right)e_{\tilde{m}_p,\tilde{j}_p}^{m_p,j_p,l_p}, 0\right\};$$

$$m_p,\tilde{m}_p \in HU^p; j_p,\tilde{j}_p \in C^p; (m_p,j_p)\in HT_p^{l_p}; (\tilde{m}_p,\tilde{j}_p)\in ZHT_p^{l_p}; l_p = 1,2,\ldots,L_p. \tag{5.131}$$

The presence (or absence) of each augment unit can be determined with the binary variables in the following logic constraints:

$$X_{\tilde{i}_p,\tilde{j}_p,k}^{y_{i_p,j_p}} - \Lambda_{\tilde{i}_p,\tilde{j}_p}\sigma_{\tilde{i}_p,\tilde{j}_p,k}^{y_{i_p,j_p}} \le 0;$$

$$i_p,\tilde{i}_p \in H^p; j_p,\tilde{j}_p \in C^p; (i_p,j_p)\in Y^p; (\tilde{i}_p,\tilde{j}_p)\in Z_p; k\in ST; y_{i_p,j_p} = 1,2,\ldots,N_{i_p,j_p}. \tag{5.132}$$

$$X_{\tilde{i}_p,\tilde{n}_p}^{i_p,n_p,w_p} - \Lambda_{\tilde{i}_p,\tilde{n}_p}\sigma_{\tilde{i}_p,\tilde{n}_p}^{i_p,n_p,w_p} \le 0;$$

$$i_p,\tilde{i}_p \in H^p; n_p,\tilde{n}_p \in CU^p; (i_p,n_p)\in CL_p^{w_p}; (\tilde{i}_p,\tilde{n}_p)\in ZCL_p^{w_p}; w_p = 1,2,\ldots,W_p. \tag{5.133}$$

$$X_{\tilde{m}_p,\tilde{j}_p}^{m_p,j_p,l_p} - \Lambda_{\tilde{m}_p,\tilde{j}_p}\sigma_{\tilde{m}_p,\tilde{j}_p}^{m_p,j_p,l_p} \le 0;$$

$$m_p,\tilde{m}_p \in HU^p; j_p,\tilde{j}_p \in C^p; (m_p,j_p)\in HT_p^{l_p}; (\tilde{m}_p,\tilde{j}_p)\in ZHT_p^{l_p}; l_p = 1,2,\ldots,L_p. \tag{5.134}$$

Finally, the presence (or absence) of bypasses can also be determined with the binary variables in equations (5.105)–(5.108).

5.3.4 Objective Function

The overall saving achieved by retrofitting and building the multi-plant HEN can be used as the objective function to be maximized. This function (*TACS*) should be expressed as follows

$$TACS = TUC - TUC' - af\left(ATCC + NTCC_1 + NTCC_2 + NTCC_3\right) \quad (5.135)$$

where *TUC* denotes the sum of utility costs of all single-plant HENs which is a given constant in the corresponding MINLP models; *TUC′* denotes the total utility cost of the multi-plant HEN after retrofit; *af* is the annualization factor which is another given constant; *ATCC* is the total capital cost of all augmented units; $NTCC_1$ denotes the total capital cost of all new units purchased for interplant matches; $NTCC_2$ denotes the total capital cost of all bypasses; $NTCC_3$ denotes the total capital cost of all new units purchased for inner-plant matches. Other than the aforementioned two constants (i.e., *TUC* and *af*), the detailed expressions of the remaining cost models are listed in the subsequent subsections and, for the sake of brevity, all embedded model parameters (or cost coefficients) are first defined below:

Parameters:

BY_{i_p} Cost of a single bypass on hot stream i_p

BY_{j_p} Cost of a single bypass on cold stream j_p

$CA_{i_q,j_{q'}}$ Variable cost coefficient in the cost model of heat exchanger between hot stream i_p and cold stream $j_{q'}$ $(q,q' = 1,2,\ldots,P)$

$CA_{i_p,n_{q'}}$ Variable cost coefficient in the cost model of cooler between hot stream i_q and cold utility $n_{q'}$ $(q' = 1,2,\ldots,P)$

CA_{m_q,j_p} Variable cost coefficient in the cost model of heater between hot utility m_q $(q = 1,2,\ldots,P)$ and cold stream j_p

$CF_{i_q,j_{q'}}$ Fixed cost in the cost model of heat exchanger between hot stream i_q and cold stream $j_{q'}$ $(q,q' = 1,2,\ldots,P)$

$CF_{i_p,n_{q'}}$ Fixed cost in the cost model of cooler between hot stream i_q and cold utility $n_{q'}$ $(q' = 1,2,\ldots,P)$

CF_{m_q,j_p} Fixed cost in the cost model of heater between hot utility m_q $(q = 1,2,\ldots,P$) and cold stream j_p

$CQ_{i_p,n_{q'}}$ Unit cost of cold utility $n_{q'}$ $(q' = 1,2,\ldots,P)$ for cooling hot stream i_p

CQ_{m_q,j_p} Unit cost of hot utility m_q $(q = 1,2,\ldots,P)$ for heating cold stream j_p

B Exponent of heater transfer areas in variable cost terms in the cost models of heat exchanger, cooler, and heater

5.3.4.1 Cost Models for Applying Strategy 1

For Strategy 1, all four cost items embedded in equation (5.135) are presented in detail as follows

$$TUC' = \sum_{p=1}^{P} \sum_{(i_p,n_p)\in CL_p} CQ_{i_p,n_p} q_{i_p,n_p} + \sum_{p=1}^{P} \sum_{(m_p,j_p)\in HT_p} CQ_{m_p,j_p} q_{m_p,j_p} + \sum_{p=1}^{P} \sum_{\substack{q'=1 \\ q'\neq p}}^{P} \sum_{(i_p,n_{q'})\in GC_p} CQ_{i_p,n_{q'}} q_{i_p,n_{q'}} + \sum_{p=1}^{P} \sum_{\substack{q=1 \\ q\neq p}}^{P} \sum_{(m_q,j_p)\in GH_p} CQ_{m_q,j_p} q_{m_q,j_p} \tag{5.136}$$

$$\begin{aligned} ATCC = {} & \sum_{p=1}^{P} \sum_{(i_p,j_p)\in Y^P} \sum_{y_{i_p,j_p}=1}^{N_{i_p,j_p}} \sum_{k\in ST} \left[CF_{i_p,j_p} \sigma_{i_p,j_p,k}^{y_{i_p,j_p}} + CA_{i_p,j_p} \left(X_{i_p,j_p,k}^{y_{i_p,j_p}} \right)^{\beta} \right] \\ & + \sum_{p=1}^{P} \sum_{(i_p,n_p)\in CL_p} \left[CF_{i_p,n_p} \sigma_{i_p,n_p} + CA_{i_p,n_p} \left(X_{i_p,n_p} \right)^{\beta} \right] \\ & + \sum_{p=1}^{P} \sum_{(m_p,j_p)\in HT_p} \left[CF_{m_p,j_p} \sigma_{m_p,j_p} + CA_{m_p,j_p} \left(X_{m_p,j_p} \right)^{\beta} \right] \end{aligned} \tag{5.137}$$

$$\begin{aligned} NTCC_1 = {} & \sum_{p=1}^{P} \sum_{\substack{q'=1 \\ q'\neq p}}^{P} \sum_{(i_p,j_{q'})\in G_p} \sum_{k\in ST} \left[CF_{i_p,j_{q'}} \xi_{i_p,j_{q'},k} + CA_{i_p,j_{q'}} \left(A_{i_p,j_{q'}} \right)^{\beta} \right] \\ & + \sum_{p=1}^{P} \sum_{\substack{q=1 \\ q\neq p}}^{P} \sum_{(i_q,j_p)\in G_p} \sum_{k\in ST} \left[CF_{i_q,j_p} \xi_{i_q,j_p,k} + CA_{i_q,j_p} \left(A_{i_q,j_p} \right)^{\beta} \right] \\ & + \sum_{p=1}^{P} \sum_{\substack{q'=1 \\ q'\neq p}}^{P} \sum_{(i_p,n_{q'})\in GC_p} \left[CF_{i_p,n_{q'}} \xi_{i_p,n_{q'}}^{CU} + CA_{i_p,n_{q'}} \left(A_{i_p,n_{q'}} \right)^{\beta} \right] \\ & + \sum_{p=1}^{P} \sum_{\substack{q=1 \\ q\neq p}}^{P} \sum_{(m_q,j_p)\in GH_p} \left[CF_{m_q,j_p} \xi_{m_q,j_p}^{HU} + CA_{m_q,j_p} \left(A_{m_q,j_p} \right)^{\beta} \right] \end{aligned} \tag{5.138}$$

$$\begin{aligned} NTCC_2 = {} & \sum_{p=1}^{P} \sum_{i_p\in H^P} \left[BY_{i_p} \left(rzhu_{i_p} + \sum_{k\in ST} rz_{i_p,k} \right) \right] \\ & + \sum_{p=1}^{P} \sum_{j_p\in C^P} \left[BY_{j_p} \left(rzcu_{j_p} + \sum_{k\in ST} rz_{j_p,k} \right) \right] \end{aligned} \tag{5.139}$$

$$NTCC_3 = 0 \tag{5.140}$$

5.3.4.2 Cost Models for Applying Strategy 2

The above cost models are all applicable in the present scenario except *ATCC*. In particular, equation (5.137) should be replaced by the following formula

$$
\begin{aligned}
ATC = &\sum_{p=1}^{P} \sum_{(i_p,j_p)\in Y^P} \sum_{y_{i_p,j_p}=1}^{N_{i_p,j_p}} \sum_{(\tilde{i}_p,\tilde{j}_p)\in Y^P} \sum_{k\in ST} \left[CF_{\tilde{i}_p,\tilde{j}_p}\, \sigma^{y_{i_p,j_p}}_{\tilde{i}_p,\tilde{j}_p,k} + CA_{\tilde{i}_p,\tilde{j}_p} \left(X^{y_{i_p,j_p}}_{\tilde{i}_p,\tilde{j}_p,k} \right)^{\beta} \right] \\
&+ \sum_{p=1}^{P} \sum_{w_p=1}^{W_p} \sum_{(i_p,n_p)\in CL_p^{w_p}} \sum_{(\tilde{i}_p,\tilde{n}_p)\in CL_p^{w_p}} \left[CF_{\tilde{i}_p,\tilde{n}_p}\, \sigma^{i_p,n_p,w_p}_{\tilde{i}_p,\tilde{n}_p} + CA_{\tilde{i}_p,\tilde{n}_p} \left(X^{i_p,n_p,w_p}_{\tilde{i}_p,\tilde{n}_p} \right)^{\beta} \right] \\
&+ \sum_{p=1}^{P} \sum_{l_p=1}^{L_p} \sum_{(m_p,j_p)\in HT_p^{l_p}} \sum_{(\tilde{m}_p,\tilde{j}_p)\in HT_p^{l_p}} \left[CF_{\tilde{m}_p,\tilde{j}_p}\, \sigma^{m_p,j_p,l_p}_{\tilde{m}_p,\tilde{j}_p} + CA_{\tilde{m}_p,\tilde{j}_p} \left(X^{m_p,j_p,l_p}_{\tilde{m}_p,\tilde{j}_p} \right)^{\beta} \right]
\end{aligned}
\tag{5.141}
$$

5.3.4.3 Cost Models for Applying Strategy 3

In this case, the total capital cost of all new units purchased for interplant matches ($NTCC_1$) and the total capital cost of all bypasses ($NTCC_2$) can be determined according to equations (5.138) and (5.139) respectively. The other cost models are presented below:

$$
\begin{aligned}
TUC' = &\sum_{p=1}^{P} \sum_{w_p=1}^{W_p} \sum_{(i_p,n_p)\in ZCL_p^{w_p}} CQ_{i_p,n_p} q_{i_p,n_p} + \sum_{p=1}^{P} \sum_{l_p=1}^{L_p} \sum_{(m_p,j_p)\in ZHT_p^{l_p}} CQ_{m_p,j_p} q_{m_p,j_p} \\
&+ \sum_{p=1}^{P} \sum_{\substack{q'=1 \\ q'\neq p}}^{P} \sum_{(i_p,n_{q'})\in GC_p} CQ_{i_p,n_{q'}} q_{i_p,n_{q'}} + \sum_{p=1}^{P} \sum_{\substack{q=1 \\ q\neq p}}^{P} \sum_{(m_q,j_p)\in GH_p} CQ_{m_q,j_p} q_{m_q,j_p}
\end{aligned}
\tag{5.142}
$$

$$
\begin{aligned}
ATC = &\sum_{p=1}^{P} \sum_{(i_p,j_p)\in Y^P} \sum_{y_{i_p,j_p}=1}^{N_{i_p,j_p}} \sum_{(\tilde{i}_p,\tilde{j}_p)\in Z^P} \sum_{k\in ST} \left[CF_{\tilde{i}_p,\tilde{j}_p}\, \sigma^{y_{i_p,j_p}}_{\tilde{i}_p,\tilde{j}_p,k} + CA_{\tilde{i}_p,\tilde{j}_p} \left(X^{y_{i_p,j_p}}_{\tilde{i}_p,\tilde{j}_p,k} \right)^{\beta} \right] \\
&+ \sum_{p=1}^{P} \sum_{w_p=1}^{W_p} \sum_{(i_p,n_p)\in CL_p^{w_p}} \sum_{(\tilde{i}_p,\tilde{n}_p)\in ZCL_p^{w_p}} \left[CF_{\tilde{i}_p,\tilde{n}_p}\, \sigma^{i_p,n_p,w_p}_{\tilde{i}_p,\tilde{n}_p} + CA_{\tilde{i}_p,\tilde{n}_p} \left(X^{i_p,n_p,w_p}_{\tilde{i}_p,\tilde{n}_p} \right)^{\beta} \right] \\
&+ \sum_{p=1}^{P} \sum_{l_p=1}^{L_p} \sum_{(m_p,j_p)\in HT_p^{l_p}} \sum_{(\tilde{m}_p,\tilde{j}_p)\in ZHT_p^{l_p}} \left[CF_{\tilde{m}_p,\tilde{j}_p}\, \sigma^{m_p,j_p,l_p}_{\tilde{m}_p,\tilde{j}_p} + CA_{\tilde{m}_p,\tilde{j}_p} \left(X^{m_p,j_p,l_p}_{\tilde{m}_p,\tilde{j}_p} \right)^{\beta} \right]
\end{aligned}
\tag{5.143}
$$

$$NTCC_3 = \sum_{p=1}^{P} \sum_{(i_p,j_p)\in Z^P} \sum_{k\in ST} \left[CF_{i_p,j_p} u_{i_p,j_p,k} + CA_{i_p,j_p} \left(A_{i_p,j_p,k} \right)^{\beta} \right]$$

$$+ \sum_{p=1}^{P} \sum_{w_p=1}^{W_p} \sum_{(i_p,n_p)\in ZCL_p^{w_p}} \left[CF_{i_p,n_p} u_{i_p,n_p}^{CU,w_p} + CA_{i_p,n_p} \left(A_{i_p,n_p} \right)^{\beta} \right] \quad (5.144)$$

$$+ \sum_{p=1}^{P} \sum_{l_p=1}^{L_p} \sum_{(m_p,j_p)\in ZHT_p^{l_p}} \left[CF_{m_p,j_p} u_{m_p,j_p}^{HU,l_p} + CA_{m_p,j_p} \left(A_{m_p,j_p} \right)^{\beta} \right]$$

5.3.5 Example 5.2

As an illustrative example, let us again consider three chemical plants (P1, P2, and P3) and their stream data given in Table 5.1. The corresponding utility data can be found in Table 5.7. Let us further assume that the existing single-plant HENs were synthesized according to the conventional simultaneous optimization strategy (Yee and Grossmann, 1990) and they are presented in Figure 5.8. The minimum TACs of these HENs were found to be 306,994 USD/yr (P1), 47,614 USD/yr (P2), and 279,177 USD/yr (P3) respectively. The sum of utility costs of all three single-plant HENs, i.e., *TUC*, was determined to be 633,785 USD/yr.

TABLE 5.7
Utility Data of Illustrative Example

Plant	Utility	Inlet Temp. (°C)	Outlet Temp. (°C)	Unit Cost (USD/kW·yr)
P1	Cooling water	25	30	200
	LP steam	200	200	375
	MP steam	250	250	575
	HP steam	300	300	775
	Hot oil	500	475	900
P2	Cooling water	25	30	250
	LP steam	200	200	400
	MP steam	250	250	600
	HP steam	300	300	800
	Hot oil	500	475	1000
P3	Cooling water	25	30	150
	LP steam	200	200	350
	MP steam	250	250	550
	HP steam	300	300	750
	Hot oil	500	475	850

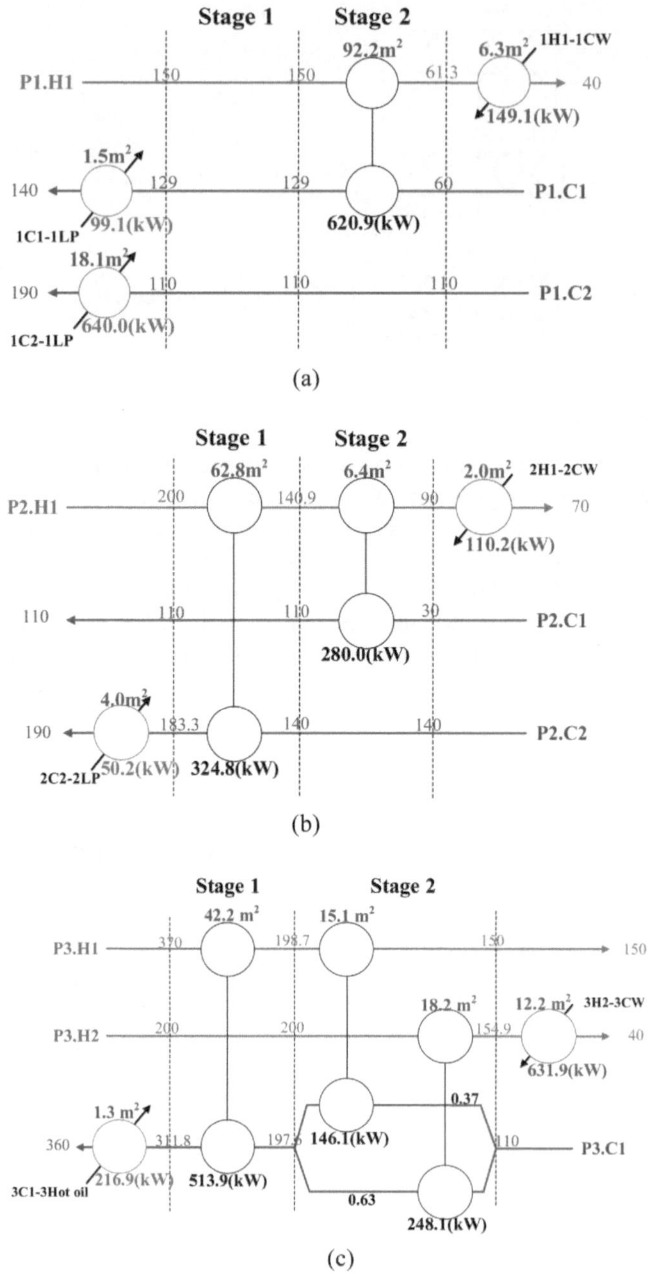

FIGURE 5.8 Single-plant HENs of (a) P1, (b) P2, and (c) P3. (Adapted with permission from *Ind. Eng. Chem. Res.* 2020, 59: 18088–18105. Copyright 2020 American Chemical Society.)

The cost model parameters used in the present example are listed below:

- Based on a life length of 10 years and a yearly interest rate of 5.85%, an annualization factor (*af*) of 0.1349 is adopted.
- The exponent of area in the variable cost term (β) is 0.83.
- The variable cost coefficients of the heat exchangers (CA_{i_p,j_p}, $CA_{i_p,j_{q'}}$, and CA_{i_q,j_p}) and those of coolers (CA_{i_p,n_p} and $CA_{i_p,n_{q'}}$) are all set to be: 380 $/m$^{1.66}$; the variable cost coefficients of the heaters (CA_{m_p,j_p} and CA_{m_q,j_p}) are set to be: 700 USD/m$^{1.66}$.
- The fixed costs of interplant heat exchangers, coolers, and heaters ($CF_{i_p,j_{q'}}$, CF_{i_q,j_p}, $CF_{i_p,n_{q'}}$, and CA_{m_q,j_p}) are set to be: 30,000 USD; the fixed costs of all other inner-plant units (CF_{i_p,j_p}, CF_{i_p,n_p}, and CA_{m_p,j_p}) are all assumed to be: 10,000 USD.
- The re-piping cost of every bypass (BY_{i_p} and BY_{j_p}) is 500 USD.
- For computation convenience, the overall heat transfer coefficients of all heat exchangers, coolers, and heaters are taken to be 1 kW/m^2°C.

Finally, it should be noted that the optimal solutions presented in the sequel were all obtained with solver COUENNE in GAMS 24.0 on a personal computer (Intel Core i7 4750; 16G).

5.3.5.1 Optimal Solution Obtained by Applying Strategy 1

An MINLP model can be constructed according to the formulations described in Sections 5.2.2, 5.3.1, and 5.3.4.1. The actual numbers of real and integer variables in the GAMS code were 1563 and 306 respectively, while those of equality and inequality constraints were 824 and 642 respectively. After solving this model, an optimal revamped design of the three-plant HEN can be generated (see Figure 5.9). The computation time in this case was around 20,000 s. Notice that the interplant matches in Figure 5.9 are represented with vertically connected circles in which different colors are filled and each match should be housed in a purchased new heat exchanger. All inner-plant matches are represented with vertically connected circles without color, and they are housed (at least partially) in the existing heat exchangers. The inner-plant matches that require augmented units are indicated with double circles, while the others are marked with single circles. The net saving of this design, i.e., *TACS*, was determined to be 75,126 USD/yr, while the sum of the yearly utility cost and annualized total capital investment of the retrofit design, i.e., $TUC' + af\left(ATCC + NTCC_1 + NTCC_2\right)$, was 558,659 USD/yr.

5.3.5.2 Optimal Solution Obtained with Strategy 2

Another MINLP model was constructed according to the formulations described in Sections 5.2.2, 5.3.2, and 5.3.4.2. The actual numbers of real and integer variables in the GAMS code were 1625 and 378 respectively, while those of equality and inequality constraints were 920 and 662 respectively. After solving this model, an optimal revamped design of the three-plant HEN can be generated and

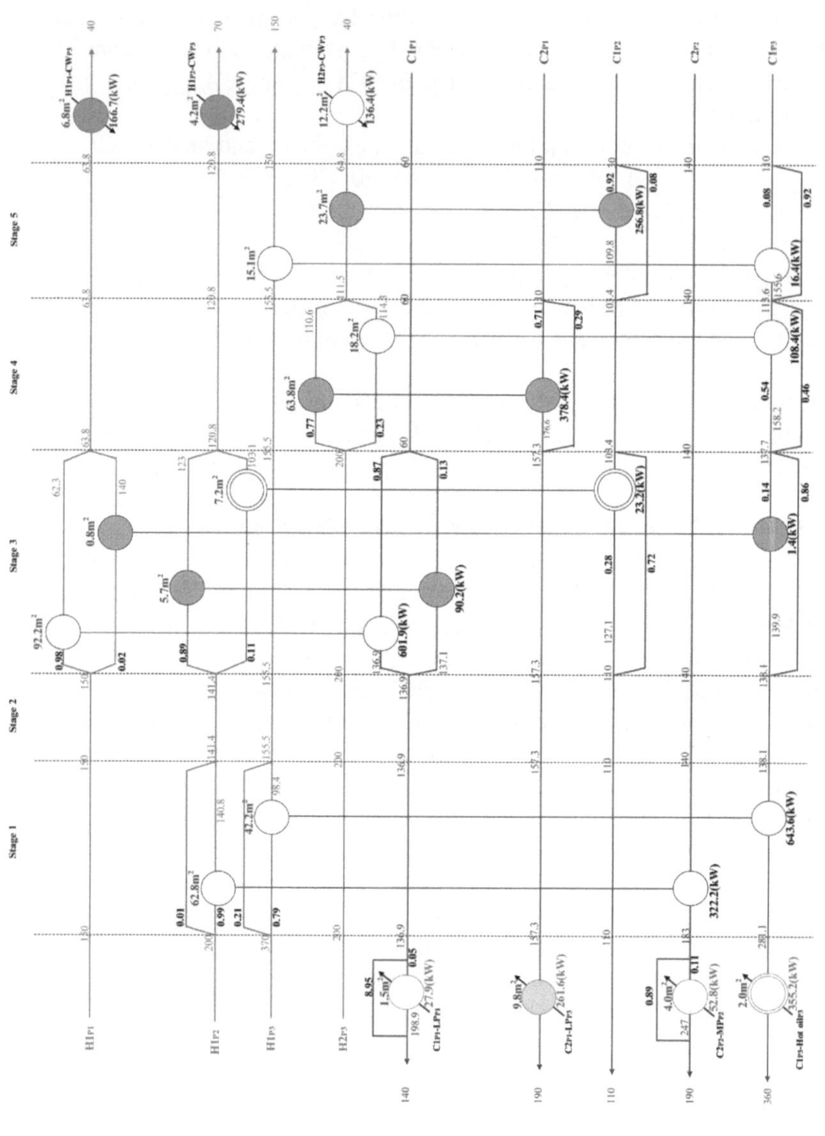

FIGURE 5.9 Interplant heat integration scheme obtained by revamping the single-plant HENs of P1, P2, and P3 with Strategy I. (Adapted with permission from *Ind. Eng. Chem. Res.* 2020, 59: 18088–18105. Copyright 2020 American Chemical Society.)

this design is presented in Figure 5.10. The computation time in this case was also around 20,000 s. The symbols used in Figure 5.10 follow exactly the same conventions described in the previous subsection. Since Strategy 2 allows the assignment of every existing heat exchanger, cooler, or heater to any existing match of the same type within the same plant, additional information can be extracted from the optimal solution. Table 5.8 shows the placement scheme of the existing units in the retrofit design, while Table 5.9 shows the arrangements of new units. Notice from Table 5.8 that the two existing heat exchangers used to house two matches $\left(H1_{P2},C1_{P2}\right)$ and $\left(H1_{P2},C2_{P2}\right)$, respectively, in the original single-plant design of P2 are switched in the revamp design. Similarly, a switch between the two existing heat exchangers for $\left(H1_{P3},C1_{P3}\right)$ and $\left(H2_{P3},C1_{P3}\right)$ in the original single-plant design of P3 is called for in the revamped multi-plant HEN. Notice also from Table 5.8 and Table 5.9 that the matches $\left(H1_{P1},CW_{P1}\right)$ and $\left(H1_{P2},CW_{P2}\right)$ are not housed in their original coolers in the retrofit design and, instead, the hot streams $H1_{P1}$ and $H1_{P2}$ are matched with the cheapest cooling water from plant P3 in new coolers. On the other hand, the existing heaters for original matches $\left(LP_{P1},C1_{P1}\right)$, $\left(LP_{P1},C2_{P1}\right)$, and $\left(LP_{P2},C2_{P2}\right)$ are not needed in the retrofit design. The first match is actually not present, while the cold streams in the remaining two matches, i.e., $C2_{P1}$ and $C2_{P2}$, are matched with the cheapest low-pressure steam from P3 in new heaters. Finally, the net saving of this second design, i.e., *TACS*, was determined to be 84,328 USD/yr, while the sum of the yearly utility cost and annualized total capital investment of the retrofit design, i.e.,$TUC' + af\left(ATCC + NTCC_1 + NTCC_2\right)$, was 549,457 USD/yr.

5.3.5.3 Optimal Solution Obtained with Strategy 3

A third MINLP model was constructed according to the formulations described in Sections 5.2.2, 5.3.3, and 5.3.4.3. The actual numbers of real and integer variables in the GAMS code were 1689 and 432 respectively, while those of equality and inequality constraints were 920 and 662 respectively. After solving this model, an optimal revamped design of the three-plant HEN can be generated according to Strategy 3 and this design is presented in Figure 5.11. The computation time in this case was around 50,000 s. Again, the symbols used in Figure 5.11 follow exactly the same conventions described in the previous two subsections. Since Strategy 3 allows the assignment of every existing heat exchanger, cooler, or heater to any match of the same type within the same plant, additional information can be extracted from the optimal solution. Table 5.10 shows the assignments of the existing units in the retrofit design, while Table 5.11 shows the arrangements of new units. Notice from Table 5.10 that the existing heat exchanger that originally housed the match $\left(H1_{P2},C2_{P2}\right)$ is used on another match $\left(H1_{P2},C1_{P2}\right)$ in plant P2 and, in addition, the two existing heat exchangers used to house two matches $\left(H1_{P3},C1_{P3}\right)$ and $\left(H2_{P3},C1_{P3}\right)$, respectively, in the original single-plant design of P3 are switched in the revamp design. On the other hand, the existing coolers for original matches $\left(H1_{P1},CW_{P1}\right)$ and $\left(H1_{P2},CW_{P2}\right)$ and the existing heater for original match $\left(LP_{P1},C2_{P1}\right)$ are not needed in the retrofit

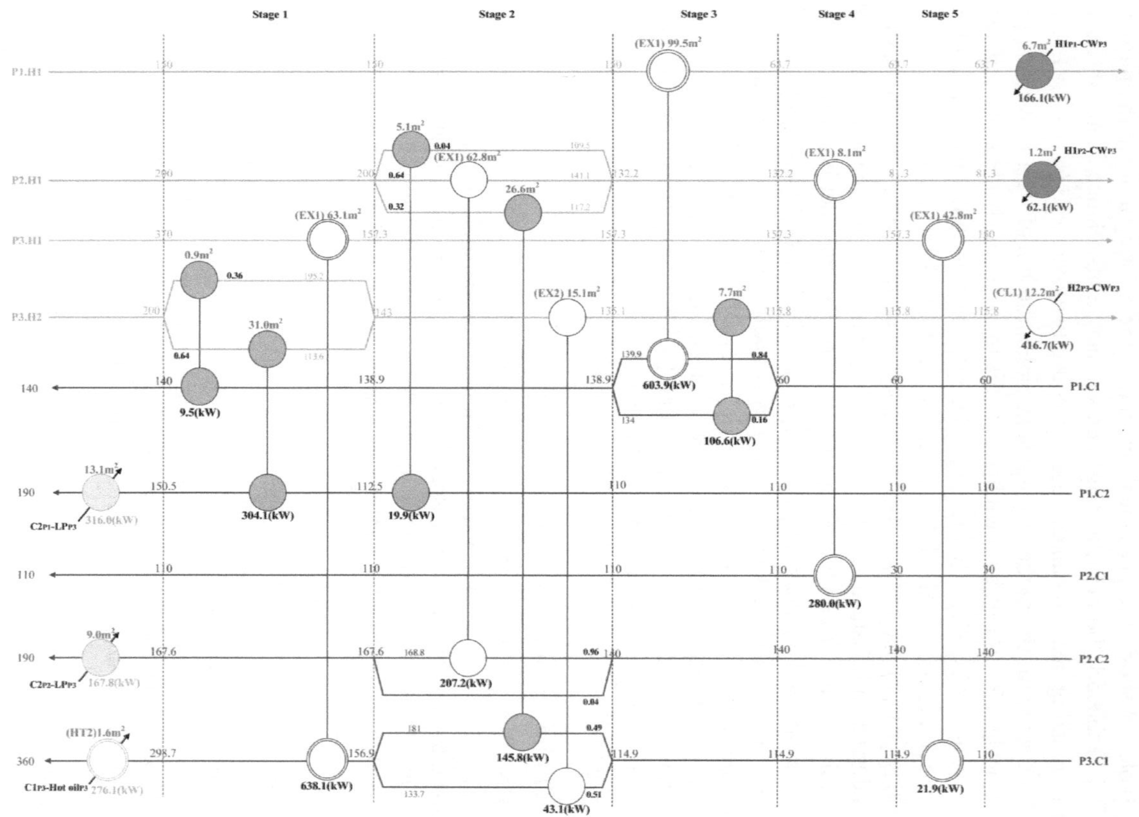

FIGURE 5.10 Multi-plant HEN design obtained by revamping the single-plant HENs of P1, P2, and P3 with Strategy 2. (Adapted with permission from *Ind. Eng. Chem. Res.* 2020, 59: 18088–18105. Copyright 2020 American Chemical Society.)

TABLE 5.8
Assignments of Existing Units Using Strategy 2

Existing Matches/Exchangers before Retrofit (i_p, j_p)	Assignments of Exchanger Matches in Retrofit Design $(\tilde{i}_p, \tilde{j}_p)$	Exchanger Order (y_{i_p,j_p})	Areas of Existing Heat Exchangers (m²)
$(H1_{P1}, C1_{P1})$	$(H1_{P1}, C1_{P1})$	1	92.2
$(H1_{P2}, C1_{P2})$	$(H1_{P2}, C2_{P2})$	1	6.4
$(H1_{P2}, C2_{P2})$	$(H1_{P2}, C1_{P2})$	1	62.8
$(H1_{P3}, C1_{P3})$	$(H1_{P3}, C1_{P3})$	1	42.2
$(H1_{P3}, C1_{P3})$	$(H2_{P3}, C1_{P3})$	2	15.1
$(H2_{P3}, C1_{P3})$	$(H1_{P3}, C1_{P3})$	1	18.2
Existing Matches/Coolers before Retrofit (i_p, n_p)	**Assignments of Cooler Matches in Retrofit Design $(\tilde{i}_p, \tilde{n}_p)$**	**Cooler Type (w_p)**	**Areas of Existing Coolers (m²)**
$(H1_{P1}, CW_{P1})$	/	/	6.3
$(H1_{P2}, CW_{P2})$	/	/	2.0
$(H2_{P3}, CW_{P3})$	$(H2_{P3}, CW_{P3})$	1	12.2
Existing Matches/Heaters before Retrofit (m_p, j_p)	**Assignments of Heater Matches in Retrofit Design $(\tilde{m}_p, \tilde{j}_p)$**	**Heater Type (l_p)**	**Areas of Existing Heaters (m²)**
$(LP_{P1}, C1_{P1})$	/	/	1.5
$(LP_{P1}, C2_{P1})$	/	/	18.1
$(LP_{P2}, C2_{P2})$	/	/	4.0
$(Hot\,oil_{P3}, C1_{P3})$	$(Hot\,oil_{P3}, C1_{P3})$	2	1.3

design. Cheaper utilities are taken from plant P3 for cooling and heating in all three cases (see Table 5.11). Finally, the net saving of the last design, i.e., *TACS*, was found to be the highest among all three strategies considered so far (i.e., 95,634 USD/yr), while the sum of the yearly utility cost and annualized total capital investment of the retrofit design, i.e.,$TUC' + af\left(ATCC + NTCC_1 + NTCC_2\right)$, was the lowest at 538,151 USD/yr.

5.3.5.4 Benefit Allocation with Shapley Values

According to the discussions in Section 5.1, the function value $\nu(\Sigma)$, where $\Sigma \in 2^{\Theta}$ and $\Sigma \subseteq \Theta$, should be viewed as the total cost saving (*TACS*) realized by the corresponding coalition. In the present case, $\Theta = \{P1, P2, P2\}$ and let us further define the following subsets of Θ, i.e., $\Sigma_1^1 = \{P1\}$, $\Sigma_2^1 = \{P2\}$, $\Sigma_3^1 = \{P3\}$; $\Sigma_1^2 = \{P1, P2\}$,

TABLE 5.9
Assignments of New Units Using Strategy 2

Interplant Matches Using New Heat Exchangers in Retrofit Design $(i_p, j_{q'})$ or (i_q, j_p)	Areas of New Heat Exchanger (m²)
$(H2_{P3}, C1_{P1})$	0.9
$(H2_{P3}, C2_{P1})$	31.0
$(H1_{P2}, C2_{P1})$	5.1
$(H1_{P2}, C1_{P3})$	26.6
$(H2_{P3}, C1_{P1})$	7.7
Interplant Matches Using New Coolers in Retrofit Design $(i_p, n_{q'})$	**Areas of New Coolers (m²)**
$(H1_{P1}, CW_{P3})$	6.7
$(H1_{P2}, CW_{P3})$	1.2
Interplant Matches Using New Heaters in Retrofit Design (m_q, j_p)	**Areas of New Heaters (m²)**
$(LP_{P3}, C2_{P1})$	13.1
$(LP_{P3}, C2_{P2})$	9.0

$\Sigma_2^2 = \{P2, P3\}$, $\Sigma_3^2 = \{P3, P1\}$, for the sake of clarity. Thus, the traditional Shapley-value vector can be written as

$$\varphi_\Theta = \begin{bmatrix} \varphi_{\Theta,P1} \\ \varphi_{\Theta,P2} \\ \varphi_{\Theta,P3} \end{bmatrix} = \begin{bmatrix} \dfrac{2\nu(\Psi)+\nu(\Sigma_3^2)+\nu(\Sigma_1^2)+2\nu(\Sigma_1^1)-2\nu(\Sigma_2^2)-\nu(\Sigma_2^1)-\nu(\Sigma_3^1)}{6} \\ \dfrac{2\nu(\Psi)+\nu(\Sigma_1^2)+\nu(\Sigma_2^2)+2\nu(\Sigma_2^1)-2\nu(\Sigma_3^2)-\nu(\Sigma_3^1)-\nu(\Sigma_1^1)}{6} \\ \dfrac{2\nu(\Psi)+\nu(\Sigma_2^2)+\nu(\Sigma_3^2)+2\nu(\Sigma_3^1)-2\nu(\Sigma_1^2)-\nu(\Sigma_1^1)-\nu(\Sigma_2^1)}{6} \end{bmatrix} \tag{5.145}$$

The allocation schemes of cost savings achieved with different retrofit strategies can all be devised according to equation (5.145), if there are no risks of plant shutdowns.

By repeatedly solving the MINLP models described in Sections 5.2.2, 5.3.1, and 5.3.4.1 for applying Strategy 1 to coalitions Σ_1^2, Σ_2^2, Σ_3^2, and Θ, one can obtain the following function values

- $\nu(\varnothing) = \nu(\Sigma_1^1) = \nu(\Sigma_2^1) = \nu(\Sigma_3^1) = 0$;
- $\nu(\Sigma_1^2) = 47368$, $\nu(\Sigma_2^2) = 6878$, $\nu(\Sigma_3^2) = 44009$;
- $\nu(\Theta) = 75126$.

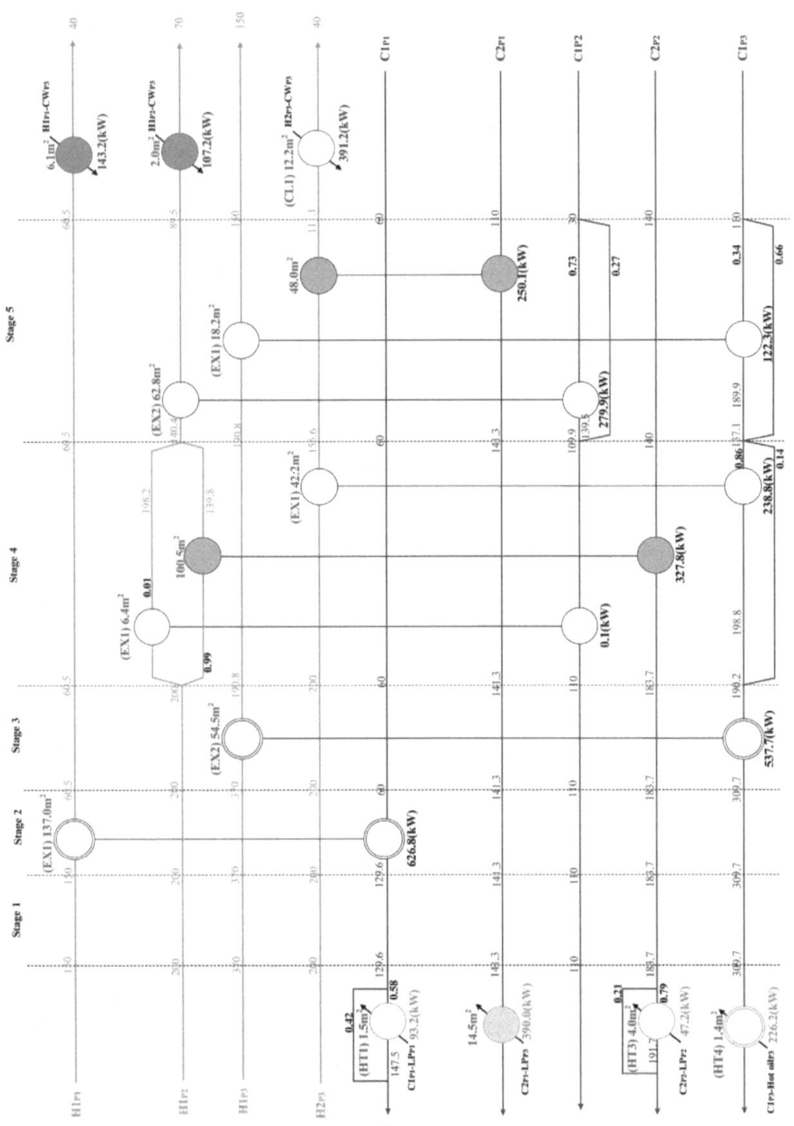

FIGURE 5.11 Multi-plant HEN design obtained by revamping the single-plant HENs of P1, P2, and P3 with Strategy 3. (Adapted with permission from *Ind. Eng. Chem. Res.* 2020, 59: 18088–18105. Copyright 2020 American Chemical Society.)

TABLE 5.10
Assignments of Existing Units Using Strategy 3

Existing Matches/Exchangers before Retrofit (i_p, j_p)	Assignments of Exchanger Matches in Retrofit Design $(\tilde{i}_p, \tilde{j}_p)$	Exchanger Order $\left(y_{i_p,j_p}\right)$	Areas of Existing Heat Exchanger (m²)
$\left(H1_{P1}, C1_{P1}\right)$	$\left(H1_{P1}, C1_{P1}\right)$	1	92.2
$\left(H1_{P2}, C1_{P2}\right)$	$\left(H1_{P2}, C1_{P2}\right)$	1	6.4
$\left(H1_{P2}, C2_{P2}\right)$	$\left(H1_{P2}, C1_{P2}\right)$	1	62.8
$\left(H1_{P3}, C1_{P3}\right)$	$\left(H2_{P3}, C1_{P3}\right)$	1	42.2
$\left(H1_{P3}, C1_{P3}\right)$	$\left(H1_{P3}, C1_{P3}\right)$	2	15.1
$\left(H2_{P3}, C1_{P3}\right)$	$\left(H1_{P3}, C1_{P3}\right)$	1	18.2
Existing Matches/Coolers before Retrofit (i_p, n_p)	**Assignments of Cooler Matches in Retrofit Design** $(\tilde{i}_p, \tilde{n}_p)$	**Cooler Type** (w_p)	**Areas of Existing Coolers (m²)**
$\left(H1_{P1}, CW_{P1}\right)$	/	/	6.3
$\left(H1_{P2}, CW_{P2}\right)$	/	/	2.0
$\left(H2_{P3}, CW_{P3}\right)$	$\left(H2_{P3}, CW_{P3}\right)$	1	12.2
Existing Matches/Heaters before Retrofit (m_p, j_p)	**Assignments of Heater Matches in Retrofit Design** $(\tilde{m}_p, \tilde{j}_p)$	**Heater Type** (l_p)	**Areas of Existing Heaters (m²)**
$\left(LP_{P1}, C1_{P1}\right)$	$\left(LP_{P1}, C1_{P1}\right)$	1	1.5
$\left(LP_{P1}, C2_{P1}\right)$	/	/	18.1
$\left(LP_{P2}, C2_{P2}\right)$	$\left(LP_{P2}, C2_{P2}\right)$	1	4.0
$\left(Hot\,oil_{P3}, C1_{P3}\right)$	$\left(Hotoil_{P3}, C1_{P3}\right)$	2	1.3

Therefore, the cost savings allocated to P1, P2, and P3, i.e., the conventional Shapley values, can be computed respectively according to equation (5.145) (see Table 5.12). On the other hand, if plant shutdowns are possible during operation, then the risk-based Shapley values described in Sections 5.1.3–5.1.7 should be computed. It is assumed in this example that, after one or more plant dropping out of the coalition, each of the remaining players can return to its original optimal single-plant HEN operation before implementing the retrofit project. In other words, the corresponding loss of the defect coalition is simply the sum of capital costs of the additional bypasses and the interplant heat exchangers. For illustration purposes, the probabilities of plant shutdowns in this example are assumed to be 0.3, 0.35, and 0.32, respectively, and the corresponding risk-based Shapley values are also presented in Table 5.12.

TABLE 5.11
Assignments of New Units Using Strategy 3

Matches Using New Heat Exchangers in Retrofit Design $(\tilde{i}_p, \tilde{j}_p)$, $(i_p, j_{q'})$, or (i_q, j_p)	Areas of New Heat Exchangers (m²)
$(H1_{P2}, C2_{P2})$	100.5
$(H2_{P3}, C2_{P1})$	48.0
Matches Using New Coolers in Retrofit Design $(i_p, n_{q'})$	**Areas of New Coolers (m²)**
$(H1_{P1}, CW_{P3})$	6.1
$(H1_{P2}, CW_{P3})$	2.0
Matches Using New Heaters in Retrofit Design (m_q, j_p)	**Areas of New Heaters (m²)**
$(LP_{P3}, C2_{P1})$	14.5

TABLE 5.12
Conventional and Risk-Based Shapley Values ($/yr) Obtained with Strategy 1 in Example 5.2

i	1	2	3
$\varphi_{\{1,2,3\},i}$	37,979	19,413	17,734
$\hat{\varphi}_{\{1,2,3\},i}$	37,918	19,439	17,768

Secondly, by repeatedly solving the MINLP models described in Sections 5.2.2, 5.3.2, and 5.3.4.2 for applying Strategy 2 to coalitions Σ_1^2, Σ_2^2, Σ_3^2, and Θ, one can obtain the following function values

- $v(\varnothing) = v(\Sigma_1^1) = v(\Sigma_2^1) = v(\Sigma_3^1) = 0;$
- $v(\Sigma_1^2) = 57884,\ v(\Sigma_2^2) = 6930,\ v(\Sigma_3^2) = 64116;$
- $v(\Theta) = 84328.$

Therefore, the cost savings allocated to P1, P2, and P3 can be determined respectively according to the conventional or risk-based Shapley values listed in Table 5.13.

TABLE 5.13
Conventional and Risk-Based Shapley Values (\$/yr) Obtained with Strategy 2 in Example 5.2

i	1	2	3
$\varphi_{\{1,2,3\},i}$	45,515	16,922	21,891
$\hat{\varphi}_{\{1,2,3\},i}$	43,856	17,526	22,946

TABLE 5.14
Conventional and Risk-Based Shapley Values (\$/yr) Obtained with Strategy 3 in Example 5.2

i	1	2	3
$\varphi_{\{1,2,3\},i}$	53,950	13,288	28,396
$\hat{\varphi}_{\{1,2,3\},i}$	52,364	13,914	29,356

Finally, by repeatedly solving the MINLP models described in Sections 5.2.2, 5.3.3, and 5.3.4.3 for applying Strategy 3 to coalitions Σ_1^2, Σ_2^2, Σ_3^2, and Θ, one can obtain the following function values

- $v(\varnothing) = v(\Sigma_1^1) = v(\Sigma_2^1) = v(\Sigma_3^1) = 0$;
- $v(\Sigma_1^2) = 58036$, $v(\Sigma_2^2) = 6930$, $v(\Sigma_3^2) = 88254$;
- $v(\Theta) = 95634$.

Therefore, the cost savings allocated to P1, P2, and P3 can be computed according to the conventional or risk-based Shapley values given in Table 5.14.

First of all, it can be observed from Tables 5.12–5.14 that each conventional Shapley value really does not differ significantly from the corresponding risk-based value. This is due to the fact that the probabilities of plant shutdowns adopted in the present example vary only slightly. It can also be observed from the aforementioned Shapley values that, although the percentages of cost savings allocated to P1 and P3 rise as the model restriction gradually relaxes from Strategy 1 to 3, the share of P2 dwindles. On the other hand, despite the fact that the cost savings achieved by the two-plant HENs for $\Sigma_1^2 = \{P1, P2\}$ and $\Sigma_3^2 = \{P3, P1\}$ both increase in the corresponding three cases, that by $\Sigma_2^2 = \{P2, P3\}$ still remains unaffected by the different revamp strategies under consideration. Thus, the above observations seem to suggest that the contribution levels to overall cost savings of the three-plant heat integration scheme can be ranked in the order of P1, P3, and P2 in the present example.

5.4 CONCLUDING REMARKS

This chapter is presented in three main segments. Firstly, the primary components of cooperative game theory are reviewed in Section 5.1, i.e., the core and the conventional and risk-based Shapley values.

A rigorous synthesis procedure is introduced in Section 5.2 to produce a realistic cost-sharing plan for interplant heat integration in the spirit of a cooperative game. Specifically, to ensure feasibility and fairness, a novel computation strategy is devised to determine the core and also the risk-based Shapley values of all players. As a result, a definite portion of the total annual cost of the multi-plant HEN can be calculated for assignment to each plant by considering both its cost contribution level and also the probabilities of potential fallouts of unexpected plant shutdowns. The proposed methodology is illustrated in detail in a simple example.

Finally, a comprehensive design procedure is proposed in Section 5.3 to *revamp* the multi-plant HENs for lowering the overall utility consumption level and to divide and distribute the resulting cost saving fairly to all members of the interplant heat integration scheme. Three retrofit strategies are devised to satisfy practical concerns, such as safety issues, space limitations, and/or operability problems, etc. The corresponding MINLP models can be constructed by augmenting the superstructure-based formulation presented in Section 5.2.2 with different versions of additional constraints imposed on the new and original matches, on the re-piping and reuse of existing units in the multi-plant HEN, and on the placement of purchased heat exchangers, coolers, and heaters. As expected, it can be observed from Example 5.2 that a greater financial gain can always be realized with a less-constrained MINLP model. The allocation plan of overall cost saving is drawn up according to the well-established Shapley values given in Section 5.1. From the allocation results in the same example, the contribution levels of the players in the corresponding cooperative game can also be easily identified.

REFERENCES

Chen, J.J.J. 1987. Comments on improvements on a replacement for the logarithmic mean. *Chemical Engineering Science* 42(10): 2488–2489.

Ciric, A.R., Floudas, C.A. 1990. A mixed integer nonlinear programming model for retrofitting heat-exchanger networks. *Industrial and Engineering Chemistry Research* 29: 239–252.

Hiete, M., Ludwig, J., Schultmann, F. 2012. Intercompany energy integration – adaptation of thermal pinch analysis and allocation of savings. *Journal of Industrial Ecology* 16: 689–698.

Høyland, A., Rausand, M. 1994. *System Reliability Theory: Models and Statistical Methods*. New York, NY: John Wiley & Sons, Inc.

Jin, Y., Chang, C.T., Li, S., Jiang, D. 2018. On the use of risk-based Shapley values for cost sharing in interplant heat integration programs. *Applied Energy* 211: 904–920.

Yee, T.F., Grossmann, I.E. 1990. Simultaneous optimization models for heat integration—II. Heat exchanger network synthesis. *Computers and Chemical Engineering* 14(10): 1165–1184.

6 Multi-Plant Water Network Designs for Continuous and Batch Processes

6.1 INTRODUCTION

Water is an important utility, being intensively used in the process industry. Common uses of water include thermal purposes, steam stripping, liquid-liquid extraction, and various washing operations. Rapid industrial growth has led to serious water pollution in the world, and therefore the environmental regulations for wastewater disposal are increasingly stringent. Concurrently, the scarcity of industrial water (partly due to climate change) has also led to the rise of freshwater supply and effluent treatment costs. The economic considerations, along with increased public awareness of environmental sustainability, have stimulated the recent development of systematic design tools for efficient and responsible use of water in industry. Particularly, systematic design of in-plant water recovery systems through process integration techniques (also known as water network synthesis) has been commonly accepted as an effective means in this regard, with reuse, recycling, and regeneration as options for the reduction of industrial freshwater intake and wastewater discharge (Wang and Smith, 1994).

Apart from in-plant water recovery, opportunities for interplant water integration (IPWI) may be explored to achieve further water recovery when considering an industrial complex with multiple plants or processes, hence the problem of interplant water network (IPWN) synthesis. The first process integration work addressing this issue was reported by Olesen and Polley (1996) using the conventional fixed-load model (Wang and Smith, 1994). Foo (2008) later addressed the problem from the fixed-flowrate perspective.

More recently, Chen et al. (2010) presented a mathematical model for the synthesis of IPWNs, in which central (among plants) and local water mains (within plants) are used to connect the water-using units so as to simplify the piping connection. Lovelady and El-Halwagi (2009) developed an optimization-based approach to IPWI in a common eco-industrial park (EIP) facility. In addition, Rubio-Castro et al. (2010, 2011) presented a mathematical model for water integration in EIPs considering process and environmental constraints. Apart from the efforts to optimize the entire IPWN for minimum flowrates/cost, there are also studies dedicated to maximizing the benefits of individual participating parties in an IPWN, using game theory models (Chew et al., 2009, 2011) and fuzzy optimization (Aviso et al., 2010).

While most works on IPWN synthesis were developed for continuous processes, there are cases in which the participating plants in an IPWN contain batch process units. For example, in petrochemical complexes most processes operate continuously, while some operate in batch mode. This requires a technique to handle process plants with a mix of batch and continuous units. In this chapter, the model developed by Chen et al. (2010) for IPWN synthesis for continuous processes is extended to a special case where the IPWN consists mostly of continuous units with a few batch units. Similar to the strategy proposed by Chen and Ciou (2007) for synthesizing a continuously operated batch mass exchanger network, water storage tanks may be used in IPWI to remove the time constraint of batch units to facilitate water recovery. Following this, the IPWN synthesis problem can be decomposed into two stages. The first stage concerns the IPWN configuration, and the second stage deals with the water storage policy for the batch units.

6.2 PROBLEM STATEMENT

The problem addressed in this chapter can be stated as follows.

- Given a set of process plants $p \in \mathbf{P}$, each plant has a set of water-using units $i \in \mathbf{I}_p$ and there are more continuous units ($i \in \mathbf{I}^{\mathrm{c}}$) than semi-continuous units ($i \in \mathbf{I}^{\mathrm{sc}}$).
- These units require water as a mass separating agent to remove a set of contaminants $c \in \mathbf{C}$ with fixed mass loads from the process streams. For this purpose, a set of freshwater sources $w \in \mathbf{W}$ with different water qualities is available. Effluents from the water-using units may be reused/recycled or sent to wastewater disposal systems $d \in \mathbf{D}$.
- To ease the operation of water networks, the IPWI scheme proposed by Chen et al. (2010) is adopted. This scheme uses a set of centralized and local (in-plant) water mains for indirect water integration, where no direct water transfer between water-using units is allowed.
- The objective is to synthesize an optimal IPWN that achieves the minimum freshwater consumption or network cost, while satisfying all process constraints.

6.3 SOLUTION APPROACH

The main challenge of synthesizing an IPWN involving continuous and semi-continuous water-using units is to integrate units of different operation modes. Continuous units operate uninterruptedly for long duration (e.g., 8000 h/y), while semi-continuous units are scheduled to operate within certain periods of time (as short as a few hours or days) and often in a cyclic manner. A two-stage approach is proposed to handle the case in which there are more continuous units than semi-continuous units. In the first stage, all the semi-continuous units are treated as continuous ones by using auxiliary water

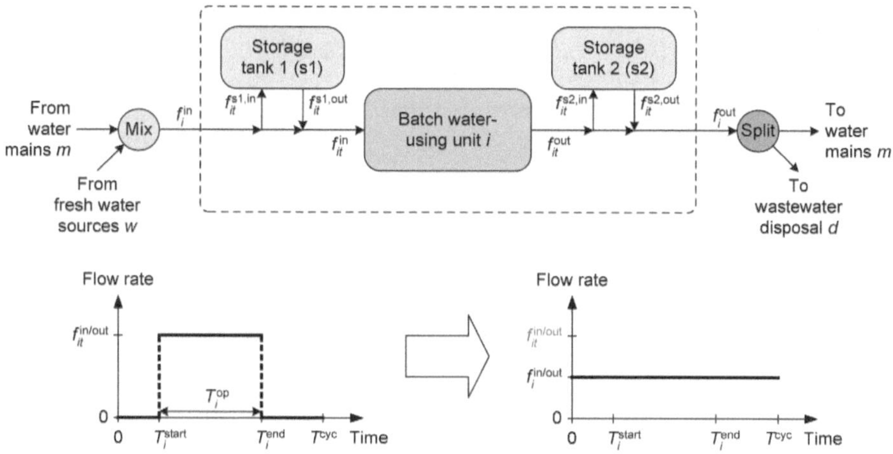

FIGURE 6.1 Use of auxiliary water storage tanks to transform the inlet/outlet water flowrates of a semi-continuous unit. (Reprinted from *Chem. Eng. Res. Des.* 92, 5: 941–953. Copyright 2013, with permission from Elsevier.)

storage tanks, and a continuously operated IPWN is synthesized. Figure 6.1 illustrates the use of two storage tanks to transform the discontinuous inlet and outlet water flowrates of a semi-continuous unit into continuous flowrates. Solving the problem of this stage entails data conversion for semi-continuous units, as will be demonstrated in detail in the illustrative examples. The water storage policy for the semi-continuous units is then determined in the second stage.

6.4 MODEL FORMULATION

With the two-stage decomposition of the problem as described in Section 6.3, the overall model consists of two parts, which are presented separately in the following subsections. Notation used in the formulation is given in the Nomenclature.

6.4.1 Stage 1: Synthesis of a Continuously Operated IPWN

By treating all semi-continuous units as continuous ones, a simplified version of the model of Chen et al. (2010) can be used in this stage to synthesize a continuously operated IPWN. The stage-1 formulation consists of mass balances and logical/IPWI constraints, based on a superstructure that incorporates all feasible network connections between water-using units (Figure 6.2) and water mains (Figure 6.3).

6.4.1.1 Mass Balances for Water-Using Units

Figure 6.2 shows a schematic diagram of a water-using unit i. Because no direct water integration between units is allowed, the inlet water to unit i may come from freshwater sources w and water mains m. The outlet water from unit i may be sent to water mains m and wastewater disposal systems d. Equations (6.1) and (6.2) describe

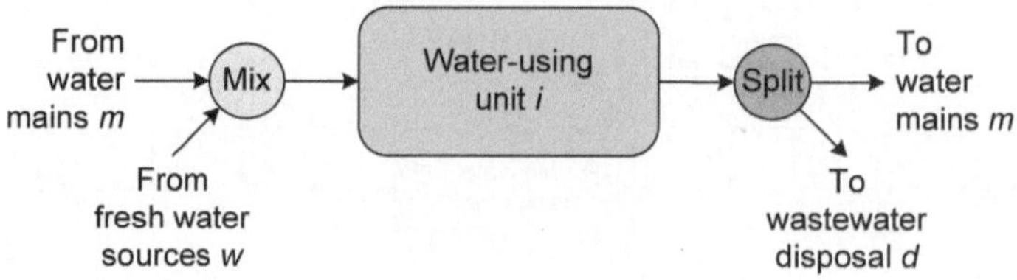

FIGURE 6.2 Schematic of a water-using unit. (Reprinted from *Chem. Eng. Res. Des.* 92, 5: 941–953. Copyright 2013, with permission from Elsevier.)

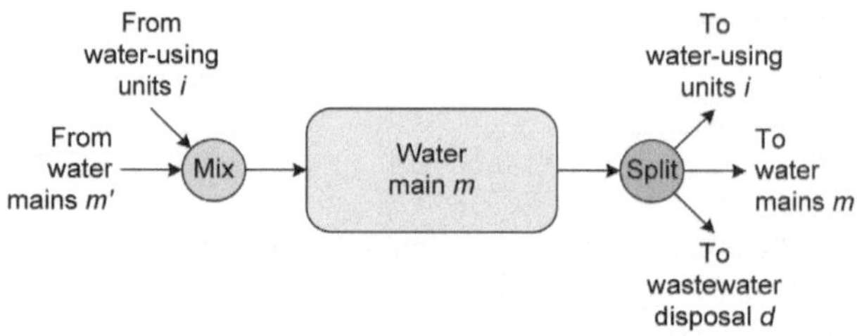

FIGURE 6.3 Schematic of a water main. (Reprinted from *Chem. Eng. Res. Des.* 92, 5: 941–953. Copyright 2013, with permission from Elsevier.)

the inlet and outlet water flowrate balances for unit i, respectively. Equation (6.3) describes the overall water flowrate balance across unit i, assuming no significant water loss or gain during the unit operation.

$$f_i^{\text{in}} = \sum_{w \in \mathbf{W}} f_{wi} + \sum_{m \in \mathbf{M}} f_{mi} \quad \forall i \in \mathbf{I} \tag{6.1}$$

$$f_i^{\text{out}} = \sum_{m \in \mathbf{M}} f_{im} + \sum_{d \in \mathbf{D}} f_{id} \quad \forall i \in \mathbf{I} \tag{6.2}$$

$$f_i^{\text{in}} = f_i^{\text{out}} \quad \forall i \in \mathbf{I} \tag{6.3}$$

Equations (6.4) and (6.5) describe the inlet and overall contaminant balances for unit i, respectively. The maximum inlet and outlet concentrations for unit i are specified using equations (6.6) and (6.7).

$$f_i^{\text{in}} c_{ic}^{\text{in}} = \sum_{w \in \mathbf{W}} f_{wi} C_{wc} + \sum_{m \in \mathbf{M}} f_{mi} c_{mc} \quad \forall c \in \mathbf{C}, i \in \mathbf{I} \tag{6.4}$$

$$f_i^{\text{in}} c_{ic}^{\text{in}} + M_{ic} = f_i^{\text{out}} c_{ic}^{\text{out}} \quad \forall c \in \mathbf{C}, i \in \mathbf{I} \tag{6.5}$$

$$c_{ic}^{\text{in}} \le C_{ic}^{\text{in,max}} \quad \forall c \in \mathbf{C}, i \in \mathbf{I} \tag{6.6}$$

$$c_{ic}^{\text{out}} \le C_{ic}^{\text{out,max}} \quad \forall c \in \mathbf{C}, i \in \mathbf{I} \tag{6.7}$$

6.4.1.2 Mass Balances for Water Mains

Figure 6.3 shows a schematic diagram of a water main m. The inlet water to water main m may come from water-using units i and other water mains m', while the outlet water may be sent to water-using units i, other water mains m', and wastewater disposal systems d. Equations (6.8) and (6.9) describe the inlet and outlet water flowrate balances for main m, respectively. Equation (6.10) describes the overall water flowrate balance across main m, assuming no water loss in the main. The contaminant balance at the inlet of main m is given by equation (6.11).

$$f_m^{\text{in}} = \sum_{i \in \mathbf{I}} f_{im} + \sum_{m' \in \mathbf{M}} f_{m'm} \quad \forall m \in \mathbf{M} \tag{6.8}$$

$$f_m^{\text{out}} = \sum_{i \in \mathbf{I}} f_{mi} + \sum_{m' \in \mathbf{M}} f_{mm'} \quad \forall m \in \mathbf{M} \tag{6.9}$$

$$f_m^{\text{in}} = f_m^{\text{out}} \quad \forall m \in \mathbf{M} \tag{6.10}$$

$$f_m^{\text{in}} c_{mc}^{\text{in}} = \sum_{i \in \mathbf{I}} f_{im} c_{ic}^{\text{out}} + \sum_{m' \in \mathbf{M}} f_{m'm} c_{m'c} \quad \forall c \in \mathbf{C}, m \in \mathbf{M} \tag{6.11}$$

6.4.1.3 Logical and IPWI Constraints

Equation (6.12) represents the lower and upper bounds of the water flowrates.

$$F_{\dagger}^{\text{L}} y_{\dagger} \le f_{\dagger} \le F_{\dagger}^{\text{U}} y_{\dagger} \quad \dagger \in \{id, im, md, mi, mm', wi\} \tag{6.12}$$

where $y\dagger$ is a binary variable indicating the presence ($y\dagger = 1$) or absence ($y\dagger = 0$) of a water flow connection.

The following two constraints are imposed to implement the IPWI scheme of Chen et al. (2010). Equation (6.13) forbids the connections between water-using units and water mains of different plants.

$$\sum_{i \in \mathbf{I}_p} \sum_{m \in \mathbf{M}'_p} \left(y_{im} + y_{mi} \right) \quad \forall p \in \mathbf{P} \tag{6.13}$$

where $\mathbf{M}'_p \equiv \mathbf{M} - \mathbf{M}^{\text{cen}} - \mathbf{M}_p$; $\mathbf{M}^{\text{cen}}$ and $\mathbf{M}_p$ are the sets of central mains and local mains (in plant p), respectively. The interconnection between central and local mains is also defined. A central main may be connected to local mains and other central mains. However, connections between local mains of different plants are forbidden. This is given in equation (6.14).

$$\sum_{m \in \mathbf{M}_p} \sum_{m' \in \mathbf{M}'_p} y_{mm'} \quad \forall p \in \mathbf{P} \tag{6.14}$$

6.4.2 Stage 2: Determination of the Storage Policy

Having dealt with the IPWN configuration in stage 1, the task of this stage is to determine the water storage policy for the batch process units. To address the time dimension of batch processes, a set of time intervals $t \in \mathbf{T}$ is defined, with index t used throughout the stage-2 formulation to reflect time dependence.

6.4.2.1 Flowrate Balance for Batch Units

As shown in Figure 6.1, two storage tanks (tank 1 and tank 2) are used for one batch unit, in order to convert its intermittent inlet and outlet water flows into continuous flowrates. Thus, the batch unit can behave as a continuous process that receives and discharges water constantly. The flowrate correlations are given in equations (6.15) and (6.16).

$$f_{it}^{\text{in}} = Z_{it}^{\text{op}} f_i^{\text{in}} \frac{T^{\text{cyc}}}{T_i^{\text{op}}} \quad \forall i \in \mathbf{I}^{\text{b}}, t \in \mathbf{T} \tag{6.15}$$

$$f_{it}^{\text{out}} = Z_{it}^{\text{op}} f_i^{\text{out}} \frac{T^{\text{cyc}}}{T_i^{\text{op}}} \quad \forall i \in \mathbf{I}^{\text{b}}, t \in \mathbf{T} \tag{6.16}$$

where Z_{it}^{op} is a binary parameter indicating whether batch unit i operates in time interval t ($Z_{it}^{\text{op}} = 1$) or not ($Z_{it}^{\text{op}} = 0$).

Equations (6.17)–(6.22) describe the water flowrate balances for batch unit i in time interval t. When batch unit i is not operating ($f_{it}^{\text{in}} = 0$), the inlet water (from water mains or freshwater sources) is all stored into tank 1 ($f_i^{\text{in}} = f_{it}^{\text{s1,in}}$), while no water is discharged from tank 1 ($f_{it}^{\text{s1,out}} = 0$). The inlet water will not be stored ($f_{it}^{\text{s1,in}} = 0$) during the operation of batch unit i. Instead, the inlet water goes directly to the unit together with the water discharged from tank 1 ($f_i^{\text{in}} + f_{it}^{\text{s1,out}} = f_{it}^{\text{in}}$). The same concept applies to tank 2. When batch unit i is operating, part of its outlet water is stored into tank 2 and the rest forms the outlet stream to be allocated to water mains or wastewater disposal systems ($f_{it}^{\text{out}} = f_{it}^{\text{s2,in}} + f_i^{\text{out}}$), while no water is discharged from tank 2 ($f_{it}^{\text{s2,out}} = 0$). The water stored in tank 2 will be discharged as the outlet stream ($f_{it}^{\text{s2,out}} = f_{it}^{\text{out}}$) when batch unit i is off ($f_{it}^{\text{out}} = f_{it}^{\text{s2,in}} = 0$).

$$f_{it}^{\text{in}} = f_i^{\text{in}} - f_{it}^{\text{s1,in}} + f_{it}^{\text{s1,out}} \quad \forall i \in \mathbf{I}^{\text{b}}, t \in \mathbf{T} \tag{6.17}$$

$$f_{it}^{\text{s1,out}} \le \overline{F}_i^{\text{s1}} Z_{it}^{\text{op}} \quad \forall i \in \mathbf{I}^{\text{b}}, t \in \mathbf{T} \tag{6.18}$$

$$f_{it}^{\text{s1,in}} \le \overline{F}_i^{\text{s1}} \left(1 - Z_{it}^{\text{op}}\right) \quad \forall i \in \mathbf{I}^{\text{b}}, t \in \mathbf{T} \tag{6.19}$$

$$f_{it}^{\text{out}} = f_i^{\text{out}} + f_{it}^{\text{s2,in}} - f_{it}^{\text{s2,out}} \quad \forall i \in \mathbf{I}^{\text{b}}, t \in \mathbf{T} \tag{6.20}$$

$$f_{it}^{\text{s2,out}} \le \overline{F}_i^{\text{s2}} \left(1 - Z_{it}^{\text{op}}\right) \quad \forall i \in \mathbf{I}^{\text{b}}, t \in \mathbf{T} \tag{6.21}$$

$$f_{it}^{\mathrm{s2,in}} \le \overline{F}_i^{\mathrm{s2}} Z_{it}^{\mathrm{op}} \quad \forall i \in \mathbf{I}^{\mathrm{b}}, t \in \mathbf{T} \tag{6.22}$$

6.4.2.2 Flow Balance for Storage Tanks

Equation (6.23) describes the overall water flow balance for both storage tanks of batch unit i. The amount of water stored at the end of time interval t is equal to that accumulated from the previous time interval (t – 1) adjusted by the inlet and outlet water flows during time interval t. Assuming cyclic operation for the batch units, the amount of water stored at the end of a cycle is taken as the initial water content for the next cycle. The amount of water stored in the tank is limited to its capacity, as given in equation (6.24).

$$q_{it}^{*} = q_{i,t-1}^{*} + \left(f_{it}^{*,\mathrm{in}} - f_{it}^{*,\mathrm{out}}\right)\delta_t \quad * \in \{\mathrm{s1}, \mathrm{s2}\}, i \in \mathbf{I}^{\mathrm{b}}, t \in \mathbf{T} \tag{6.23}$$

$$q_{it}^{*} \le \overline{q}_i^{*} \quad * \in \{\mathrm{s1}, \mathrm{s2}\}, i \in \mathbf{I}^{\mathrm{b}}, t \in \mathbf{T} \tag{6.24}$$

6.4.3 Objective Functions

If water recovery is the primary concern of IPWI, the two-stage formulations presented in Section 6.4.1 and Section 6.4.2 can be solved sequentially. In this case, the objective function of stage 1 is to minimize the freshwater consumption of water-using operations:

$$\min g_1 = \sum_{w \in \mathbf{W}} \sum_{i \in \mathbf{I}} f_{wi} \tag{6.25}$$

With binary variables in equations (6.12)–(6.14) and bilinear terms in equations (6.4), (6.5), and (6.11), the model for stage 1 (equations (6.1)–(6.14) and (6.25)) is a mixed-integer nonlinear program (MINLP).

The objective function of stage 2 is to minimize the total capacity of storage tanks:

$$\min g_2 = \sum_{i \in \mathbf{I}^{\mathrm{b}}} \left(\overline{q}_i^{\mathrm{s1}} + \overline{q}_i^{\mathrm{s2}}\right) \tag{6.26}$$

With no bilinear terms or binary variables, the model for stage 2 (equations (6.15)–(6.24) and (6.26)) is a linear program (LP).

Instead of solving the IPWN synthesis problem sequentially, the formulations of both stages can be combined and solved simultaneously for economic trade-offs. In this case, the objective may be to minimize the total network cost, which consists of costs associated with freshwater supply, wastewater disposal, water mains, piping, and storage tanks.

i. Freshwater supply cost:

$$g^{\mathrm{FW}} = AOH \sum_{w \in \mathbf{W}} \sum_{i \in \mathbf{I}} CFW_w f_{wi} \tag{6.27}$$

where AOH is the annual operating hours.

ii. Wastewater disposal cost:

$$g^{\mathrm{WD}} = AOH\sum_{d\in\mathbf{D}} CWD_d\left(\sum_{i\in\mathbf{I}} f_{id} + \sum_{m\in\mathbf{M}} f_{md}\right) \tag{6.28}$$

iii. Water main cost:

$$g^{\mathrm{main}} = AF\sum_{m\in\mathbf{M}}\left(CM_m^{\mathrm{var}} f_m^{\mathrm{in}} + CM_m^{\mathrm{fix}} y_m\right) \tag{6.29}$$

where AF is the annualization factor; y_m is a binary variable indicating the existence of main m. Equation (6.30) represents the lower and upper limit constraint correlating the binary variable with the inlet flowrate of water mains.

$$F_m^{\mathrm{L}} y_m \le f_m^{\mathrm{in}} \le F_m^{\mathrm{U}} y_m \quad \forall m \in \mathbf{M} \tag{6.30}$$

iv. Piping cost:

$$g^{\mathrm{pipe}} = AF\sum_{\dagger}\left(CP_{\dagger}^{\mathrm{var}} f_{\dagger} + CP_{\dagger}^{\mathrm{fix}} y_{\dagger}\right) \quad \dagger \in \{id, im, md, mi, mm', wi\} \tag{6.31}$$

v. Storage tank cost:

$$g^{\mathrm{tank}} = AF\sum_{i\in\mathbf{I}^{\mathrm{b}}}\left(CST_i^{\mathrm{var}}\overline{q}_i^{*} + CST_i^{\mathrm{fix}} y_i^{*}\right) \quad * \in \{\mathrm{s1}, \mathrm{s2}\} \tag{6.32}$$

where y_i^* is a binary variable defined to indicate the presence of storage tanks of batch unit i. However, since both storage tanks are needed for the batch units to behave as continuous ones, all y_i^* can be set to one.

Taking into account the above cost functions (equations (6.27)–(6.29), (6.31), and (6.32)), the objective function is to minimize the total annualized network cost:

$$\min g_3 = g^{\mathrm{FW}} + g^{\mathrm{WD}} + g^{\mathrm{main}} + g^{\mathrm{pipe}} + g^{\mathrm{tank}} \tag{6.33}$$

The combined model for cost minimization (equations (6.1)–(6.24) and (6.33)) is an MINLP.

6.5 ILLUSTRATIVE EXAMPLES

Two literature examples are solved to illustrate the proposed approach. In both examples, the optimization models are implemented and solved in GAMS, using BARON as the MINLP solver and CPLEX as the LP solver.

6.5.1 Example 1

The first example is a single-contaminant problem adapted from Olesen and Polley (1996). This example involves 3 process plants and 15 water-using units, 4 of which operate in batch mode. Table 6.1 shows the operating data for the water-using units. Figure 6.4 shows the Gantt chart for the batch operations. According to the scheduled operating periods of units 3, 5, 6, and 14, the batch cycle time of ten hours is divided into 4 time intervals. In this example, a single pure freshwater source is available for use ($C_{wc} = 0$).

As shown in Table 6.1, the mass loads of continuous units are instantaneous values (in kg/h), while those of batch units are given as total amounts (in kg). When batch units are treated as continuous ones, their mass loads need to be converted to instantaneous values. The converted data for the batch units are given in Table 6.2, where the total mass loads are averaged over the 10-h cycle time. For instance, the mass load in unit 3 is converted from 50 kg to 5 kg/h (= 50 kg/10 h). After this data conversion, the models for both stages of the problem can be solved either sequentially or simultaneously.

Sequential optimization is carried out first for maximum water recovery. With the use of three local mains and a central main for in-plant and interplant water integration, the model for stage 1 (MINLP formulation) involves 468 constraints, 243 continuous variables, and 170 binary variables. The minimum freshwater consumption of 315.31 t/h was obtained in 35 CPU s. Figure 6.5 shows the optimal IPWN configuration. Note that all the process plants send water to and also receive water from the

TABLE 6.1
Operating Data for the Water-Using Units for Example 1

Plant	Unit	Duration (h)	Maximum Inlet Concentration (ppm)	Maximum Outlet Concentration (ppm)	Mass Load
A	1	-	0	100	2 kg/h
	2	-	50	80	2 kg/h
	3	2–7	50	100	50 kg
	4	-	80	800	30 kg/h
	5	2–6	400	800	40 kg
B	6	0–2	0	100	20 kg
	7	-	50	80	2 kg/h
	8	-	80	400	5 kg/h
	9	-	100	800	30 kg/h
	10	-	400	1000	4 kg/h
C	11	-	0	100	2 kg/h
	12	-	25	50	2 kg/h
	13	-	25	125	5 kg/h
	14	6–10	50	800	300 kg
	15	-	100	150	15 kg/h

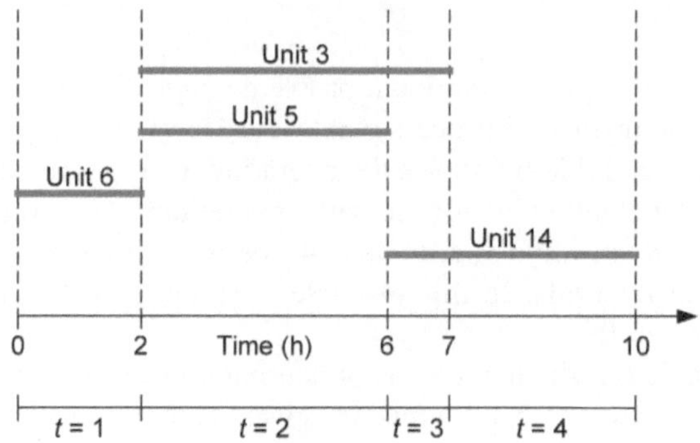

FIGURE 6.4 Gantt chart for the batch operations for Example 1. (Reprinted from *Chem. Eng. Res. Des.* 92, 5: 941–953. Copyright 2013, with permission from Elsevier.)

TABLE 6.2
Converted Data for the Batch Units for Example 1

Plant	Unit	Mass Load (kg/h)	Maximum Inlet Concentration (ppm)	Maximum Outlet Concentration (ppm)
A	3	5	50	100
	5	4	400	800
B	6	2	0	100
C	14	30	50	800

central main (M4). The concentrations of water mains M1–M4 are determined to be 86.96, 400, 90, and 125.98 ppm, respectively.

To evaluate the effect of water integration, two reference cases are identified for comparison. Before exploring any opportunities for water reuse/recycling, the minimum freshwater requirement is calculated to be 474 t/h, which is given by the sum of the minimum freshwater flowrate for each water-using unit (i.e., $\sum_{i \in \mathbf{I}} \max_{c \in \mathbf{C}} \left\{ M_{ic} / \left(C_{ic}^{\text{out,max}} - 0 \right) \right\}$). When only in-plant water integration is allowed, in which case each plant uses a local main, the minimum freshwater consumption is determined to be 343.42 t/h. This value can be obtained by solving the stage-1 model, while setting the inlet and outlet flowrates of the central main to zero. The comparison shows that a 27.5% reduction in freshwater consumption is achieved through in-plant water integration, and IPWI offers a further water reduction of 6%.

Using the IPWN flowrates determined in stage 1 (f_i^{in} and f_i^{out}) as parameters, the model for stage 2 is solved to determine the water storage policy for the batch units. The LP formulation entails 129 constraints and 137 continuous variables. It took 0.2 CPU s to solve the model, and the minimum total storage capacity is determined

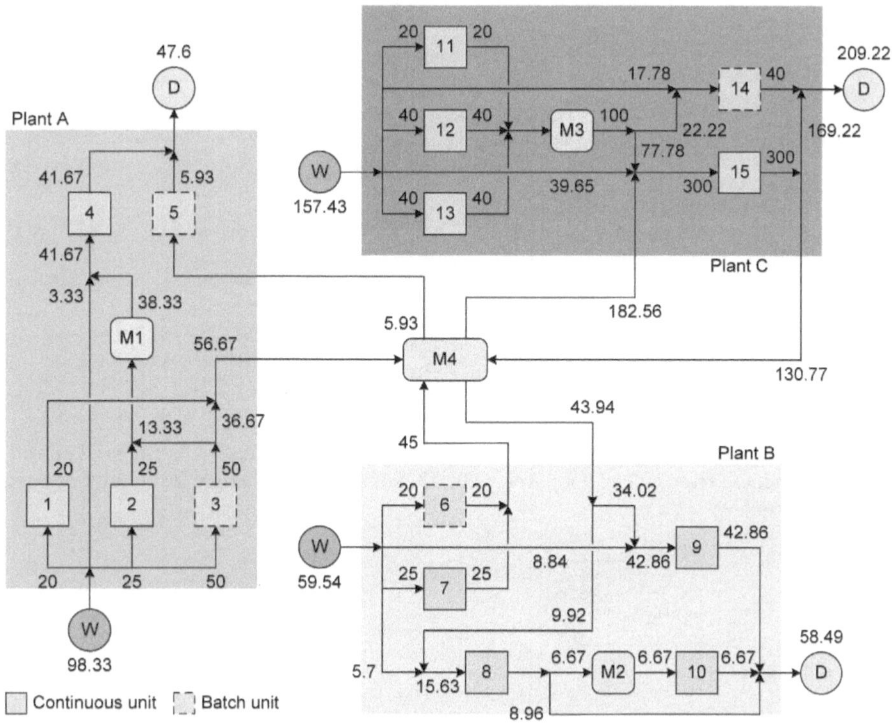

FIGURE 6.5 Optimal IPWN configuration for Example 1 (maximum water recovery). (Reprinted from *Chem. Eng. Res. Des.* 92, 5: 941–953. Copyright 2013, with permission from Elsevier.)

to be 1371.22 t. Figure 6.6 shows the water storage profiles of the auxiliary tanks. Although it is proposed to use eight storage tanks, in fact four of them are dispensable. As shown in Figure 6.5, the inlet water to units 3 and 6 comes directly from the freshwater source, while the outlet water from units 5 and 14 is sent directly to wastewater disposal. Because these water streams have no connection with other units, the flowrate transformation is not necessary. Thus, the inlet tanks of units 3 and 6 can be removed along with the outlet tanks of units 5 and 14. Figure 6.7(a)–(d) shows the modified operating conditions of units 3, 5, 6, and 14, respectively. Note that it is valid to assume freshwater supply to be intermittent.

Simultaneous optimization is then carried out for economic evaluation. The cost data are given as follows: freshwater, \$0.2/t; wastewater disposal, \$0.1/t; the variable and fixed costs for a water main are assumed to be \$240 h/t and \$8000, respectively; the variable and fixed costs for a storage tank, \$120/t and \$10,000, respectively. An equal distance of 200 m is assumed for all cross-plant pipelines, which in this work refer to the connections between process plants and central mains. Piping cost within the individual plants is assumed to be negligible, as it is normally much smaller than

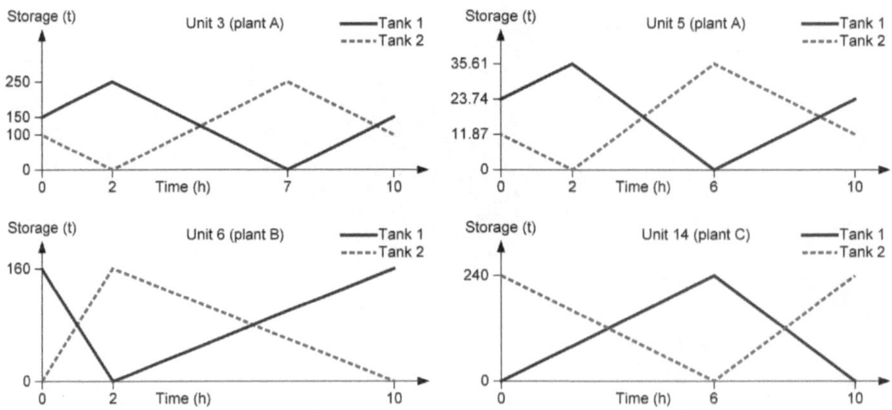

FIGURE 6.6 Water storage profiles of the auxiliary tanks for Example 1 (sequential optimization). (Reprinted from *Chem. Eng. Res. Des.* 92, 5: 941–953. Copyright 2013, with permission from Elsevier.)

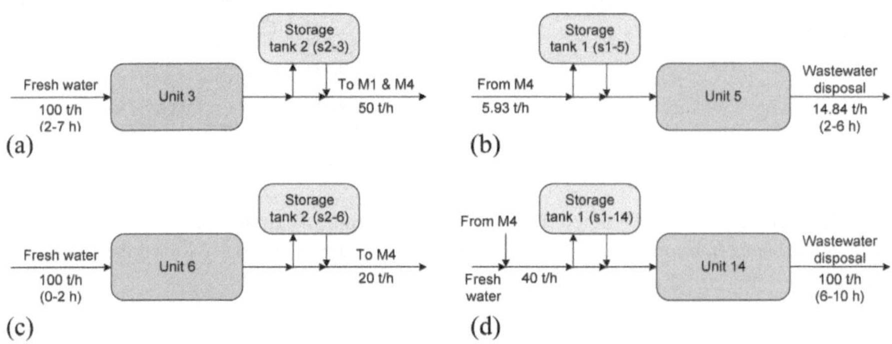

FIGURE 6.7 Modified operating conditions for (a) unit 3, (b) unit 5, (c) unit 6, and (d) unit 14. (Reprinted from *Chem. Eng. Res. Des.* 92, 5: 941–953. Copyright 2013, with permission from Elsevier.)

the cross-plant piping cost. Therefore, the variable and fixed costs for cross-plant piping are estimated at \$200 h/t and \$25,000, respectively, and those for in-plant pipes are all taken as zero. In addition, all plants are assumed to operate for 8600 h annually with an annualization factor of 0.4/y.

With the same number of water mains available for use (three locals and one central), the combined model involves 600 constraints, 379 continuous variables, and 174 binary variables. It was solved in 65 CPU s and the minimum total network cost is determined to be \$1,000,805/y. Figure 6.8 shows the optimal IPWN configuration. Compared to the maximum water recovery design in Figure 6.5, this design features fewer cross-plant streams; only plant A sends water to the central main (M4) while plants B and C receive the water. Therefore, the throughput of M4

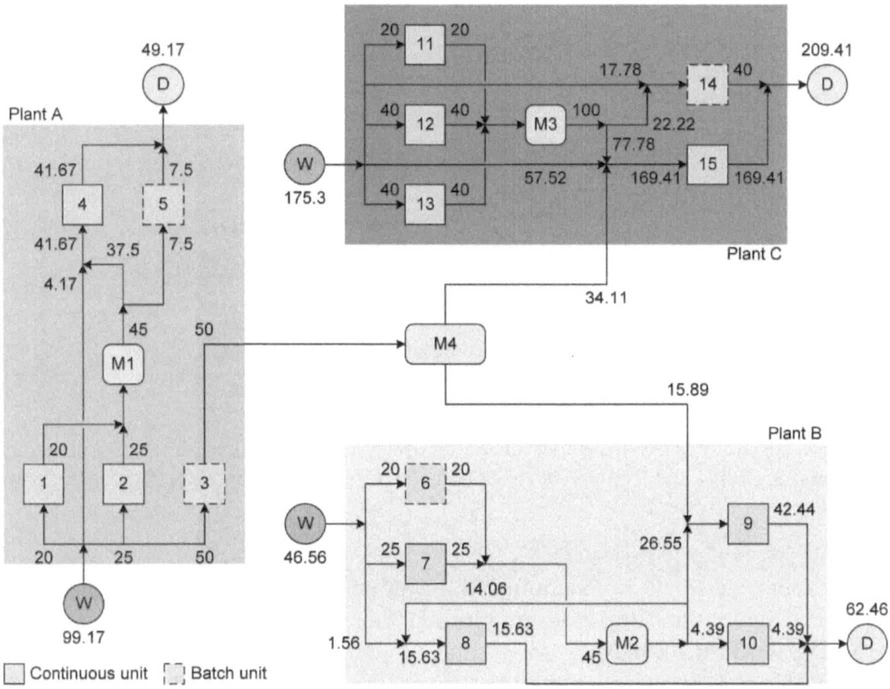

FIGURE 6.8 Optimal IPWN configuration for Example 1 (minimum network cost). (Reprinted from *Chem. Eng. Res. Des.* 92, 5: 941–953. Copyright 2013, with permission from Elsevier.)

is reduced from 232.44 to 50 t/h. On the other hand, there is a slight increase in the freshwater consumption from the minimum of 315.31 to 321.03 t/h. These contrasts indicate the trade-offs between the costs of freshwater supply, water mains, and piping. The concentrations of water mains M1–M4 are determined to be 88.89, 88.89, 90, and 100 ppm, respectively. As with the case of sequential optimization, four of the storage tanks can be removed (i.e., the inlet tanks of units 3 and 6 as well as the outlet tanks of units 5 and 14). With the removal of these tanks, the total network cost is reduced to $951,445/y. Figure 6.9 shows the water storage profiles of the tanks needed.

6.5.2 Example 2

The second example is adapted from Ma et al. (2007), involving two process plants, ten water-using units (three operating in batch mode), and three contaminants. Table 6.3 shows the operating data for the water-using units. Figure 6.10 shows the Gantt chart for the batch operations. According to the scheduled operating periods of units 3, 7, and 10, the batch cycle time of 8 h is divided into three time intervals.

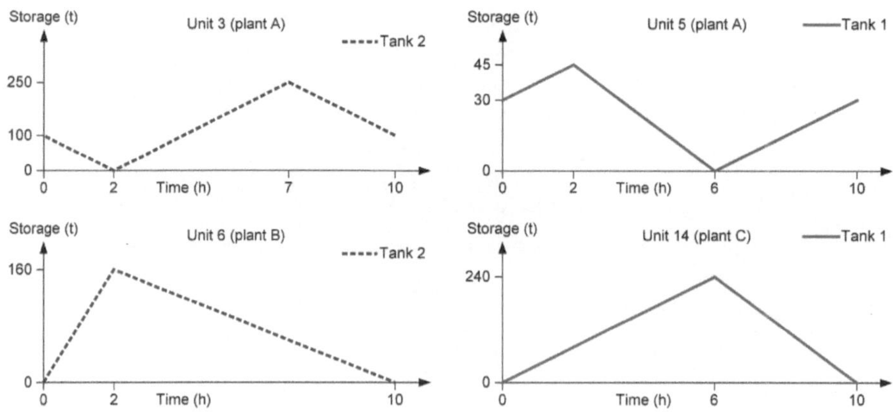

FIGURE 6.9 Water storage profiles of the auxiliary tanks for Example 1 (simultaneous optimization). (Reprinted from *Chem. Eng. Res. Des.* 92, 5: 941–953. Copyright 2013, with permission from Elsevier.)

In this example, a single uncontaminated freshwater source is available for use. The converted data for the batch units are given in Table 6.4, where the total mass loads are averaged over the 8-h cycle time.

For sequential optimization with the use of two local mains and a central main, the model for stage 1 involves 356 constraints, 188 continuous variables, and 92 binary variables. It took 235 CPU s to solve the model and the minimum freshwater consumption is determined to be 393.28 t/h. This corresponds to a 16.34% reduction in freshwater use compared to the case without water reuse/recycling, and a further 7.39% reduction compared to the case with only in-plant water integration. Figure 6.11 shows the optimal IPWN configuration. The concentrations of water mains M1–M3 are determined to be (20, 60, 20), (11.92, 600, 14.9), and (15, 400, 35) ppm, respectively. Using the IPWN flowrates obtained in stage 1, the model for stage 2 involves 73 constraints and 79 continuous variables. It was solved in 0.2 CPU s and the minimum total storage capacity is determined to be 381.86 t. Since unit 3 uses freshwater, while the outlet water from units 7 and 10 is sent for wastewater disposal (see Figure 6.11), the inlet tank of unit 3 and the outlet tanks of units 7 and 10 can be removed. Figure 6.12(a) shows the water storage profiles of the tanks needed.

For simultaneous optimization, the cost data used are the same as in Example 1. With three water mains available for use (two locals and one central), the combined model has 431 constraints, 266 continuous variables, and 95 binary variables. The minimum total network cost of $1,124,324/y was obtained in 167 CPU s. Figure 6.13 shows the optimal IPWN configuration. The concentrations of water mains M1–M3 are determined to be (20, 60, 20), (11.92, 600, 14.9), and (15, 400, 35) ppm, respectively. Similar contrasts between the maximum water recovery and minimum network cost designs are also found in this example. The latter design features fewer

TABLE 6.3
Operating Data for the Water-Using Units for Example 2

Plant	Unit	Duration (h)	Maximum Inlet Concentration (ppm)	Maximum Outlet Concentration (ppm)	Mass Load
I	1	-	20	120	3400 g/h
			300	12,500	414,800 g/h
			45	180	4590 g/h
	2	-	120	220	5600 g/h
			20	45	1400 g/h
			200	9500	520,800 g/h
	3	2–6	0	20	1280 g
			0	60	3840 g
			0	20	1280 g
II	4	-	50	150	800 g/h
			400	8000	60,800 g/h
			60	120	480 g/h
	5	-	0	15	750 g/h
			0	400	20,000 g/h
			0	35	1750 g/h
	6	-	10	70	2000 g/h
			200	600	100,700 g/h
			20	90	2500 g/h
	7	0–2	25	150	14,400 g
			230	1000	54,400 g
			20	220	4800 g
	8	-	5	100	3000 g/h
			45	4000	102,300 g/h
			50	300	8140 g/h
	9	-	13	100	4600 g/h
			200	3000	200,000 g/h
			5	200	1900 g/h
	10	2–8	10	100	32,000 g
			90	500	82,400 g
			70	800	72,000 g

cross-plant streams and therefore reduced throughput of the central main (M3). However, a higher freshwater consumption of 403.71 t/h is required. It is observed from Figure 6.13 that the inlet tank of unit 3 and the outlet tanks of units 7 and 10 can be removed. With the removal of these unnecessary tanks, the total network cost is reduced to $1,103,313/y. Figure 6.12(b) shows the water storage profiles of the tanks needed. Note that the storage policies for both optimization cases are much the same except the capacity required.

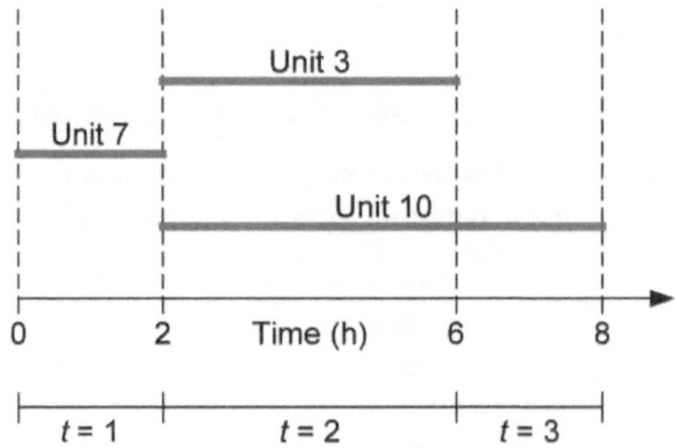

FIGURE 6.10 Gantt chart for the batch operations for Example 2. (Reprinted from *Chem. Eng. Res. Des.* 92, 5: 941–953. Copyright 2013, with permission from Elsevier.)

TABLE 6.4
Converted Data for the Batch Units for Example 2

Plant	Unit	Mass Load (g/h)	Maximum Inlet Concentration (ppm)	Maximum Outlet Concentration (ppm)
I	3	160	0	20
		480	0	60
		160	0	20
II	7	1800	25	150
		6800	230	1000
		600	20	220
	10	4000	10	100
		10,300	90	500
		9000	70	800

6.6 SUMMARY

A two-stage approach for the synthesis of IPWNs involving continuous and batch units has been presented in this chapter, with mathematical models developed for both stages. In the first stage, a continuously operated IPWN is synthesized, where all batch units are treated as continuous ones with the use of water storage tanks. Based on the IPWN configuration obtained, the water storage policy for the batch units is determined in the second stage. The models can be solved either sequentially for maximum water recovery, or simultaneously for minimum IPWN cost, taking into account water, main, piping, and storage costs. Two literature examples were solved to illustrate the proposed approach. Note however that the solution approach and optimization models developed in this chapter are suitable for the case in which process plants consist mainly of continuous units with a few batch units. For methods developed for the opposite case, readers are referred to the paper of Lee et al. (2013).

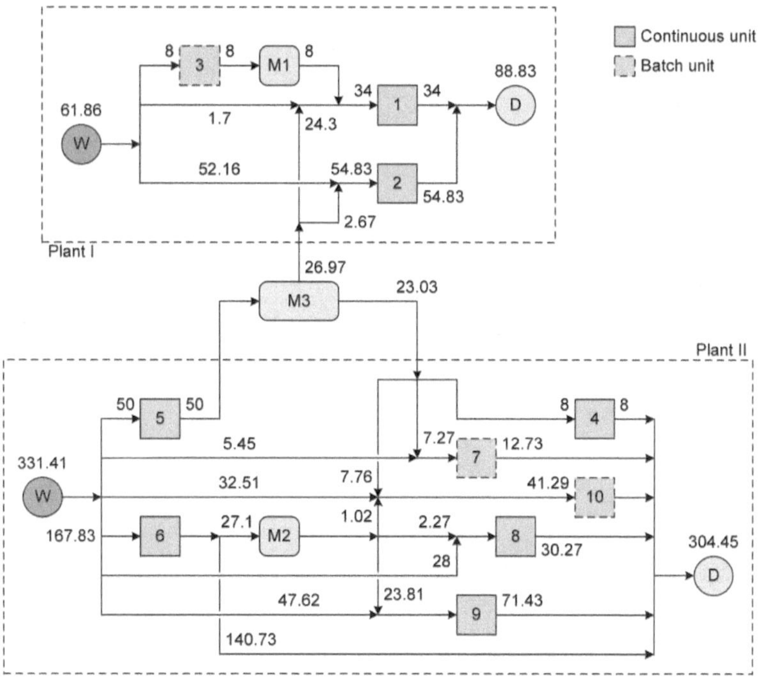

FIGURE 6.11 Optimal IPWN configuration for Example 2 (maximum water recovery). (Reprinted from *Chem. Eng. Res. Des.* 92, 5: 941–953. Copyright 2013, with permission from Elsevier.)

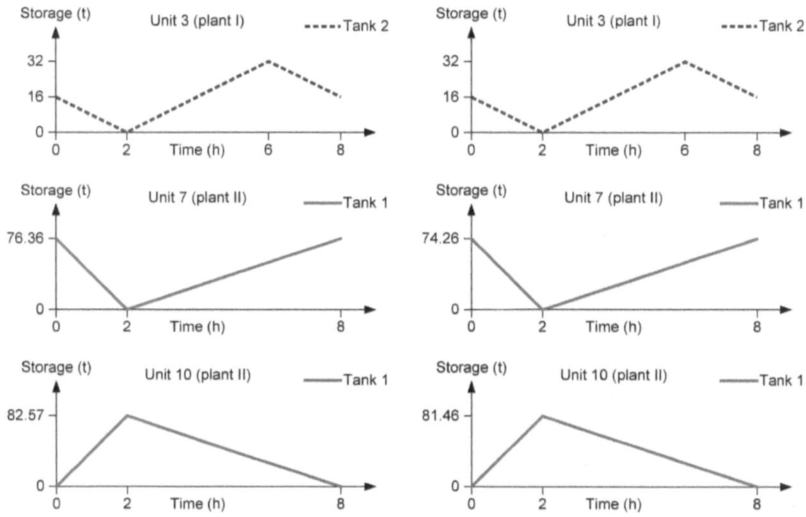

FIGURE 6.12 Water storage profiles of the auxiliary tanks for Example 2, obtained by (a) sequential and (b) simultaneous optimization. (Reprinted from *Chem. Eng. Res. Des.* 92, 5: 941–953. Copyright 2013, with permission from Elsevier.)

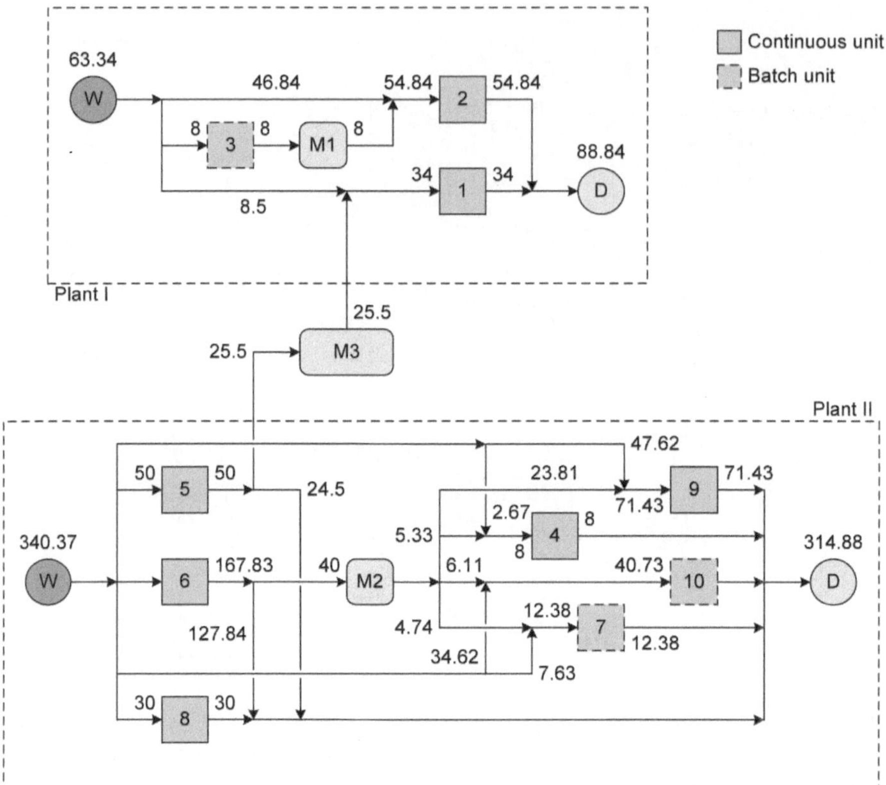

FIGURE 6.13 Optimal IPWN configuration for Example 2 (minimum network cost). (Reprinted from *Chem. Eng. Res. Des.* 92, 5: 941–953. Copyright 2013, with permission from Elsevier.)

NOMENCLATURE

Indices and Sets

$c \in \mathbf{C}$	contaminants
$d \in \mathbf{D}$	wastewater disposal systems
$i \in \mathbf{I}$	water-using units
$i \in \mathbf{I}^{b} \subset \mathbf{I}$	batch water-using units
$i \in \mathbf{I}^{c} \subset \mathbf{I}$	continuous water-using units
$i \in \mathbf{I}_{p} \subset \mathbf{I}$	water-using units in plant p
$m \in \mathbf{M}$	water mains
$m \in \mathbf{M}_{p} \subset \mathbf{M}$	water mains in plant p (local mains)
$m \in \mathbf{M}^{cen} \subset \mathbf{M}$	centralized water mains
$p \in \mathbf{P}$	process plants
$t \in \mathbf{T}$	time intervals
$w \in \mathbf{W}$	freshwater sources

Parameters

AF annualization factor
AOH annual operating hours
$C_{ic}^{\text{in,max}}$ maximum inlet concentration of contaminant c for unit i
$C_{ic}^{\text{out,max}}$ maximum outlet concentration of contaminant c for unit i
C_{wc} concentration of contaminant c in freshwater source w
CFW_w unit cost of freshwater from source w
CM_m^{fix} fixed cost for main m
CM_m^{var} variable cost for main m
$CP_\dagger^{\text{fix}}$ fixed cost for pipe connection $\dagger \in \{id, im, md, mi, mm', wi\}$
$CP_\dagger^{\text{var}}$ variable cost for pipe connection $\dagger \in \{id, im, md, mi, mm', wi\}$
CST_i^{fix} fixed cost for the storage tanks of batch unit i
CST_i^{var} variable cost for the storage tanks of batch unit i
CWD_d unit cost of wastewater disposal through system d
Δ_t length of time interval t
$F_\dagger^{\text{L}}$ lower bound for the flowrate in connection $\dagger \in \{id, im, md, mi, mm', wi\}$
$F_\dagger^{\text{U}}$ upper bound for the flowrate in connection $\dagger \in \{id, im, md, mi, mm', wi\}$
$\bar{F}_i^{\text{s1}}$ upper bound for the inlet/outlet flowrate of tank 1 of batch unit i
$\bar{F}_i^{\text{s2}}$ upper bound for the inlet/outlet flowrate of tank 2 of batch unit i
F_m^{L} lower bound for the inlet flowrate of main m
F_m^{U} upper bound for the inlet flowrate of main m
M_{ic} mass load of contaminant c in unit i
T^{cyc} batch cycle time
T_i^{op} total operating time of batch unit i in a cycle
Z_{it}^{op} binary parameter indicating whether batch unit i operates in time interval t

Variables

c_{ic}^{in} inlet concentration of contaminant c to unit i
c_{ic}^{out} outlet concentration of contaminant c from unit i
c_{mc} concentration of contaminant c in main m
f_i^{in} inlet water flowrate of unit i
f_i^{out} outlet water flowrate of unit i
f_{it}^{in} inlet water flowrate of batch unit i in time interval t
f_{it}^{out} outlet water flowrate of batch unit i in time interval t
$f_{it}^{\text{s1,in}}$ inlet water flowrate to tank 1 of batch unit i in time interval t
$f_{it}^{\text{s1,out}}$ outlet water flowrate from tank 1 of batch unit i in time interval t
$f_{it}^{\text{s2,in}}$ inlet water flowrate to tank 2 of batch unit i in time interval t
$f_{it}^{\text{s2,out}}$ outlet water flowrate from tank 2 of batch unit i in time interval t
f_{id} water flowrate from unit i to wastewater disposal system d
f_{im} water flowrate from unit i to main m

f_m^{in} inlet water flowrate of main m
f_m^{out} outlet water flowrate of main m
f_{md} water flowrate from main m to wastewater disposal system d
f_{mi} water flowrate from main m to unit i
$f_{mm'}$ water flowrate from main m to main m'
f_{wi} water flowrate from freshwater source w to unit i
q_{it}^{s1} amount of water stored in tank 1 of batch unit i at the end of time interval t
q_{it}^{s2} amount of water stored in tank 2 of batch unit i at the end of time interval t
$\overline{q}_i^{s1}$ storage capacity of tank 1 of batch unit i
$\overline{q}_i^{s2}$ storage capacity of tank 2 of batch unit i
$y\dagger$ binary variable indicating the existence of connection $\dagger \in \{id, im, md, mi, mm', wi\}$
y_m binary variable indicating the existence of main m

REFERENCES

Aviso, K.B., Tan, R.R., Culaba, A.B. 2010. Designing eco-industrial water exchange networks using fuzzy mathematical programming. *Clean Technologies and Environmental Policy* 12(4): 353–363.

Chen, C.-L., Ciou, Y.-J. 2007. Synthesis of a continuously operated mass-exchanger network for a semiconsecutive process. *Industrial and Engineering Chemistry Research* 46(22): 7136–7151.

Chen, C.-L., Hung, S.-W., Lee, J.-Y. 2010. Design of inter-plant water network with central and decentralized water mains. *Computers and Chemical Engineering* 34(9): 1522–1531.

Chew, I.M.L., Tan, R.R., Foo, D.C.Y., Chiu, A.S.F. 2009. Game theory approach to the analysis of inter-plant water integration in an eco-industrial park. *Journal of Cleaner Production* 17(18): 1611–1619.

Chew, I.M.L., Thillaivarrna, S.L., Tan, R.R., Foo, D.C.Y. 2011. Analysis of inter-plant water integration with indirect integration schemes through game theory approach: Pareto optimal solution with interventions. *Clean Technologies and Environmental Policy* 13(1): 49–62.

Foo, D.C.Y. 2008. Flowrate targeting for threshold problems and plant-wide integration for water network synthesis. *Journal of Environmental Management* 88(2): 253–274.

Lee, J.-Y., Chen, C.-L., Lin, C.-Y. 2013. A mathematical model for water network synthesis involving mixed batch and continuous units. *Industrial and Engineering Chemistry Research* 52(22), 7047–7055.

Lovelady, E.M., El-Halwagi, M.M. 2009. Design and integration of eco-industrial parks for managing water resources. *Environmental Progress and Sustainable Energy* 28(2): 265–272.

Ma, H., Feng, X., Cao K. 2007. A rule-based design methodology for water networks with internal water mains. *Chemical Engineering Research and Design* 85(4): 431–444.

Olesen, S.G., Polley, G.T. 1996. Dealing with plant geography and piping constraints in water network design. *Process Safety and Environmental Protection* 74(4): 273–276.

Rubio-Castro, E., Ponce-Ortega, J.M., Nápoles-Rivera, F., El-Halwagi, M.M., Serna-González, M., Jiménez-Gutiérrez, A. 2010. Water integration of eco-industrial parks using a global optimization approach. *Industrial and Engineering Chemistry Research* 49(20): 9945–9960.

Rubio-Castro, E., Ponce-Ortega, J.M., Serna-González, M., Jiménez-Gutiérrez, A., El-Halwagi, M.M. 2011. A global optimal formulation for the water integration in eco-industrial parks considering multiple pollutants. *Computers and Chemistry Engineering* 35(8): 1558–1574.

Wang, Y.P., Smith, R. 1994. Wastewater minimisation. *Chemical Engineering Science* 49(7): 981–1006.

7 Total-Site Water Integration Based on a Cooperative-Game Model

Against the backdrop of extreme climate change, efficient water conservation has become an important issue in modern process design. In order to solve the optimization problem of single-plant water network design, Huang et al. (1999) first built an NLP model to minimize freshwater consumption. Wang et al. (2003) later introduced water main(s) into the superstructure for model building so as to generate optimal multi-contaminant water networks. Gunaratnam et al. (2005) also developed an MINLP model to produce the cost-optimal solution without limiting the number of water-treatment units. From the brief review mentioned above and many other published studies, it can be observed that the solution strategies of single-plant water-network design problem are already matured.

On the other hand, the total-site water integration (TSWI) has gradually become a popular research issue in recent years. Chew et al. (2008) proposed a methodology to generate a multi-plant water network (MPWN) indirectly via a central hub. Chen et al. (2010) subsequently designed MPWNs with the local and central water mains. In these studies, the TSWI program has been proven to be a viable solution to the water conservation problem. However, the TSWI arrangements obtained with the aforementioned approaches may not always be implementable. This is because of the fact that these designs only focused upon cost minimization and, as a result, the benefit allocation plan can be unacceptable to one or more participating party. Thus, the key to a successful TSWI project should be to develop, in addition to a cost-optimal MPWN design, a fair and stable benefit allocation scheme under uncertainties.

Based on the above arguments, it is clear that there is a need to develop a TSWI synthesis strategy that facilitates a feasible benefit allocation plan. Chew et al. (2009) utilized the concepts of Nash equilibrium and Pareto optimum to secure maximum benefit for every plant individually in a TSWI program. Although satisfactory results were reported, there is an obvious weakness in these methods, i.e., for total-site water integration, the assumption of non-cooperative game may not be valid. On the other hand, Hiete et al. (2012) first treated the above benefit distribution issue as a cooperative game in multi-plant heat exchanger network design, but placed the required interplant heat exchangers based on heuristic judgments. In a later work, Jin et al. (2018) developed a rigorous model-based procedure to handle this allocation problem for total-site heat integration in the spirit of a cooperative game. However, it should be noted that the counterpart studies in TSWI are still missing.

Similar to the TSHI design strategy described in Chapter 5, the present chapter presents a rigorous procedure for constructing a TSWI framework that facilitates a fair and stable benefit allocation plan according to the cooperative game theory. The proposed approach is also applied basically in two stages. The minimum total annual cost of each and every potential coalition on an industrial park is first determined with two different design strategies. One is to synthesize the MPWN directly based on the process data of all water-using units under consideration, while the other is to produce the optimal single-plant water networks first and then combine them into an MPWN. The problem of distributing cost saving is dealt with in the second stage on the basis of three criteria derived from the cooperative game theory, i.e., the core, the Shapley value, and the risk-based Shapley value (see Chapter 5). In particular, a definite portion of the total annual cost-saving of the MPWN can be determined at designated time intervals for assignment to each plant by considering both its benefit contribution level and also the risk of potential fallouts of unexpected plant shutdowns in a coalition (Jin et al., 2018). On the other hand, the lowest and the highest expected savings achieved by each player are described according to the core solution. As a result, by taking both core and risk-based Shapley value into account, the stability and the fairness of the benefit allocation scheme can be ensured.

7.1 MULTI-PLANT WATER NETWORK DESIGN

As mentioned previously, the aim of this chapter is to develop the benefit allocation plan for facilitating the implementation of interplant water integration programs. To this end, it is necessary to first minimize the total annual cost (TAC) in any given MPWN synthesis problem. For the sake of brevity, its scope is limited to grass-root designs only in this chapter. The corresponding mathematical programming model is delineated below.

7.1.1 Unit Models

The units in a water network are divided into five categories, and their respective models are reviewed in the following subsections:

7.1.1.1 Water Sources

A chemical plant may obtain process water from different primary and secondary sources. Their labels are collected in two separate sets:

$$WA = \{\, w \mid w \text{ is the label of a primary water source}\} \tag{7.1}$$

$$WB = \{\, w \mid w \text{ is the label of a secondary water source}\} \tag{7.2}$$

To facilitate later discussions, let us define

$$W = WA \cup WB \tag{7.3}$$

and also introduce the following set to identify all contaminants in the given system.

$$I = \{i \mid i \text{ is the label of a contaminant}\} \tag{7.4}$$

The contaminant concentrations in each water source ($C_{w,i}$) can be expressed as below:

$$C_{w,i} \approx 0, \forall w \in WA, \forall i \in I \tag{7.5}$$

$$C_{w,i} = Const_{w,i}, \forall w \in WB, \forall i \in I \tag{7.6}$$

where $Const_{w,i}$ is a given constant representing the concentration of contaminant i at the secondary source w.

7.1.1.2 Water-Using Units

The labels of water-using operations in the mathematical programming models can be collected in another set as follows:

$$U = \{u \mid u \text{ is the label of a water using unit}\} \tag{7.7}$$

The equality and inequality constraints associated with water-using units can thus be written as

$$X_u^{in} C_{u,i}^{in} + M_{u,i}^{load} = X_u^{out} C_{u,i}^{out} \quad \forall u \in U, \forall i \in I \tag{7.8}$$

$$C_{u,i}^{in} \le C_{u,i}^{in,max} \quad \forall u \in U, \forall i \in I \tag{7.9}$$

$$C_{u,i}^{out} \le C_{u,i}^{out,max} \quad \forall u \in U, \forall i \in I \tag{7.10}$$

where X_u^{in} and X_u^{out} denote the water flowrates (ton/hr) at the inlet and outlet of water-using unit u, respectively; $M_{u,i}^{load}$ is a given constant representing the mass load of contaminant i (g/hr) of water-using unit u; $C_{u,i}^{in}$ and $C_{u,i}^{out}$ denote the concentrations of contaminant i (ppm) at the inlet and outlet of water-using unit u, respectively, and $C_{u,i}^{in,max}$ and $C_{u,i}^{out,max}$ represent the corresponding upper bounds (ppm).

7.1.1.3 Water-Treatment Units

In the proposed programming model, the collection of water-treatment units is represented with the following set:

$$T = \{t \mid t \text{ is the label of a water treatment unit}\} \tag{7.11}$$

Two simple treatment models can be expressed as follows

$$C_{t,i}^{in}\left(1-\gamma_{t,i}\right)=C_{t,i}^{out} \quad \forall t \in T, \forall i \in I \tag{7.12}$$

$$C_{t,i}^{out} = Const_{t,i}^{out} \quad \forall t \in T, \forall i \in I \tag{7.13}$$

where $C_{t,i}^{in}$ and $C_{t,i}^{out}$ denote the concentrations of contaminant i (ppm) at the inlet and outlet of water-treatment unit t, respectively; $\gamma_{t,i}$ is a given model parameter representing the removal ratio of contaminant i of water-treatment unit t; $Const_{t,i}^{out}$ is a given constant representing the fixed concentration of contaminant i reached at the outlet of treatment unit t. In general, a water-treatment unit can only process waters that are not too dirty, i.e.

$$C_{t,i}^{in} \le C_{t,i}^{in,max} \tag{7.14}$$

where $C_{t,i}^{in,max}$ denotes the upper bound of the concentrations of contaminant i at the inlet of water-treatment unit t (ppm).

7.1.1.4 Water Sinks

Let us define the label set of water sinks as follows:

$$S = \{s \mid s \text{ is the label of a water sink}\} \tag{7.15}$$

The concentrations at the sinks should comply with government regulations, i.e.

$$C_{s,i} \le Ce_i^{max} \quad \forall s \in S, \forall i \in I \tag{7.16}$$

where $C_{s,i}$ is the concentration of contaminant i at sink s and Ce_i^{max} is the corresponding upper bound.

7.1.1.5 Water Mains

Finally, let us define the label set of local water mains as follows:

$$M = \left\{m \mid m \text{ is the label of a local mater main}\right\} \tag{7.17}$$

The component balance around a water main can be written as

$$C_{m,i}^{in} = C_{m,i}^{out} \quad \forall m \in M, \forall i \in I \tag{7.18}$$

where $C_{m,i}^{in}$ and $C_{m,i}^{out}$ denote the concentrations of contaminant i (ppm) at the inlet and outlet of water main m, respectively.

7.1.2 Superstructure of Water Network in a Single Plant

The interconnections of all units within any single plant are assumed to be embedded in the superstructure depicted in Figure 7.1. In this figure, w_i^A $(i = 1,2,\ldots,N_a)$ and w_j^B $(j = 1,2,\ldots,N_b)$ denote the primary water source i and the secondary water sources j respectively; $u_k(k = 1,2,\ldots,N_u)$ denotes the kth water-using unit; $t_l(l = 1,2,\ldots,N_t)$ denotes the lth water-treatment unit; $m_m(m = 1,2,\ldots,N_m)$ denotes the mth water main; $S_n(n = 1,2,\ldots,N_s)$ denotes the nth water sink. A splitter is installed at the exit of every water source, and a mixer at the entrance of every sink. Finally, every water-using unit, water-treatment unit, and water main is equipped with an inlet mixer and an outlet splitter to facilitate connections among units according to the following rules:

1. The primary and secondary sources only supply waters to the water-using units.
2. Every water-using unit accepts waters from the water sources and the water mains, and sends its effluents to water sinks to be discharged to the environment and also to water mains for reuse.
3. Every water-treatment unit receives waters from water mains and other water-treatment units, and delivers the regenerated waters to water sinks, water mains, and other water-treatment units.

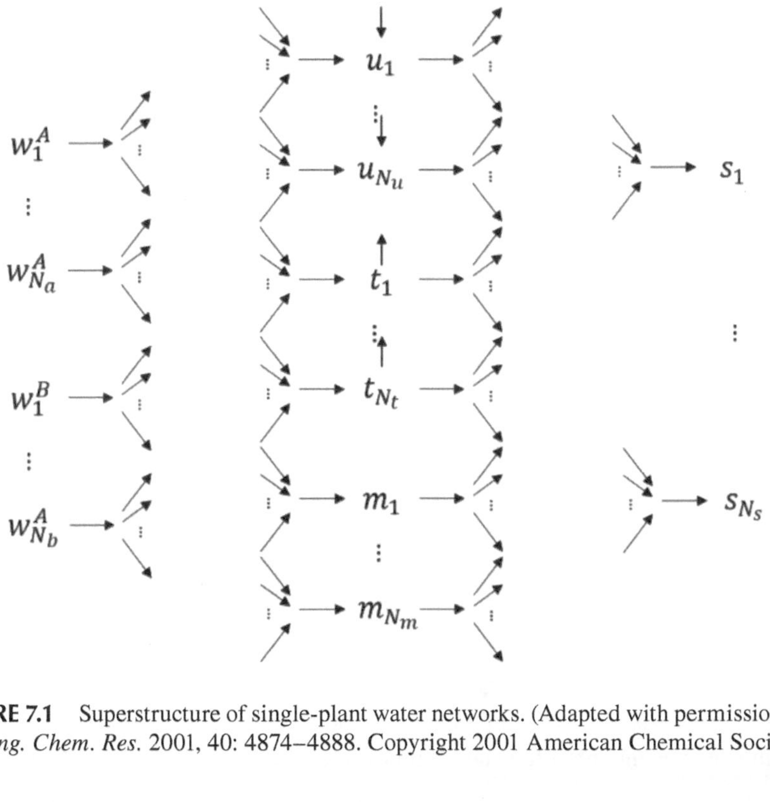

FIGURE 7.1 Superstructure of single-plant water networks. (Adapted with permission from *Ind. Eng. Chem. Res.* 2001, 40: 4874–4888. Copyright 2001 American Chemical Society.)

4. Every water main collects waters from water-using units, water-treatment units, and other water mains, and delivers waters to water-treatment units, water sinks, other water-using units, and other water mains.
5. Every water sink takes waters from water-using units, water-treatment units, and water mains, and discharges the mixed wastewater to the environment.

To facilitate the construction of a unique set of model constraints for the water network within a single plant in the TSWI scheme, let us introduce a set of additional indices to distinguish the plants:

$$P = \{p \mid p \text{ is the label of a plant}\} \tag{7.19}$$

Other than the label set of contaminants, i.e., I, every set defined in Section 7.1.1 is attached with a subscript $p \in P$ later in this chapter for illustration clarity.

7.1.2.1 Outlet Splitter of a Water Source in Plant *p*

The mass balance at the outlet splitter of a water source can be written as

$$W_w^p = \sum_{u \in U_p} f_{w,u} \quad \forall w \in W_p, \forall p \in P \tag{7.20}$$

where W_w^p denotes the water supply rate (ton/hr) of source w in plant p; $f_{w,u}$ is the water flowrate (ton/hr) from source w to water-using unit u in plant p.

7.1.2.2 Inlet Mixer and Outlet Splitter of a Water-Using Unit in Plant *p*

The mass balances at the inlet mixer and outlet splitter of a water-using unit can be expressed as

$$X_u^{in} = \sum_{w \in W_p} f_{w,u} + \sum_{m \in M_p} f_{m,u} \quad \forall u \in U_p, \forall p \in P \tag{7.21}$$

$$X_u^{out} = \sum_{m \in M_p} f_{u,m} + \sum_{s \in S_p} f_{u,s} \quad \forall u \in U_p, \forall p \in P \tag{7.22}$$

$$X_u^{in} = X_u^{out} \quad \forall u \in U_p, \forall p \in P \tag{7.23}$$

$$X_u^{in} C_{u,i}^{in} = \sum_{w \in W_p} f_{w,u} C_{w,i} + \sum_{m \in M_p} f_{m,u} C_{m,i}^{out} \quad \forall u \in U_p, \forall i \in I, \forall p \in P \tag{7.24}$$

where X_u^{in} and X_u^{out} respectively denote the water flowrates (ton/hr) at the inlet and outlet of water-using unit u in plant p; $f_{m,u}$ is the water flowrate (ton/hr) from water main m to water-using unit u in plant p; $f_{u,m}$ and $f_{u,s}$ denote the water flowrates (ton/hr) from water-using unit u to water main m and water sink s, respectively; $C_{u,i}^{in}$, $C_{w,i}$, and $C_{m,i}^{out}$ denote the concentrations of contaminant i at the inlet of water-using unit u, at water source w, and at the outlet of water main m, respectively.

7.1.2.3 Inlet Mixer and Outlet Splitter of a Water-Treatment Unit in Plant *p*

The mass balances at the inlet mixer and outlet splitter of a water-treatment unit can be written as

$$X_t^{in} = \sum_{m \in M_p} f_{m,t} + \sum_{\substack{t' \in T_p \\ t' \neq t}} f_{t',t} \quad \forall t \in T_p, \forall p \in P \tag{7.25}$$

$$X_t^{out} = \sum_{m \in M_p} f_{t,m} + \sum_{\substack{t' \in T_p \\ t' \neq t}} f_{t,t'} + \sum_{s \in S_p} f_{t,s} \quad \forall t \in T_p, \forall p \in P \tag{7.26}$$

$$X_t^{in} = X_t^{out} \quad \forall t \in T_p, \forall p \in P \tag{7.27}$$

$$X_t^{in} C_{t,i}^{in} = \sum_{m \in M_p} f_{m,t} C_{m,i}^{out} + \sum_{\substack{t' \in T_p \\ t' \neq t}} f_{t',t} C_{t',i}^{out} \quad \forall t \in T_p, \forall i \in I, \forall p \in P \tag{7.28}$$

where X_t^{in} and X_t^{out} respectively denote the input and output water flowrates (ton/hr) of water-treatment unit t in plant p; $f_{m,t}$ and $f_{t',t}$ denote the water flowrates (ton/hr) from water main m and water-treatment unit t', respectively, to water-treatment unit t in plant p; $f_{t,m}$, $f_{t,t'}$, and $f_{t,s}$ denote the water flowrates (ton/hr) from water-treatment unit t to water main m, water-treatment unit t', and water sink s, respectively; $C_{t,i}^{in}$ and $C_{t',i}^{out}$ denote the concentrations of contaminant i at the inlet of water-treatment unit t and at the outlet of water-treatment unit t', respectively.

7.1.2.4 Inlet Mixer and Outlet Splitter of a Local Water Main in Plant *p*

The mass balances at the inlet mixer and outlet splitter of a water main can be described as follows

$$X_m^{in} = \sum_{u \in U_p} f_{u,m} + \sum_{t \in T_p} f_{t,m} + \sum_{\substack{m' \in M_p \\ m' \neq m}} f_{m',m}$$

$$\forall m \in M_p, \forall p \in P \tag{7.29}$$

$$X_m^{out} = \sum_{u \in U_p} f_{m,u} + \sum_{t \in T_p} f_{m,t} + \sum_{\substack{m' \in M_p \\ m' \neq m}} f_{m,m'} + \sum_{s \in S_p} f_{m,s}$$

$$\forall m \in M_p, \forall p \in P \tag{7.30}$$

$$X_m^{in} = X_m^{out}$$

$$\forall m \in M_p, \forall p \in P \tag{7.31}$$

$$X_m^{in} C_{m,i}^{in} = \sum_{u \in U_p} f_{u,m} C_{u,i}^{out} + \sum_{t \in T_p} f_{t,m} C_{t,i}^{out} + \sum_{\substack{m' \in M_p \\ m' \neq m}} f_{m',m} C_{m',i}^{out}$$

$$\forall m \in M_p, \forall i \in I, \forall p \in P \tag{7.32}$$

where X_m^{in} and X_m^{out} respectively denote the input and output water flowrates (ton/hr) of water main m in plant p; $f_{u,m}$, $f_{t,m}$, and $f_{m',m}$ denote the water flowrates (ton/hr) from water-using unit u, water-treatment unit t, and water main m', respectively, to water main m in plant p; $f_{m,m'}$ and $f_{m,s}$ denote the water flowrates (ton/hr) from water main m to water main m' and water sink s, respectively; $C_{m,i}^{in}$ and $C_{m',i}^{out}$ denote the concentrations of contaminant i at the inlet of water main m and at the outlet of water main m', respectively.

7.1.2.5 Inlet Mixer of a Water Sink in Plant *p*

The mass balances at the inlet splitter of a water sink can be written as

$$X_s^{in} = \sum_{m \in M_p} f_{m,s} + \sum_{t \in T_p} f_{t,s} + \sum_{u \in U_p} f_{u,s}$$

$$\forall s \in S_p, \forall p \in P \tag{7.33}$$

$$X_s^{in} C_{s,i}^{in} = \sum_{m \in M_p} f_{m,s} C_{m,i}^{out} + \sum_{t \in T_p} f_{t,s} C_{t,i}^{out} + \sum_{u \in U_p} f_{u,s} C_{u,i}^{out}$$

$$\forall s \in S_p, \forall i \in I, \forall p \in P \tag{7.34}$$

where X_s^{in} denotes the input water flowrate (ton/hr) of water sink s in plant p; $C_{s,i}^{in}$, $C_{m,i}^{out}$, $C_{t,i}^{out}$, and $C_{u,i}^{out}$ denote the concentrations of contaminant i at the inlet of water sink s, at the outlet of water main m, at the outlet of water-treatment unit t, and at the outlet of water-using unit u, respectively.

7.1.3 MPWN Superstructure

Let us next assume that the interplant water exchange scheme is facilitated with a collection of central water mains. The connections in this scheme can be described with the superstructure in Figure 7.2. More specifically, every central water main $mcen_i$ $(i = 1, 2, \ldots, N_{mcen})$ receives waters from the water-using units, the water-treatment units, and the local water mains in all plants participating in the water integration program together with the waters from the other central water mains, and supplies

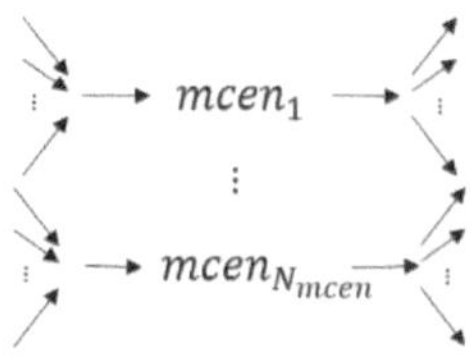

FIGURE 7.2 Superstructure for central water mains.

waters to the water-using units, the water-treatment units, the local water mains, and the water sinks in each plant mentioned above and also sends waters to the other central water mains. Based on this description, the mass balances mentioned in Section 7.1.2 can be modified by considering the extra water flows to and from the central water mains. These changes are summarized below.

7.1.3.1 Outlet Splitter of a Water Source within MPWN

There is no need to modify equation (7.20).

7.1.3.2 Inlet Mixer and Outlet Splitter of a Water-Using Unit within MPWN

Three extra terms, i.e., $\sum_{mcen \in MCEN} f_{mcen,u}$, $\sum_{mcen \in MCEN} f_{u,mcen}$, and $\sum_{mcen \in MCEN} f_{mcen,u} C^{out}_{mcen,i}$, should be added to the right side of equations (7.21), (7.22), and (7.24), respectively, where $f_{mcen,u}$ and $f_{u,mcen}$ denote the water flowrates directed from central water main *mcen* to water-using unit *u* and vice versa, $C^{out}_{mcen,i}$ is the concentration of contaminant *i* at the outlet of water main *mcen*, and *MCEN* is the label set of central water mains.

7.1.3.3 Inlet Mixer and Outlet Splitter of a Water-Treatment Unit within MPWN

Three extra terms, i.e., $\sum_{mcen \in MCEN} f_{mcen,t}$, $\sum_{mcen \in MCEN} f_{t,mcen}$, and $\sum_{mcen \in MCEN} f_{mcen,t} C^{out}_{mcen,i}$, should be added to the right sides of equations (7.25), (7.26), and (7.28), respectively, where $f_{mcen,t}$ and $f_{t,mcen}$ denote the water flowrates directed from central water main *mcen* to water-treatment unit *t* and vice versa.

7.1.3.4 Inlet Mixer and Outlet Splitter of a Local Water Main within MPWN

Three extra terms, i.e., $\sum_{mcen \in MCEN} f_{mcen,m}$, $\sum_{mcen \in MCEN} f_{m,mcen}$, and $\sum_{mcen \in MCEN} f_{mcen,m} C^{out}_{mcen,i}$, should be added to the right sides of equations (7.29), (7.30), and (7.32), respectively, where $f_{mcen,m}$ and $f_{m,mcen}$ denote the water flowrates directed from central water main *mcen* to local water main *m* and vice versa.

7.1.3.5 Inlet Mixer of a Water Sink within MPWN

Two extra terms, i.e., $\sum_{mcen \in MCEN} f_{mcen,s}$ and $\sum_{mcen \in MCEN} f_{mcen,s} C^{out}_{mcen,i}$, should be added to the right sides of equations (7.33) and (7.34), respectively, where $f_{mcen,s}$ and $f_{s,mcen}$ denote the water flowrates directed from central water main *mcen* to water sink *s* and vice versa.

7.1.4 Cost Models

Since the minimum total annual cost (TAC) of every possible MPWN in a TSWI project is considered here for benefit allocation, all cost estimation formulas are presented below to facilitate clear illustration. Specifically, TAC is treated as the sum of six cost terms:

$$TAC = Cost^{fw} + Cost^{tre} + Cost^{main} + Cost^{sink} + Cost^{pump} + Cost^{pipe} \tag{7.35}$$

where $Cost^{fw}$ denotes the total annual freshwater cost; $Cost^{tre}$ denotes the sum of the total annualized capital cost and total annual operating cost of water-treatment units; $Cost^{main}$ is the total annualized capital cost of water mains; $Cost^{sink}$ denotes the total annualized capital cost of water sinks; $Cost^{pump}$ is denotes the sum of the total annualized capital cost and the total annual operating cost of interplant pumps; $Cost^{pipe}$ is the total annualized capital cost of interplant pipelines. Specifically, the total annual freshwater cost is expressed as follows:

$$Cost^{fw} = CW \cdot H \cdot W_0 \tag{7.36}$$

$$W_0 = \sum_{p \in P} \sum_{w \in W_p} W_w^p \tag{7.37}$$

where CW is the unit cost of freshwater (\$/ton); H denotes the yearly operation time (hr/yr); W_0 is the total amount of freshwater usage (ton/hr). The sum of the total annualized capital cost and total annual operating cost of water-treatment units is determined as follows

$$Cost^{tre} = AF \sum_{p \in P} \sum_{t \in T_p} \psi_t^{inv} \left(X_t^{in}\right)^{0.7} + H \sum_{p \in P} \sum_{t \in T_p} \psi_t^{op} X_t^{in} \tag{7.38}$$

where AF is the annualization factor; ψ_t^{inv} is a model coefficient for computing the investment cost of a water-treatment unit; ψ_t^{op} is the unit cost of operating water-treatment unit t (\$/ton). The total annualized capital cost of local and central water mains can be calculated as follows:

$$\begin{aligned} Cost^{main} &= AF \sum_{p \in P} \sum_{m \in M_p} \left(\psi_m^{var} X_m^{in} + \psi_m^{fix} y_m\right) \\ &+ AF \sum_{mcen \in MCEN} \left(\psi_{mcen}^{var} X_{mcen}^{in} + \psi_{mcen}^{fix} y_{mcen}\right) \end{aligned} \tag{7.39}$$

where ψ_m^{var} and ψ_{mcen}^{var} represent the variable cost coefficients of the capital cost models of local water main m and central water main $mcen$, respectively; ψ_m^{fix} and ψ_{mcen}^{fix} respectively denote the fixed cost coefficients of the capital cost models of local water main m and central water main $mcen$; y_m and y_{mcen} are binary variables respectively used to reflect whether or not local water main m and central water main $mcen$ exist, i.e.

$$X_m^{in} - LN \cdot y_m \leq 0 \quad \forall m \in M_p, \forall p \in P \tag{7.40}$$

$$X_{mcen}^{in} - LN \cdot y_{mcen} \leq 0 \quad \forall mcen \in MCEN \tag{7.41}$$

Note that, in the above two inequalities, LN is a large positive constant number. The total annualized capital cost of water sinks can be determined according to the following constraints

$$Cost^{sink} = AF \sum_{p \in P} \sum_{s \in S_p} (\psi_s^{var} X_s^{in} + \psi_s^{fix} y_s) \tag{7.42}$$

$$X_s^{in} - LN \cdot y_s \leq 0 \quad \forall s \in S_p, \forall p \in P \tag{7.43}$$

where ψ_s^{var} and ψ_s^{fix} respectively denote the variable and fixed cost coefficients of the capital cost models of water sink s; y_s is a binary variable used to reflect whether or not local water sink s is present. The sum of the total annualized capital cost and total annual operating cost of interplant pumps can be expressed as

$$Cost^{pump} = \sum_{*} \left[H \cdot L_* \cdot \alpha (f_*)^{\beta} + AF \cdot \kappa (f_*)^{\mu} \right] \tag{7.44}$$

where the index * denotes all possible connections described in Section 7.1.3 that are between the central water mains and the units in every plant; f_* represents the water flowrate in the pipeline that facilitate the connection * (ton/hr); L_* denotes the pipeline length of connection * (m); α and β respectively denote the coefficient and exponent used in the operating cost model of a pump; κ and μ respectively denote the coefficient and exponent used in the capital cost model for a pump. The annualized total capital cost of the interplant pipelines can be estimated according to the following formula

$$Cost^{pipe} = AF \cdot L_* \cdot C \cdot \sum_{*} y_* \tag{7.45}$$

where C is the cost of pipeline per unit length (\$/m); y_* is the binary variable reflecting where or not pipeline connection * is present. In other words, the following inequality constraints may be imposed

$$f_* - UB \cdot y_* \leq 0 \tag{7.46}$$

$$f_* - LB \cdot y_* \geq 0 \tag{7.47}$$

where *UB* and *LB* are two constants denoting the upper and lower bounds of water flowrate f_*.

7.2 INTEGRATION STRATEGIES

The traditional TSWI strategy calls for minimization of the aforementioned TAC with only the needed pipeline connections (Chew et al., 2008). Specifically, the model formulations described in Sections 7.1.1, 7.1.3, and 7.1.4 are included in a mathematical programming model for generating the optimal MPWN designs. To facilitate later discussions, this approach is referred to as Strategy I in this chapter.

A slightly different approach has also been presented in this chapter. In addition to the needed pipeline connections mentioned above, this strategy tries to incorporate extra pipeline structures to allow single-plant operations when one or more plant in the TSWI scheme shuts down unexpectedly. More specifically, these MPWN designs are synthesized in two stages. The optimal single-plant network configurations are produced first with model constraints given in Sections 7.1.1 and 7.1.2. In the second stage, by fixing the connections obtained in stage 1, the overall MPWN structure can then be created by solving the programming model formulated according to Sections 7.1.1, 7.1.3, and 7.1.4. This modified water integration approach is referred to as Strategy II later in this chapter.

7.3 AN ILLUSTRATIVE EXAMPLE

To illustrate the above TSWI strategies more clearly, let us consider a simple example in this section. It is assumed that there is only one freshwater source on site, which can be acquired at the price of 0.2 $/ton. The specifications of existing water-using units in three different plants in this industrial park are listed in Table 7.1. It is also assumed

TABLE 7.1
Specifications of Water-Using Units

Plant	Unit	$C_{u,i}^{in,max}$ (ppm)	$C_{u,i}^{out,max}$ (ppm)	$M_{u,i}^{load}$ (g/hr)
A	1	0	15	750
	2	20	120	3400
	3	120	220	5600
	4	50	150	800
B	1	150	900	22500
	2	20	120	3400
	3	120	220	560
	4	0	50	1250
C	1	20	150	4550
	2	50	600	22000
	3	500	1100	30000
	4	0	15	675

TABLE 7.2
Specifications of Water-Treatment Units

Removal ratio, $\gamma_{t,i}$	0.8
Investment cost coefficient, ψ_t^{inv} ($)	12600
Operating cost coefficient, ψ_t^{op} ($/tons)	0.0067

TABLE 7.3
Specifications of Water Mains

Variable cost coefficient, ψ_m^{var} ($/tons)	116.95
Fixed cost coefficient, ψ_m^{fix} ($)	10142.16

TABLE 7.4
Specifications of Water Sinks

Maximum discharge concentration, Ce_i^{max} (ppm)	60
Variable cost coefficient, ψ_s^{var} ($/tons)	116.95
Fixed cost coefficient, ψ_s^{op} ($)	10142.16

TABLE 7.5
Specifications of Pumps

Operating cost coefficient, α ($)	$2.21*10^{-6}$
Operating cost exponent, β	1
Capital cost coefficient, κ ($)	1932
Capital cost exponent, μ	0.732

that at most one water-treatment unit, two local water mains, and one water sink are allowed to be added to each plant, and their specifications are given in Tables 7.2–7.4, respectively. The distance between the central water main station and each plant is 200 m, and the upper and lower bounds of water flowrate in each connection pipeline are 319 ton/hr and 5 ton/hr, respectively. The allowed maximum number of central water mains is two, and their specifications are the same as those listed in Table 7.3. The specifications of the corresponding pumps are presented in Table 7.5. Let us assume that the interest rate is 1%/yr and the operation life is 10 years and, thus, the annualization factor AF used in this study is a constant value of 0.106. Finally, the operation time of every plant per year (H) is taken to be 8600 hours.

7.3.1 Single-Plant Water Networks

With the above data, one can solve a programming model built according to the unit models presented in Section 7.1.1, the mass balances given in Section 7.1.2, and the cost models outlined in Section 7.1.5. The resulting minimum TACs for the three plants in this example and the corresponding single-plant water networks can be found in Table 7.6 and Figures 7.3–7.5. These results are used later in synthesizing the MPWN designs and also devising the benefit allocation schemes. Notice also that, to facilitate easy understanding of the network structures in Figures 7.3–7.5, each unit is distinguished with a three-character code. The first character in the code is used to denote the unit type, the second indicates the plant in which the unit is located, and the third is a serial number. For example, the code "MA2" represents the second water main in plant A.

TABLE 7.6
Total Annual Costs of Single-Plant Water Networks

Plant	TAC ($/year)
A	113588.0
B	123748.6
C	181060.7
Total	418397.3

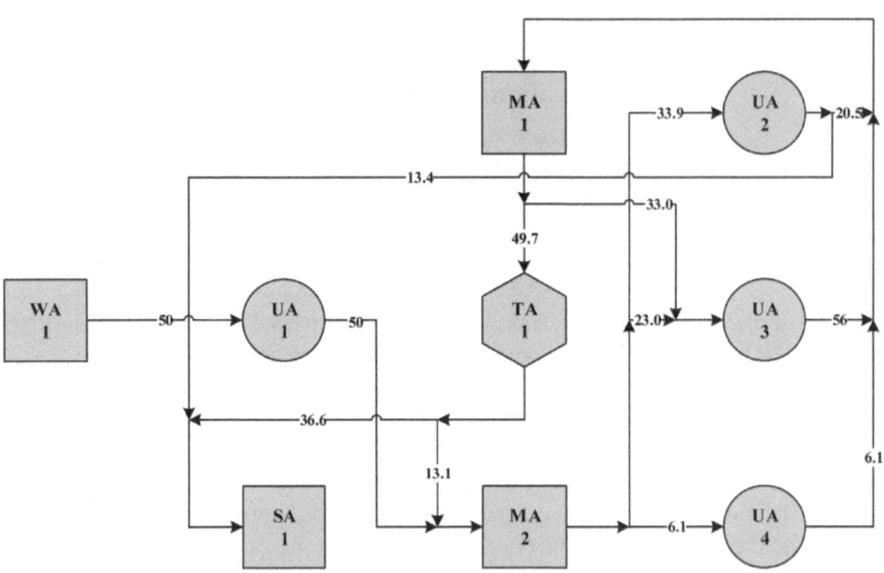

FIGURE 7.3 Cost-optimal water network of plant A.

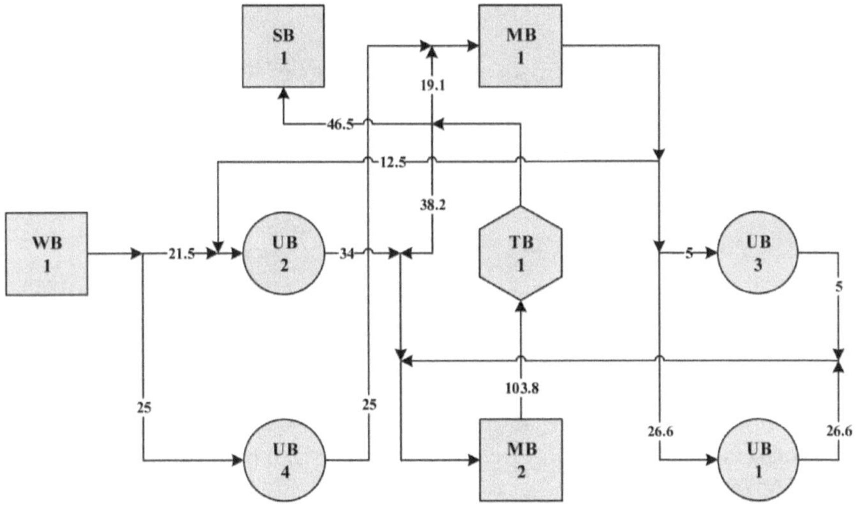

FIGURE 7.4 Cost-optimal water network of plant B.

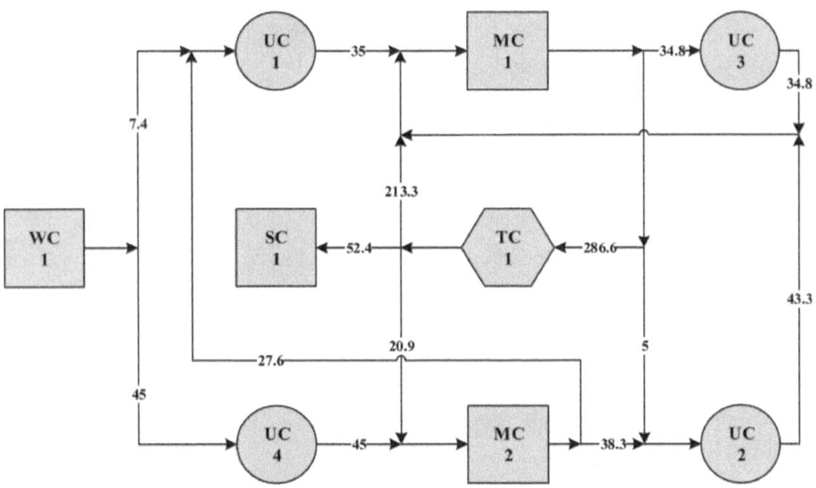

FIGURE 7.5 Cost-optimal water network of plant C.

7.3.2 MPWN Designs

As mentioned before, two integration strategies have been considered in this work. The corresponding results are presented in the sequel.

7.3.2.1 Strategy I

The MPWN designs can be obtained by solving programming models assembled by incorporating the unit models presented in Section 7.1.1, the mass balances described in Section 7.1.3, and the cost models given in Section 7.1.5. Note also

TABLE 7.7
Total Annual Costs and TAC Savings of MPWNs Obtained by Strategy I

Integration Scenarios	TAC ($/year)	TAC Saving ($/year)
A,B	216130.0	21206.6 (5.1%)
A,C	265514.4	29134.3 (7.0%)
B,C	257679.4	47129.9 (7.1%)
A,B,C	368787.0	49610.3 (7.5%)

that $* \in \{(u,mcen),(t,mcen),(m,mcen),(mcen,u),(mcen,t),(mcen,m),(mcen,s)\}$ in equations (7.44)–(7.47). The resulting minimum TACs and TAC savings for four integration scenarios and the corresponding MPWNs can be found in Table 7.7 and Figures 7.6–7.9.

7.3.2.2 Strategy II

As mentioned before, Strategy II is implemented in two stages. The single-plant water networks are produced in the first stage. In the present example, these networks have already been obtained in Section 7.3.1. In the second stage, the overall MPWN design can be generated by solving the programming model formulated according to Sections 7.1.1–7.1.4 under additional constraints imposed according to the aforementioned single-plant water networks. More specifically, this programming model can be constructed on the basis of the following principles:

- Unit models are formulated according to equations (7.1)–(7.18) in Section 7.1.1.
- Mass balances are formulated according to equations (7.19)–(7.34) in Section 7.1.2 and the modifications introduced in Section 7.1.3.
- Other than the inequalities in equations (7.46) and (7.47), additional constraints are imposed on the binary variables in order to keep the structures of the optimal single-plant water networks intact. These constraints can be written as

$$y_{\#} \leq y_{\#}^{fix} \tag{7.48}$$

where $\# \in \{(w,u),(m,u),(u,m),(u,s),(t,m),(m,m'),(m,t),(m,s),(t,t'),(t,s)\}$.

Notice that $y_{\#}^{fix}$ is a binary parameter fixed according to the optimal single-plant water networks obtained in Section 7.3.1. Its value is 1 if the connection is present, 0 if otherwise.

- The unit capacities in each plant of the MPWN design should also be kept below those in the single-plant water networks originally obtained in Section 7.3.1, i.e.

$$X_t^{in} \leq X_t^{org} \tag{7.49}$$

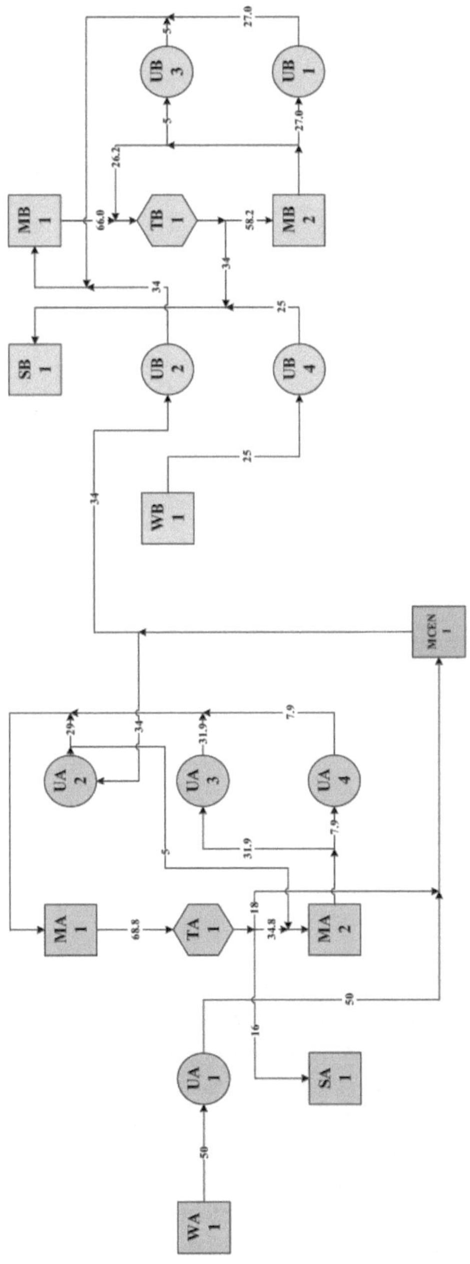

FIGURE 7.6 MPWN design of plants A and B using Strategy I.

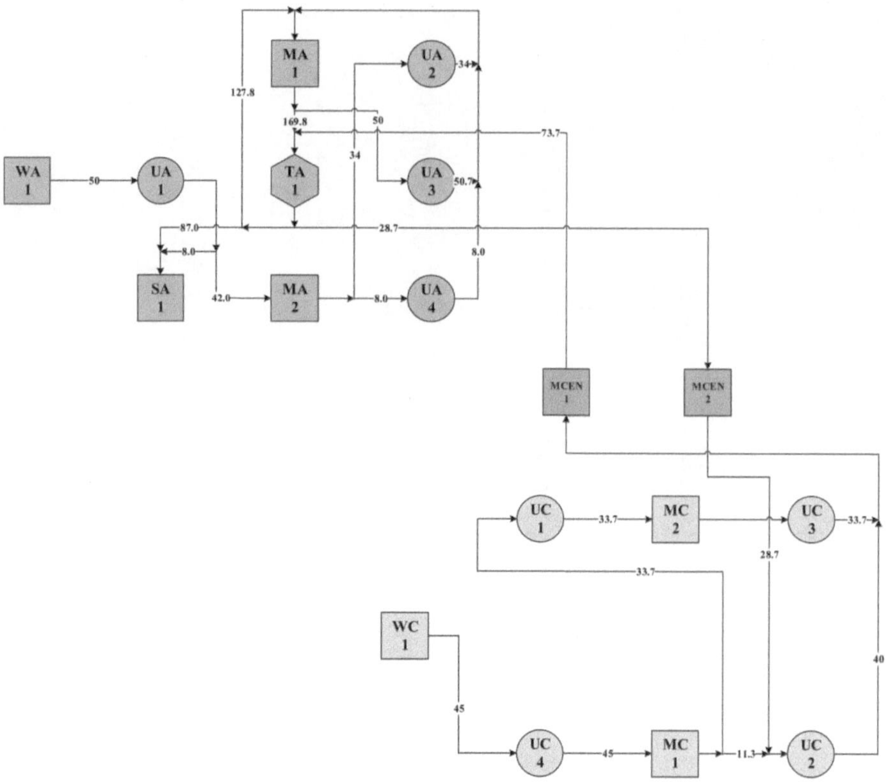

FIGURE 7.7 MPWN design of plants A and C using Strategy I.

$$X_m^{in} \leq X_m^{org} \tag{7.50}$$

$$X_s^{in} \leq X_s^{org} \tag{7.51}$$

where X_t^{org}, X_m^{org}, and X_s^{org} denote the inlet water flowrates of the water-treatment unit, local water main, and water sink obtained in the original single-plant water networks, respectively.

- The cost models should in general be the same as equations (7.35)–(7.47) in Section 7.1.4. However, since the optimal single-plant network structure must be kept unchanged in applying Strategy II, the cost estimates of water-treatment units, local water mains, and water sinks in each plant of an MPWN should be determined on the basis of the unit capacities obtained in the single-plant scenarios. More specifically, equations (7.38), (7.39), and (7.42) should be modified respectively as follows:

$$Cost^{tre} = AF\sum_{t\in T}\psi_t^{inv}\left(X_t^{org}\right)^{0.7} + H\sum_{t\in T}\psi_t^{op}X_t \tag{7.52}$$

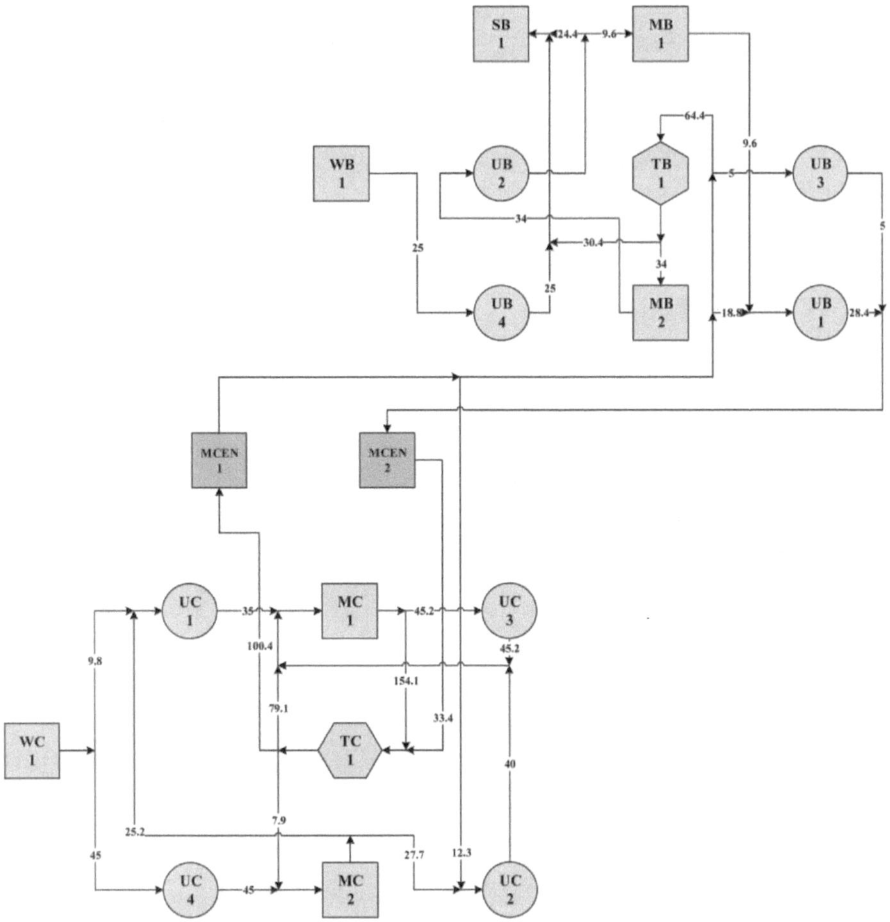

FIGURE 7.8 MPWN design of plants B and C using Strategy I.

$$Cost^{main} = AF\sum_{m\in M}\left(\psi_m^{var}X_m^{org} + \psi_m^{fix}y_m^{org}\right) + AF\sum_{mcen\in MCEN}\left(\psi_{mcen}^{var}X_{mcen}^{in} + \psi_{mcen}^{fix}y_{mcen}\right) \tag{7.53}$$

$$Cost^{sink} = AF\sum_{s\in S}(\psi_s^{var}X_s^{org} + \psi_s^{fix}y_s^{org}) \tag{7.54}$$

Notice that y_m^{org} and y_s^{org} denote binary parameters determined according to the optimal single-plant water networks in Section 7.3.1. The value of y_m^{org} is 1 if local water main m is present, 0 if otherwise. Similarly, the value of y_s^{org} is 1 if water sink s is present, 0 if otherwise.

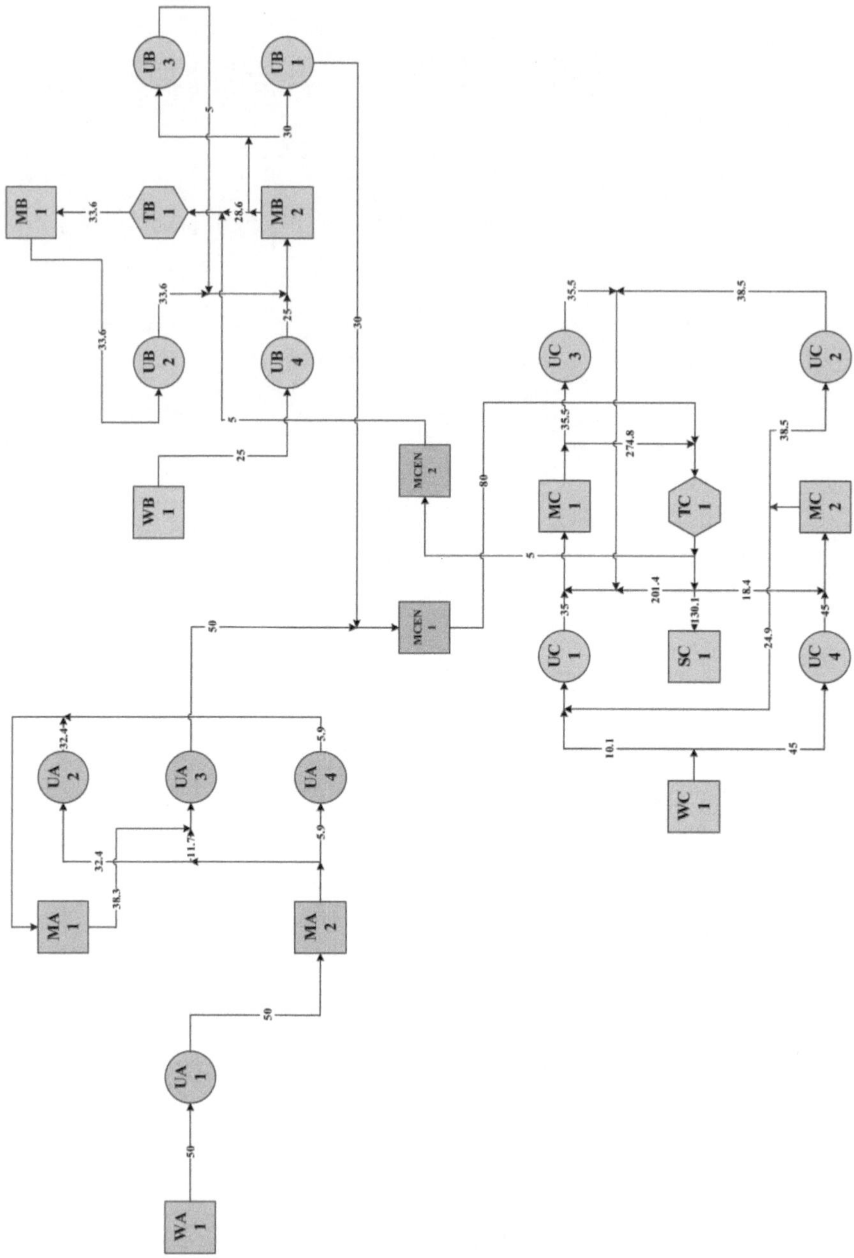

FIGURE 7.9 MPWN design of plants A, B and C using Strategy I.

TABLE 7.8
Total Annual Costs and TAC Savings of MPWNs Obtained by Strategy II

Integration Scenario	TAC ($/year)	TAC Saving ($/year)
A,B	230905.9	6430.7 (1.5%)
A,C	294648.7	0 (0.0%)
B,C	279410.3	25399 (3.6%)
A,B,C	389096.9	29300.4 (4.1%)

The MPWN designs can be obtained by solving the above programming models in four integration scenarios. The resulting minimum TACs and TAC savings and the corresponding MPWNs can be found in Table 7.8 and Figures 7.10–7.13. Note that the units in these figures are connected with either solid or dashed lines. The former represent the pipelines adopted in both MPWN and single-plant water networks, while the latter denote the pipelines used only in single-plant operations.

7.4 BENEFIT ALLOCATION BASED ON COOPERATIVE GAME MODEL

Although it has been shown that the optimal MPWN designs can be obtained with the aforementioned mathematical models, the corresponding benefit allocation scheme still has not been discussed. The fair benefit distribution plan used to facilitate smooth execution of any given TSWI project is detailed in the sequel.

7.4.1 CORE

The tolerable lower limits of the benefit allocated to every member in a coalition are characterized with the so-called "core" (see Section 5.1.1 in Chapter 5). To be specific, let us first assign each member in a coalition a numerical label and group them in a set N, i.e.

$$N = \{1,2,3,\ldots,n\} \tag{7.55}$$

A characteristic function $v(\cdot)$ can be defined accordingly to represent the revenue generated collectively by a sub-coalition of N. Specifically,

$$v : 2^N \rightarrow R \tag{7.56}$$

where 2^N is the power set of N and $v(\varnothing) = 0$. In addition, a vector $\boldsymbol{DP}_S = (dp_{1,S} \quad dp_{2,S} \quad \cdots \quad dp_{m,S})$, where $m \leq n$, is adopted in this work to represent the benefit distribution in sub-coalition $S \subseteq N$, i.e., $dp_{i,S}$ $(i = 1,2,\ldots,m)$ denotes a portion of the total

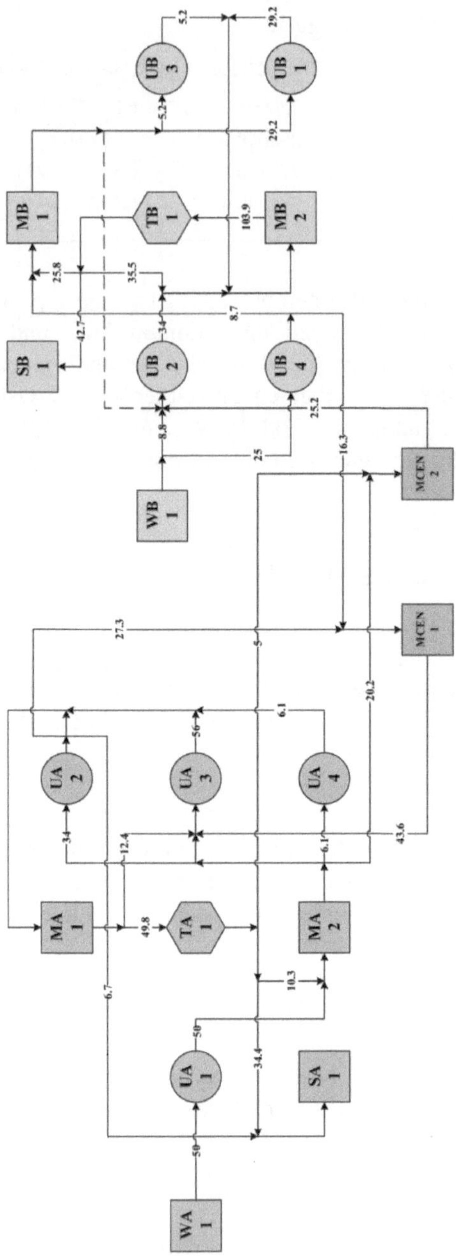

FIGURE 7.10 MPWN design of plants A and B using Strategy II.

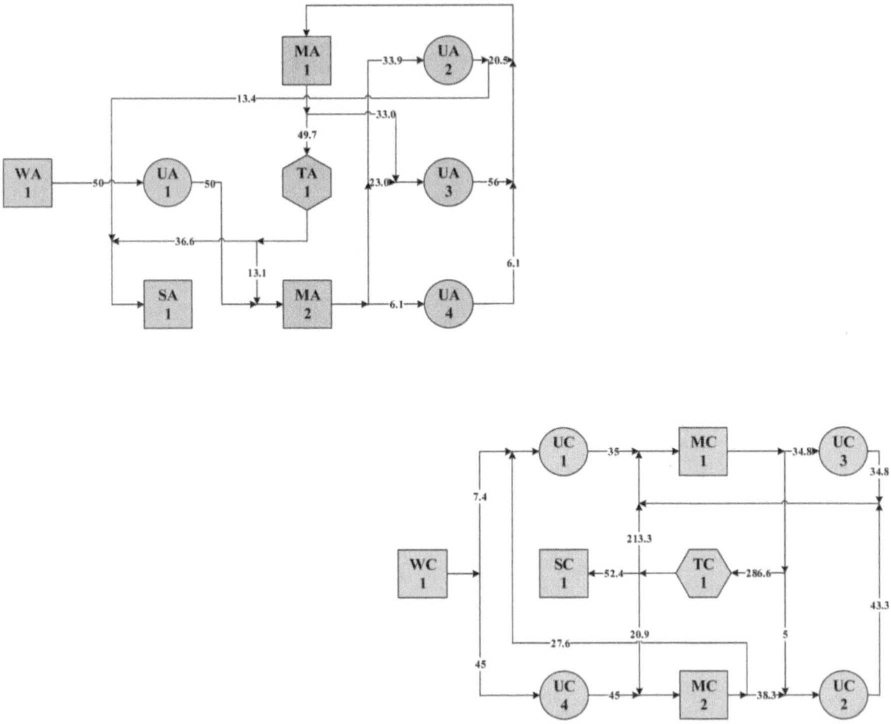

FIGURE 7.11 MPWN design of plants A and C using Strategy II.

revenue allocated to participant i of S. On the basis of the above definitions, the four basic rules of core can be repeated as follows.

- Individual rationality. A participant is not willing to join a coalition from which the allocated revenue is higher than or equal to the individual income earned by itself, i.e.

$$dp_{i,N} \geq v(i) \quad \forall i \in N \tag{7.57}$$

- Group rationality. The total revenue generated by a coalition must be distributed to all its members, i.e.

$$\sum_{i \in N} dp_{i,N} = v(N) \tag{7.58}$$

- Coalitional rationality. Forming the coalition N is not worse than forming any of its sub-coalition $S \subseteq N$. In other words, the sum of revenues

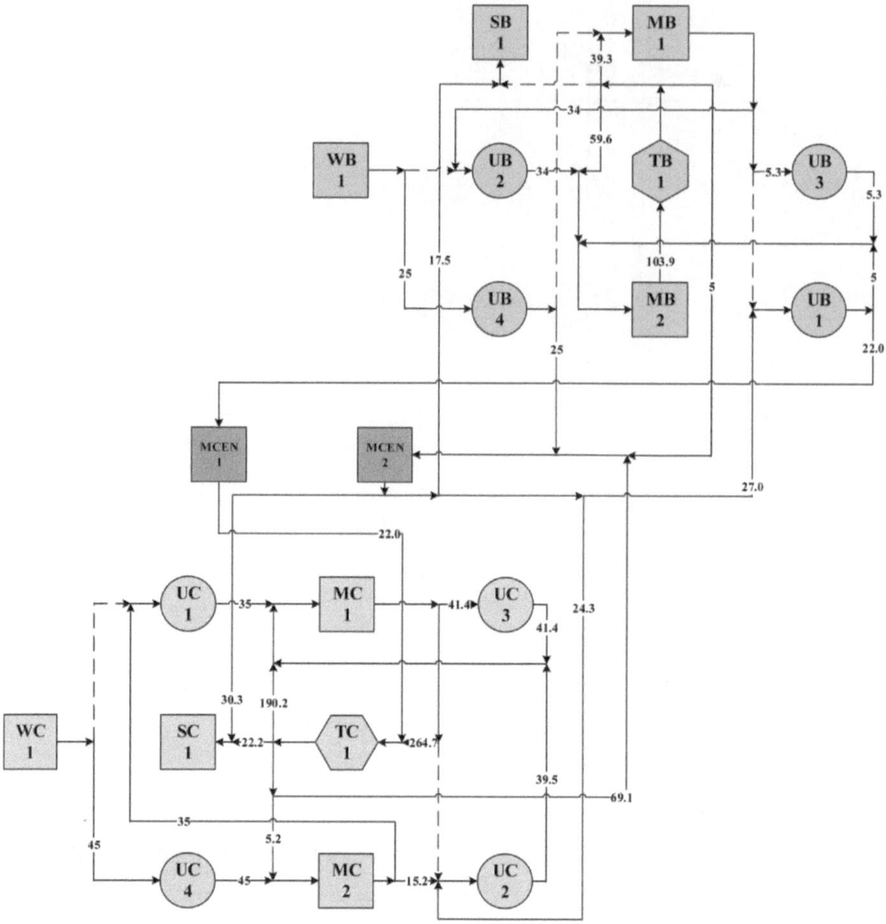

FIGURE 7.12 MPWN design of plants B and C using Strategy II.

allocated to members of S in coalition N is greater than or equal to the total revenue generated individually by S, i.e.

$$\sum_{i \in S} dp_{i,N} \geq v(S) \quad \forall S \subseteq N \tag{7.59}$$

- No-subsidy principle. The total revenue of the coalition as a whole may be increased by accepting a new member. In order to ensure the other member agrees to such an addition, this increase in total revenue must be larger than or equal to the individual revenue allocated to the new member in the resulting coalition. The following inequality can be used to characterize the present principle:

$$dp_{i,N} \leq v(N) - v(N \setminus i) \quad \forall i \in N \tag{7.60}$$

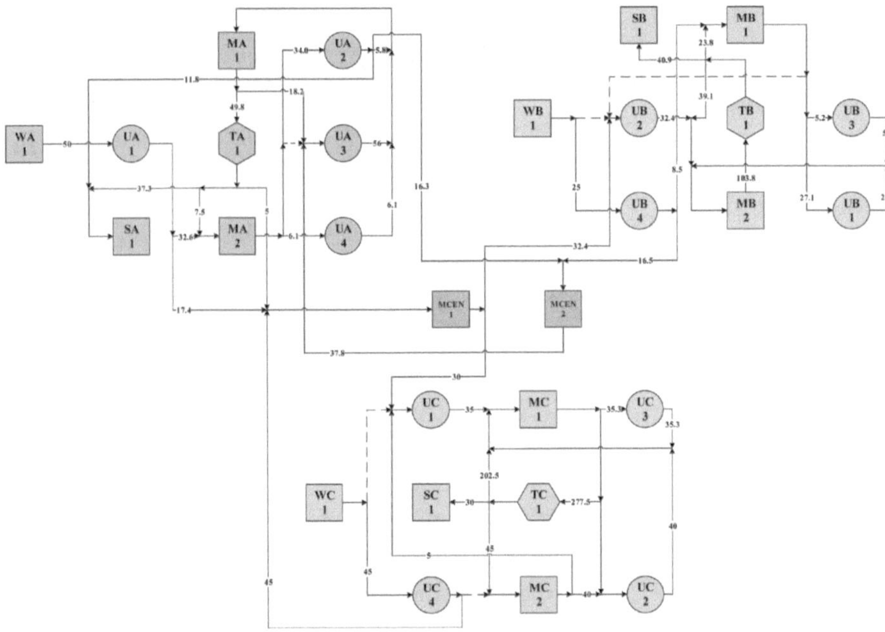

FIGURE 7.13 MPWN design of plants A, B, and C using Strategy II.

Let us next revisit the illustrative example in Section 7.3 to further illustrate the concept of core. Note first that the TAC savings listed in Tables 7.7 and 7.8 are viewed as the results of characteristic mapping defined in equation (7.56) for Strategies I and II respectively. In addition, equations (7.57)–(7.59) can be translated in this example as follows

$$dp_{A,\{A,B,C\}} \geq v(\{A\}) \tag{7.61}$$

$$dp_{B,\{A,B,C\}} \geq v(\{B\}) \tag{7.62}$$

$$dp_{C,\{A,B,C\}} \geq v(\{C\}) \tag{7.63}$$

$$dp_{A,\{A,B.C\}} + dp_{B,\{A,B.C\}} \geq v(\{A,B\}) \tag{7.64}$$

$$dp_{A,\{A,B.C\}} + dp_{C,\{A,B.C\}} \geq v(\{A,C\}) \tag{7.65}$$

$$dp_{B,\{A,B.C\}} + dp_{C,\{A,B.C\}} \geq v(\{B,C\}) \tag{7.66}$$

$$dp_{A,\{A,B.C\}} + dp_{B,\{A,B.C\}} + dp_{C,\{A,B.C\}} = v(\{A,B,C\}) \tag{7.67}$$

Substituting equation (7.67) into equations (7.61)–(7.66) yields

$$dp_{B,\{A,B.C\}} + dp_{C,\{A,B.C\}} \leq v(\{A,B,C\}) - v(\{A\}) \tag{7.68}$$

$$dp_{A,\{A,B.C\}} + dp_{C,\{A,B.C\}} \leq v(\{A,B,C\}) - v(\{B\}) \tag{7.69}$$

$$dp_{A,\{A,B.C\}} + dp_{B,\{A,B.C\}} \leq v(\{A,B,C\}) - v(\{C\}) \tag{7.70}$$

$$dp_{C,\{A,B.C\}} \leq v(\{A,B,C\}) - v(\{A,B\}) \tag{7.71}$$

$$dp_{B,\{A,B.C\}} \leq v(\{A,B,C\}) - v(\{A,C\}) \tag{7.72}$$

$$dp_{A,\{A,B.C\}} \leq v(\{A,B,C\}) - v(\{B,C\}) \tag{7.73}$$

Notice that equations (7.71)–(7.73) can also be obtained from equation (7.60). Dividing both sides of the inequalities in equations (7.61), (7.62), (7.64), (7.70), (7.72), and (7.73) by $v(\{A,B,C\})$ yields the lower and upper bounds of TAC savings of the three plants under consideration, which are shown graphically in Figures 7.14 and 7.15 for Strategies I and II respectively. The right triangles in these two figures are associated with equations (7.61), (7.62), and (7.70), while the dashed lines are constructed according to equations (7.64), (7.72), and (7.73). Note that the intersection of the above six inequalities is the so-called "core" of the cooperative game, and the cores in Figures 7.14 and 7.15 are the regions bordered by red lines.

On the other hand, the core for a two-player game is independent of the characteristic function values. Under the assumption that the TAC saving allocated to each participant in a coalition formed by two plants (denoted as X and Y) is greater than

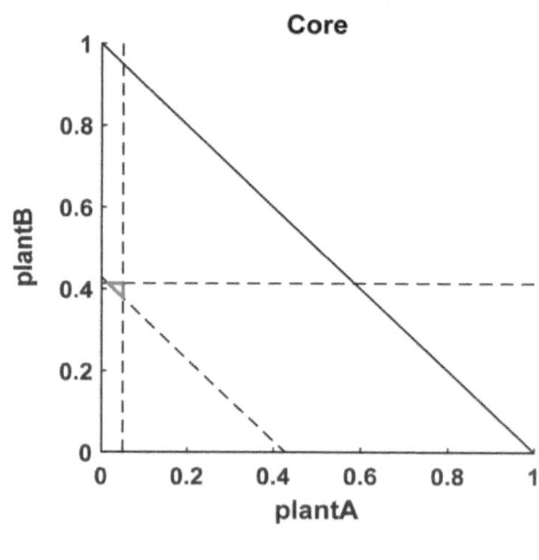

FIGURE 7.14 Core of three-plant cooperative game under Strategy I (Table 7.7).

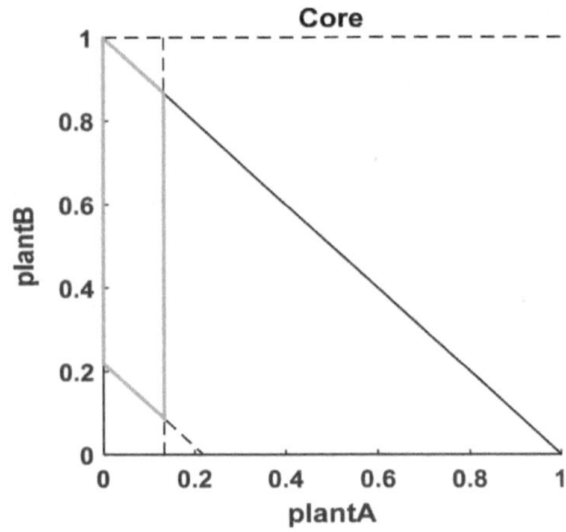

FIGURE 7.15 Core of three-plant cooperative game under Strategy II (Table 7.8).

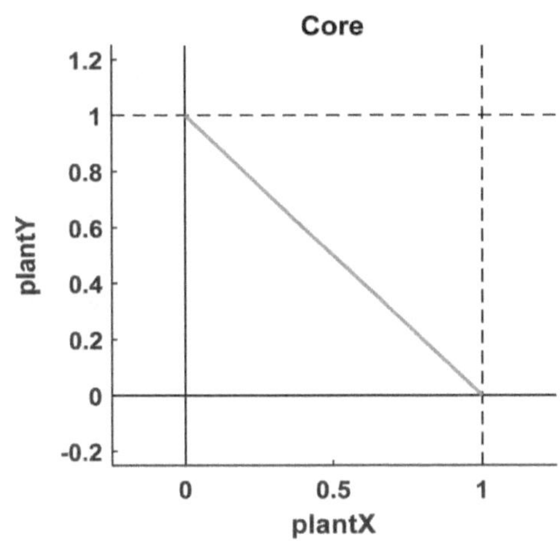

FIGURE 7.16 Core of two-plant cooperative game.

or equal to zero, the corresponding core can always be described with the following constraints

$$dp_{X,\{X,Y\}} \geq 0 \tag{7.74}$$

$$dp_{Y,\{X,Y\}} \geq 0 \tag{7.75}$$

$$dp_{X,\{X,Y\}} + dp_{Y,\{X,Y\}} = v\left(\{X,Y\}\right) \tag{7.76}$$

The corresponding graphical representation is shown in Figure 7.16, and the red line segment denotes the core. Finally, notice that $\{X,Y\} \in \{\{A,B\},\{B,C\},\{C,A\}\}$ in the present example.

7.4.2 Traditional Shapley Values

Shapley (1952) proposed to allocate the overall benefit generated by a coalition to its members according to their average contribution levels. Specifically, the marginal benefit of a member joining the coalition was regarded as its contribution, and the actual allocated amounts have been referred to as Shapley values.

In the present application, the TAC saving realized by the TSWI project is treated as the overall benefit of the coalition. Let us again revisit the example in Section 7.3 for illustration convenience. The formulas for computing the Shapley values for three possible two-player coalitions are shown below:

$$\left(\Phi_{A,\{A,B\}},\Phi_{B,\{A,B\}}\right)=\left(\frac{v(\{A,B\})-v(\{B\})+v(\{A\})}{2},\frac{v(\{A,B\})-v(\{A\})+v(\{B\})}{2}\right) \quad (7.77)$$

$$\left(\Phi_{B,\{B,C\}},\Phi_{C,\{B,C\}}\right)=\left(\frac{v(\{B,C\})-v(\{C\})+v(\{B\})}{2},\frac{v(\{B,C\})-v(\{B\})+v(\{C\})}{2}\right) \quad (7.78)$$

$$\left(\Phi_{C,\{C,A\}},\Phi_{A,\{C,A\}}\right)=\left(\frac{v(\{C,A\})-v(\{A\})+v(\{C\})}{2},\frac{v(\{C,A\})-v(\{C\})+v(\{A\})}{2}\right) \quad (7.79)$$

Since the TAC saving (characteristic function value) of each single-plant "coalition" should be zero, i.e., $v(\{A\})=v(\{B\})=v(\{C\})=0$, the Shapley values in the above three scenarios essentially suggest a 50/50 split of the benefits obtained due to interplant water integration schemes. This outcome essentially ensures the stability of every two-plant coalition according to the core in Figure 7.16. On the other hand, the Shapley values for the three-plant coalition in the illustrative example can be written as

$$\begin{gathered}\left(\phi_{A,\{A,B,C\}},\phi_{B,\{A,B,C\}},\phi_{C,\{A,B,C\}}\right)= \\ \left(\frac{2\times v(\{A,B,C\})+v(\{A,B\})+v(\{A,C\})+2\times v(\{A\})-v(\{B\})-v(\{C\})}{6},\right. \\ \frac{2\times v(\{A,B,C\})+v(\{A,B\})+v(\{B,C\})+2\times v(\{B\})-v(\{A\})-v(\{C\})}{6}, \\ \left.\frac{2\times v(\{A,B,C\})+v(\{A,C\})+v(\{B,C\})+2\times v(\{C\})-v(\{A\})-v(\{B\})}{6}\right)\end{gathered} \quad (7.80)$$

From the TAC savings listed in Tables 7.7 and 7.8 and the above formula, the Shapley values obtained with the two proposed integration strategies are plotted respectively in Figures 7.17 and 7.18 against the backdrops of corresponding cores. It can be

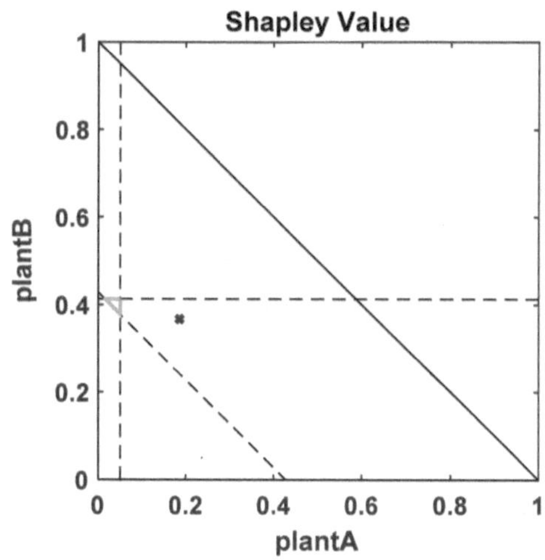

FIGURE 7.17 Point corresponding to Shapley values obtained with Strategy I.

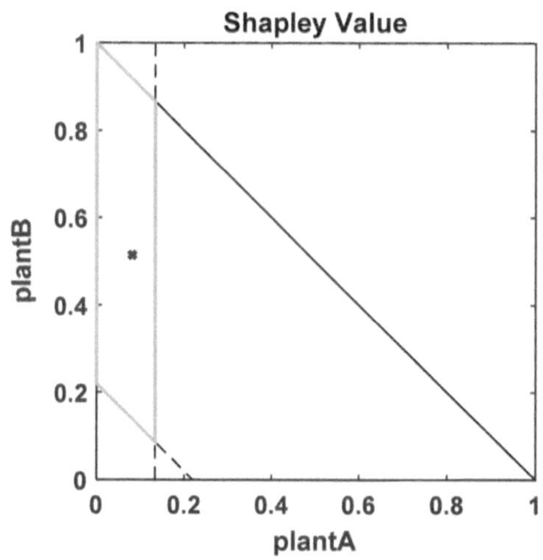

FIGURE 7.18 Point corresponding to Shapley values obtained with Strategy II.

observed from Figure 7.17 that the allocated TAC savings are not located inside the core under Strategy I. This outcome implies that, although the corresponding benefit distribution plan is "fair," the resulting coalition may not be stable. On the other hand, Figure 7.18 shows that the benefit allocation plan drawn up according to Strategy II not only yields a stable coalition but also ensures fair distribution of the total TAC saving.

7.4.3 Risk-Based Shapley Values

As a general rule, the reliability of any engineering system with interconnected components should be treated as an important design issue. If a TSWI scheme can be viewed as a coalition in a cooperative game, then the potential risk of its collapse should be considered seriously. An unplanned plant shutdown, which may be due to a wide variety of equipment failures and/or human errors, halts the water flows to/from some of the water-using units in the MPWN and forces the other plants in coalition, which are still working, to take emergency measures. Thus, it is necessary to adjust the aforementioned conventional Shapley values by assessing penalties for the potential fallouts of the TSWI system. Since the characteristic function $v(S)$ mentioned previously yields the maximum TAC saving of a *functional* MPWN for coalition $S \subseteq N$, the shutdown risk of plant $i \in S$ can be expressed with its unreliability (denoted as α_i). Note that the corresponding failure rate (denoted as λ_i) may be extracted from the historical maintenance data, e.g., see Hoyland and Rausand (1994), which should be viewed as an intrinsic parameter of the given plant. For simplicity, let us assume that the time-to-failure of plant i follows the exponential distribution, i.e.

$$\alpha_i = 1 - e^{-\lambda_i} \tag{7.81}$$

To enumerate all precedence order of members joining the three-player coalition in the illustrative example exhaustively, let us first enumerate the 3! permutations and define the corresponding set as

$$\begin{aligned} \pi_{\{A,B,C\}} &= \{\pi_1, \pi_2, \pi_3, \pi_4, \pi_5, \pi_6\} \\ &= \{(A,B,C),(A,C,B),(B,C,A),(B,A,C),(C,B,A),(C,B,A)\} \end{aligned} \tag{7.82}$$

Let $\pi_\sigma = (X,Y,Z) \in \grave{A}_{\{A,B,C\}}$ and $\sigma = 1,2,\cdots,6$. The sum of marginal TAC saving and expected loss for every plant in each precedence order defined in equation (7.82) can be stored in a vector and expressed as follows:

$$\boldsymbol{h}_{\pi_\sigma} = \begin{bmatrix} h_X & h_Y & h_Z \end{bmatrix} \tag{7.83}$$

$$h_X = \left[v(\{X\}) - v(\varnothing)\right] + E(X:\{X\}) \tag{7.84}$$

$$h_Y = \left[v(\{X,Y\}) - v(\{X\})\right] + E(Y:\{X,Y\}) \tag{7.85}$$

$$h_Z = \left[v(\{X,Y,Z\}) - v(\{X,Y\})\right] + E(Z:\{X,Y,Z\}) \tag{7.86}$$

where $\sigma = 1,2,\ldots,6$; $E\left(Z:\{X,Y,Z\}\right)$ in equation (7.86) can be interpreted as the expected loss of plant Z should the coalition $\{X, Y, Z\}$ break down due to failure(s) of X and/or Y; $E\left(Y:\{X,Y\}\right)$ in equation (7.85) can be interpreted as the expected loss of plant Y should the coalition $\{X, Y\}$ break down due to failure of X; $E\left(X:\{X\}\right)=0$ in equation (7.84) since there are no other members in coalition $\{X\}$. It should be noted that a more general derivation of equations (7.83)–(7.86) is available in Section 5.1 in Chapter 5.

To calculate the expected losses defined above, it is necessary to first determine the unreliability of each plant (α_A, α_B, and α_C) and the financial penalties to the working plants in a coalition if one or more plant fails. For computational simplicity, an approximation method is adopted in all case studies presented in this chapter. Specifically, it is assumed that every interplant water usage that utilizes the water stream from a disabled plant can always be compensated with freshwater from the counterpart plant which is still working. The detailed configurations of resulting water networks can be found in the Appendix.

As mentioned before, sequence $\pi_\sigma \in \pi_{\{A,B,C\}}$ and the entries of corresponding vector $\boldsymbol{h}_{\pi_\sigma}$ in equation (7.83) are arranged accordingly. Since there are six different sequences in $\pi_{\{A,B,C\}}$, it is necessary to first rearrange the entries in every $\boldsymbol{h}_{\pi_\sigma}$ $(\sigma = 1,2,\cdots,6)$ so as to make them follow the same original sequence (A, B, C) and then store them in another vector Θ_{π_σ} for computing the risk-based Shapley values (see Section 5.1 in Chapter 5). Specifically,

$$\Psi_{\{A,B,C\}} = \frac{1}{3!}\sum_{\sigma=1}^{6}\Theta_{\pi_\sigma} \tag{7.87}$$

$$\hat{\Phi}_{\{A,B,C\}} = \frac{\Psi_{\{A,B,C\}}}{\Psi_{\{A,B,C\}}\cdot 1_{3\times1}}\cdot v\left(\{A,B,C\}\right) \tag{7.88}$$

where $\hat{\Phi}_{\{A,B,C\}} = \left[\hat{\phi}_{A,\{A,B,C\}} \quad \hat{\phi}_{B,\{A,B,C\}} \quad \hat{\phi}_{C,\{A,B,C\}}\right]$ and $\hat{\phi}_{A,\{A,B,C\}}$, $\hat{\phi}_{B,\{A,B,C\}}$, and $\hat{\phi}_{C,\{A,B,C\}}$ respectively denote the risk-based Shapley values of plant A, plant B, and plant C (or the TAC cost savings allocated to these plants); $1_{3\times1}$ is a column vector in which all three entries equal one.

To demonstrate the use of risk-based Shapley values in the illustrative example, let us set the failure rates of plants A, B, and C to be 0.5, 0.2, and 0.05 (1/year), respectively. For a time horizon of 10 years, the trajectories of time-dependent risk-based Shapley values are plotted against the backdrops of cores in various scenarios in Figures 7.19–7.25. It can be observed from these figures and Tables 7.7 and 7.8 that TSWI projects almost always yield considerable overall TAC saving (except in the case of coalition $\{A,C\}$ under Strategy II) and a plant that bears a higher risk usually gets a larger share of this saving. Note that $\lambda_A > \lambda_B >> \lambda_C$ in the present example and the above trend can be clearly identified at the end of a ten-year span in

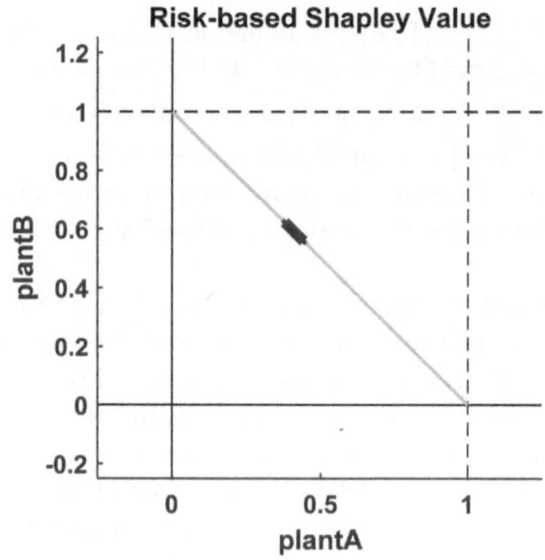

FIGURE 7.19 Trajectory of risk-based Shapley values for coalition {A,B} under Strategy I.

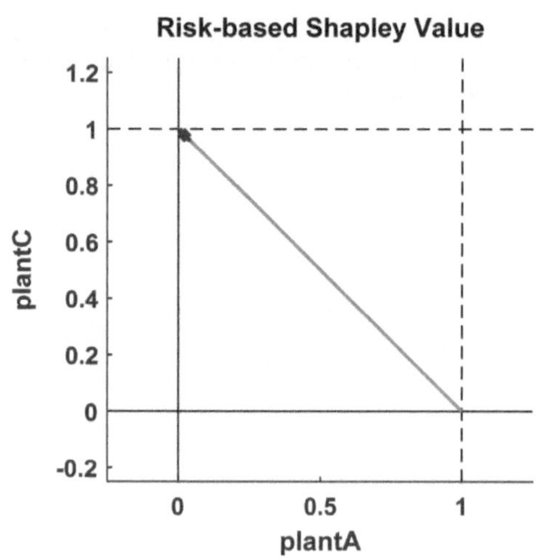

FIGURE 7.20 Trajectory of risk-based Shapley values for coalition {C, A} under Strategy I.

Figures 7.19–7.25. Although, as mentioned previously in Section 7.4.2, the traditional Shapley values in a two-plant coalition suggest a 50/50 split of the benefit obtained due to interplant water integration, almost the entire TAC saving is allocated to plant C in coalition {A,C} under Strategy I according to the risk-based scheme (see Figure 7.20). Finally, despite Tables 7.7 and 7.8 showing that coalition {A,B,C} under Strategy I results in the largest overall TAC saving, Figure 7.22 indicates that this

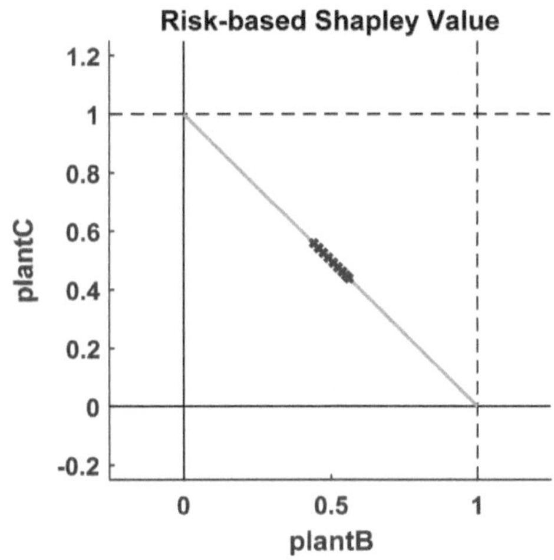

FIGURE 7.21 Trajectory of risk-based Shapley values for coalition {B, C} under Strategy I.

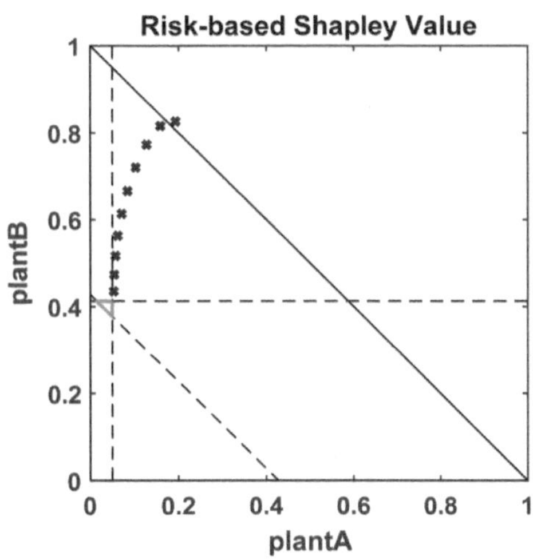

FIGURE 7.22 Trajectory of risk-based Shapley values for coalition {A, B, C} under Strategy I.

coalition is unstable. Thus, coalition {B,C} under Strategy I should be selected as the final candidate for the TSWI project in the illustrative example.

7.5 CONCLUDING REMARKS

A systematic methodology has been developed in this study to produce the distribution scheme of TAC saving realized by any TSWI project. The coalition stability is

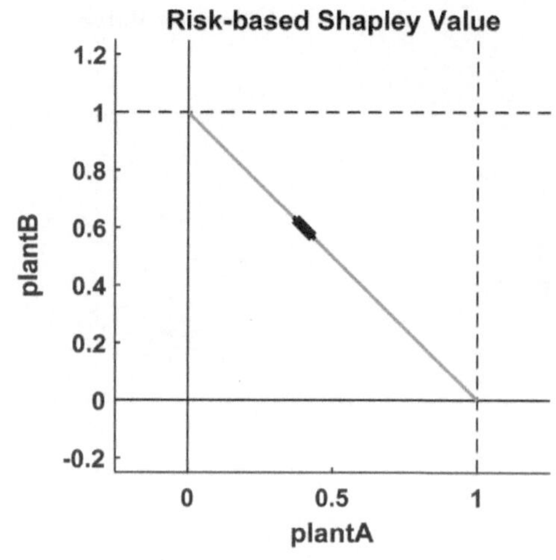

FIGURE 7.23 Trajectory of risk-based Shapley values for coalition $\{A, B\}$ under Strategy II.

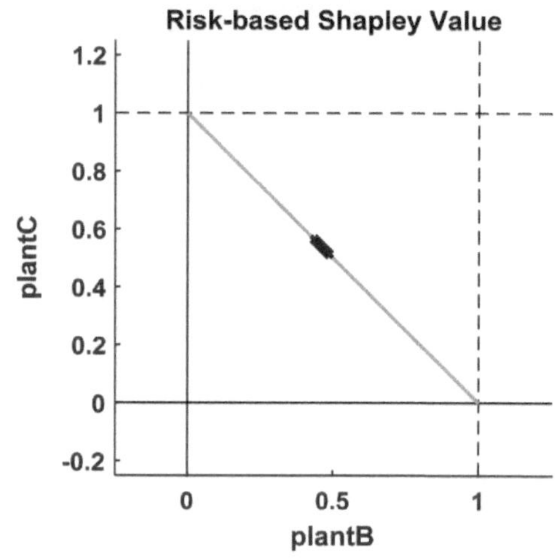

FIGURE 7.24 Trajectory of risk-based Shapley values for coalition $\{B, C\}$ under Strategy II.

ensured on the basis of the corresponding core, while the fairness of benefit allocation plan can be facilitated according to the conventional Shapley values. If there is a need to assess the impacts of unexpected plant shutdowns in a coalition, the time-dependent risk-based Shapley values can be adopted for use as an evaluation tool. Finally, it should be noted that, although the illustration example presented is concerned with a single-contaminant system, this same approach has been applied successfully to the multi-contaminant counterparts (Liu, 2019). These case studies are omitted for the sake of brevity.

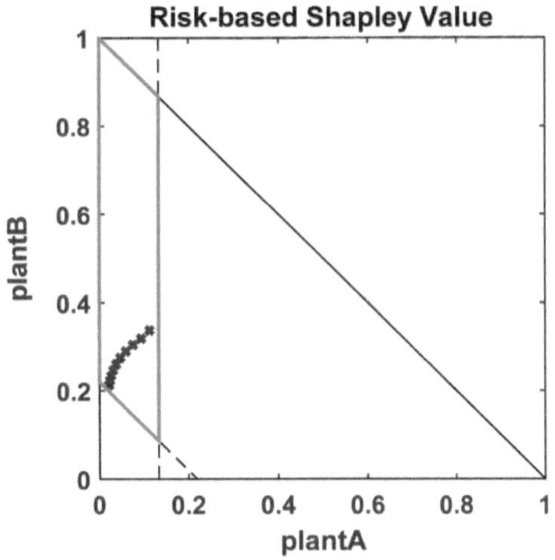

FIGURE 7.25 Trajectory of risk-based Shapley values for coalition {*A*, *B*, *C*} under Strategy II.

REFERENCES

Chen, C.L., Hung, S.W., Lee, J.Y. 2010. Design of inter-plant water network with central and decentralized water mains. *Computers and Chemical Engineering* 34(9): 1522–1531.

Chew, I.M.L., Tan, R.R., Foo, D.C.Y., Chiu, A.S.F. 2009. Game theory approach to the analysis of inter-plant water integration in an eco-industrial park. *Journal of Cleaner Production* 17(18): 1611–1619.

Chew, I.M.L., Tan, R.R., Ng, D.K.S., Foo, D.C.Y., Majozi, T., Gouws, J. 2008. Synthesis of direct and indirect interplant water network. *Industrial and Engineering Chemistry Research* 47(23): 9485–9496.

Gunaratnam, M., Alva-Argáez, A., Kokossis, A., Kim, J.K., Smith, R. 2005. Automated design of total water systems. *Industrial and Engineering Chemistry Research* 44(3): 588–599.

Hiete, M., Ludwig, J., Schultmann, F. 2012. Intercompany energy integration – adaptation of thermal pinch analysis and allocation of savings. *Journal of Industrial Ecology* 16(5): 689–698.

Hoyland, A., Rausand, M. 1994. *System Reliability Theory – Models and Statistical Method.* New York: John Wiley & Sons, Inc.

Huang, C.H., Chang, C.T., Ling, H.C., Chang, C.C. 1999. A mathematical programming model for water usage and treatment network design. *Industrial and Engineering Chemistry Research* 38(7): 2666–2679.

Jin, Y., Chang, C.T., Li, S., Jiang, D. 2018. On the use of risk-based Shapley values for cost sharing in interplant heat integration programs. *Applied Energy* 211: 904–920.

Liu, W.T. 2019. *On the Application of Cooperative Game Theory for Interplant Water Indirect Integration.* MS Thesis, NCKU, Tainan, Taiwan.

Shapley, L.S. 1952. *A Value for n-Person Games.* Santa Monica, CA: Rand Corp.

Wang, B., Feng, X., Zhang, Z. 2003. A design methodology for multiple-contaminant water networks with single internal water main. *Computers and Chemical Engineering* 27(7): 903–911.

APPENDIX: DEFECTIVE MPWN DESIGNS IN ILLUSTRATIVE EXAMPLE

Defective MPWN designs obtained in all scenarios considered in the illustrative example are shown in this appendix (Figures A.1–A.9). Notice that in all figures presented here are three types of connecting lines:

- The blue dashed lines are used to represent extra pipelines introduced to cope with the unexpected failure(s) of the other plant(s) in the coalition.
- The red solid lines denote the originally existing pipelines which are also used in the defective MPWNs.
- The black solid lines denote the originally existing pipelines which are not used in the defective MPWNs.

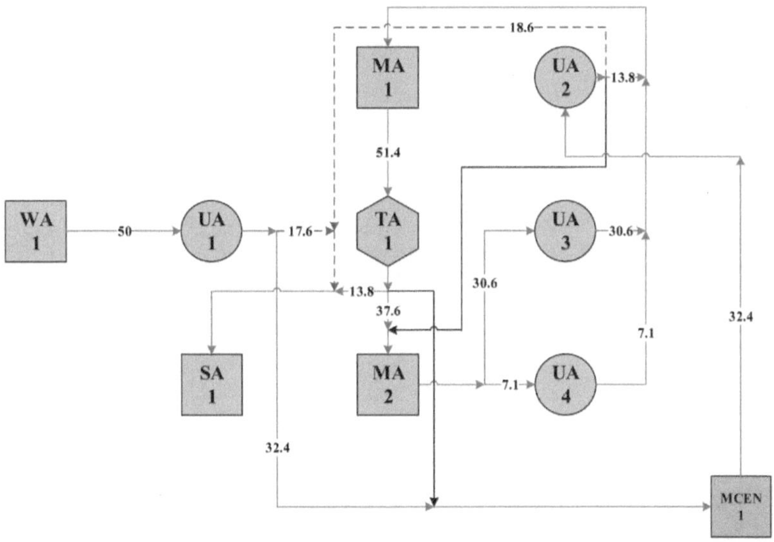

(a)

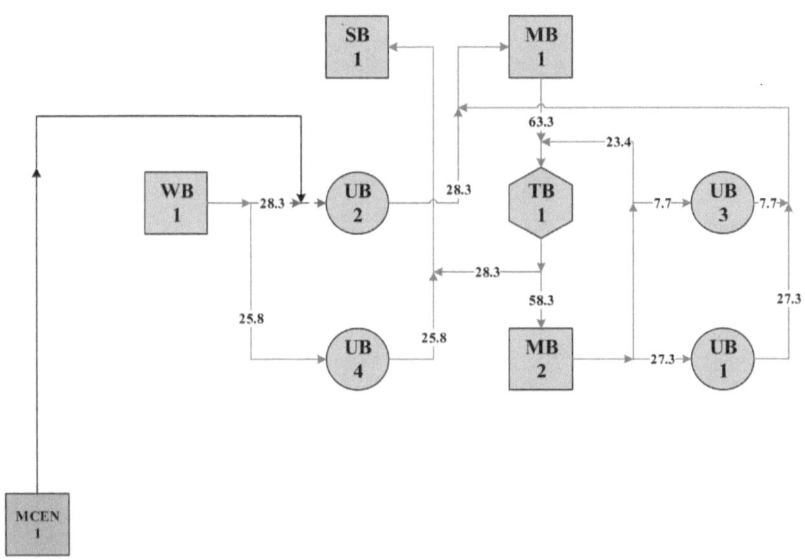

(b)

FIGURE A.1 Defective MPWNs of coalition $\{A, B\}$ using Strategy I. (a) Plant A runs alone after plant B shuts down; (b) plant B runs alone after plant A shuts down.

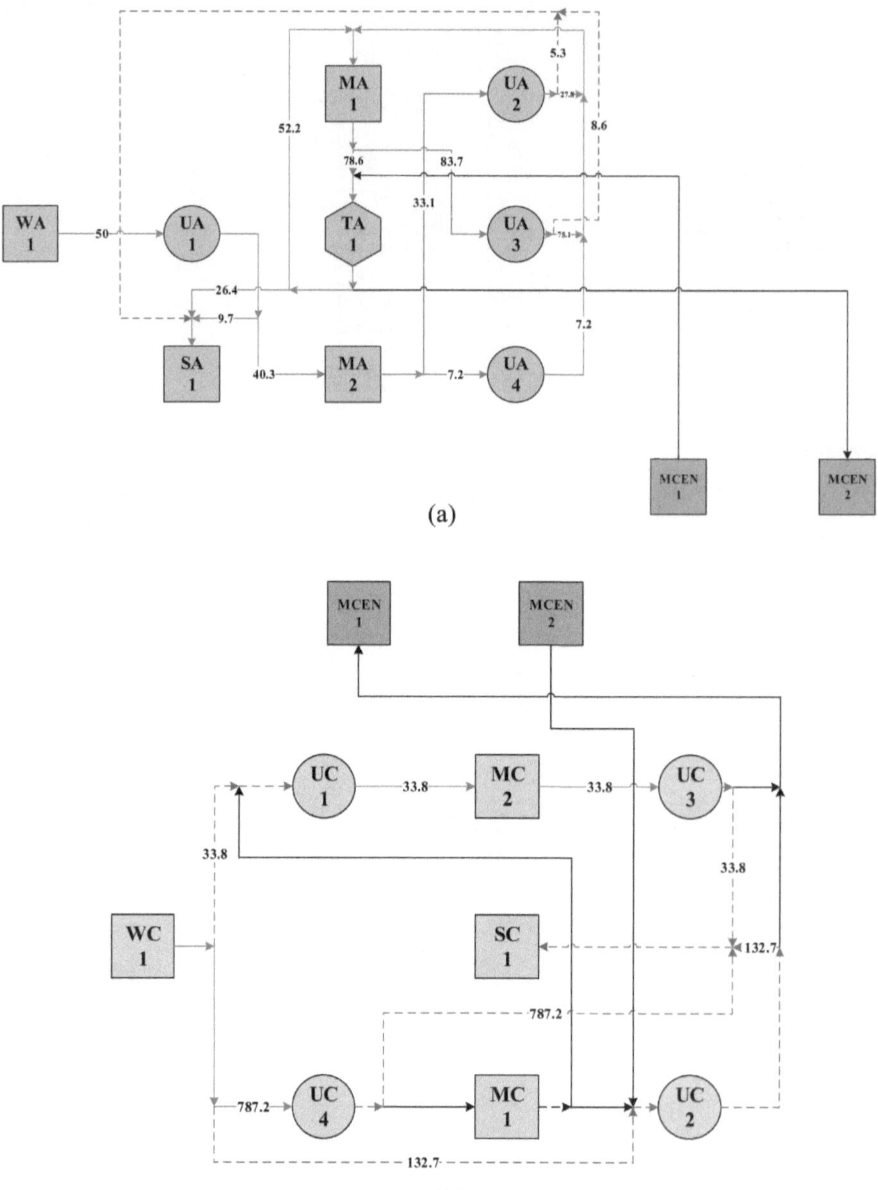

FIGURE A.2 Defective MPWNs of coalition $\{C, A\}$ using Strategy I. (a) Plant A runs alone after plant C shuts down; (b) plant C runs alone after plant A shuts down.

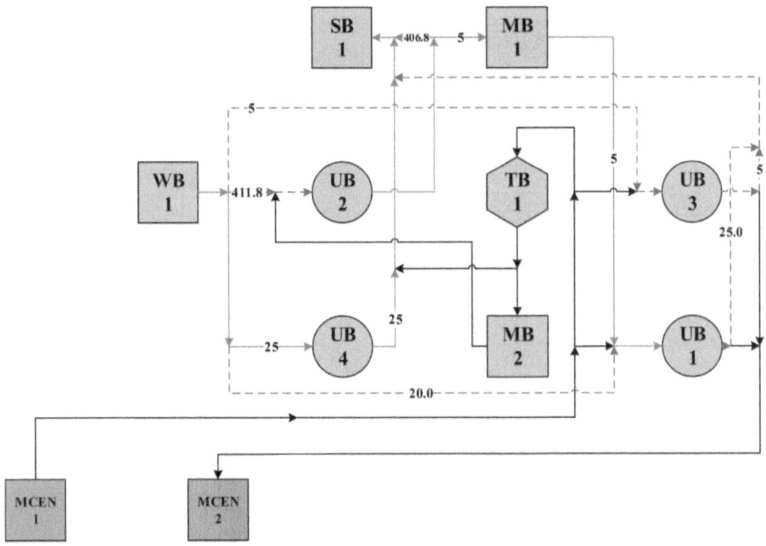

(a)

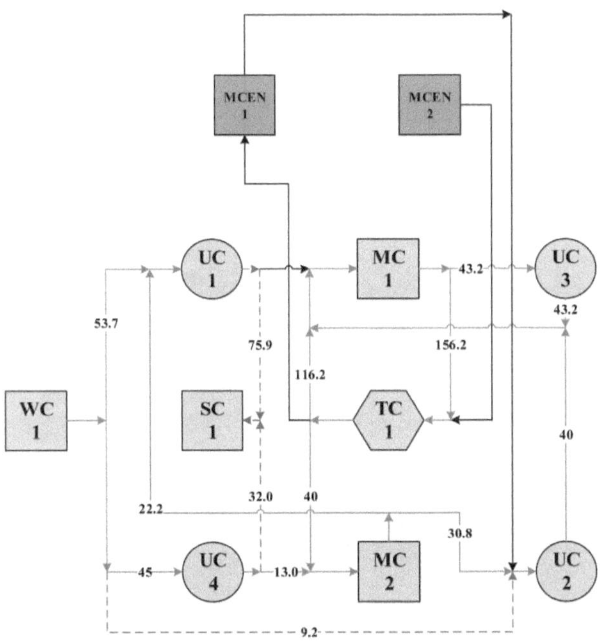

(b)

FIGURE A.3 Defective MPWNs of coalition {B, C} using Strategy I. (a) Plant B runs alone after plant C shuts down; (b) plant C runs alone after plant B shuts down.

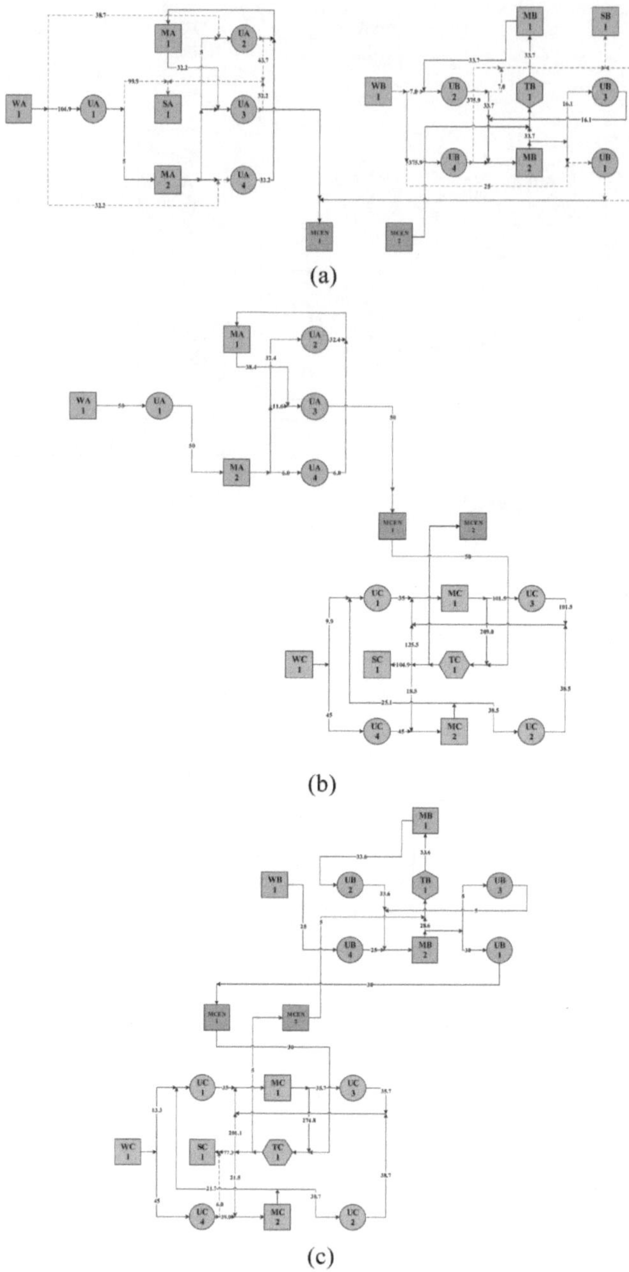

FIGURE A.4 Defective MPWNs of coalition {*A*, *B*, *C*} using Strategy I. (a) Plants A and B run together after plant C shuts down; (b) plants C and A run together after plant B shuts down; (c) plants B and C run together after plant A shuts down.

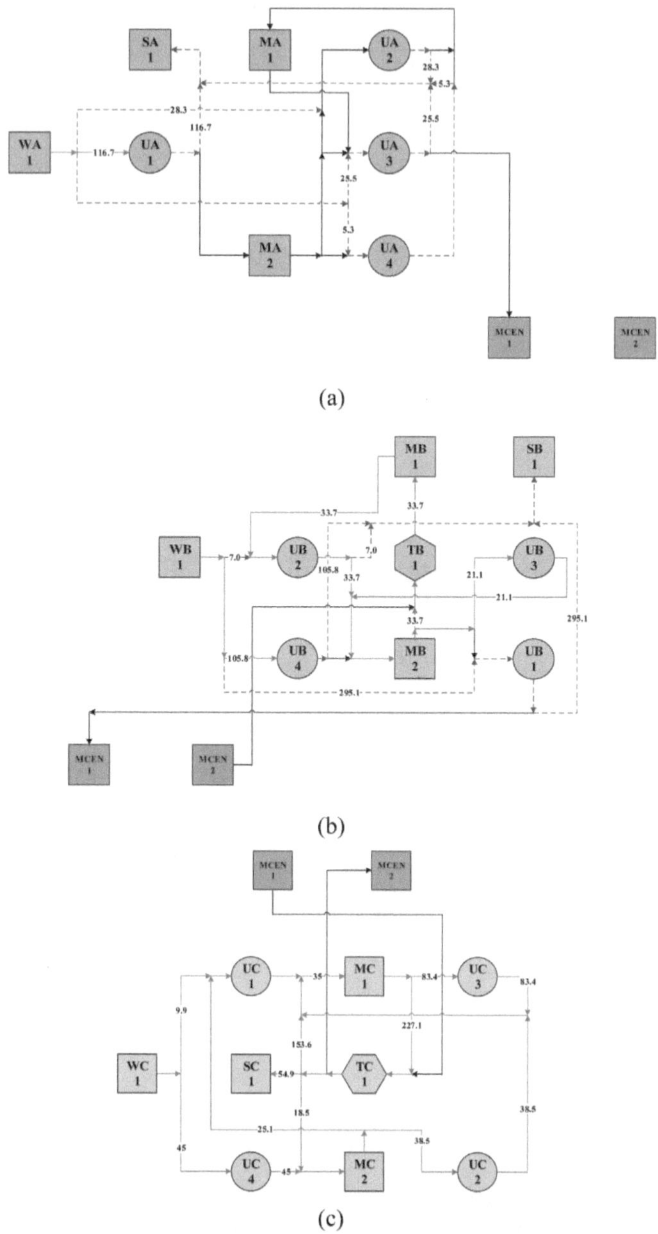

FIGURE A.5 Defective MPWNs of coalition {*A*, *B*, *C*} using Strategy I. (a) Plant A runs alone after plants B and C both shut down; (b) plant B runs alone after plants C and A both shut down; (c) plant C runs alone after plants A and B both shut down.

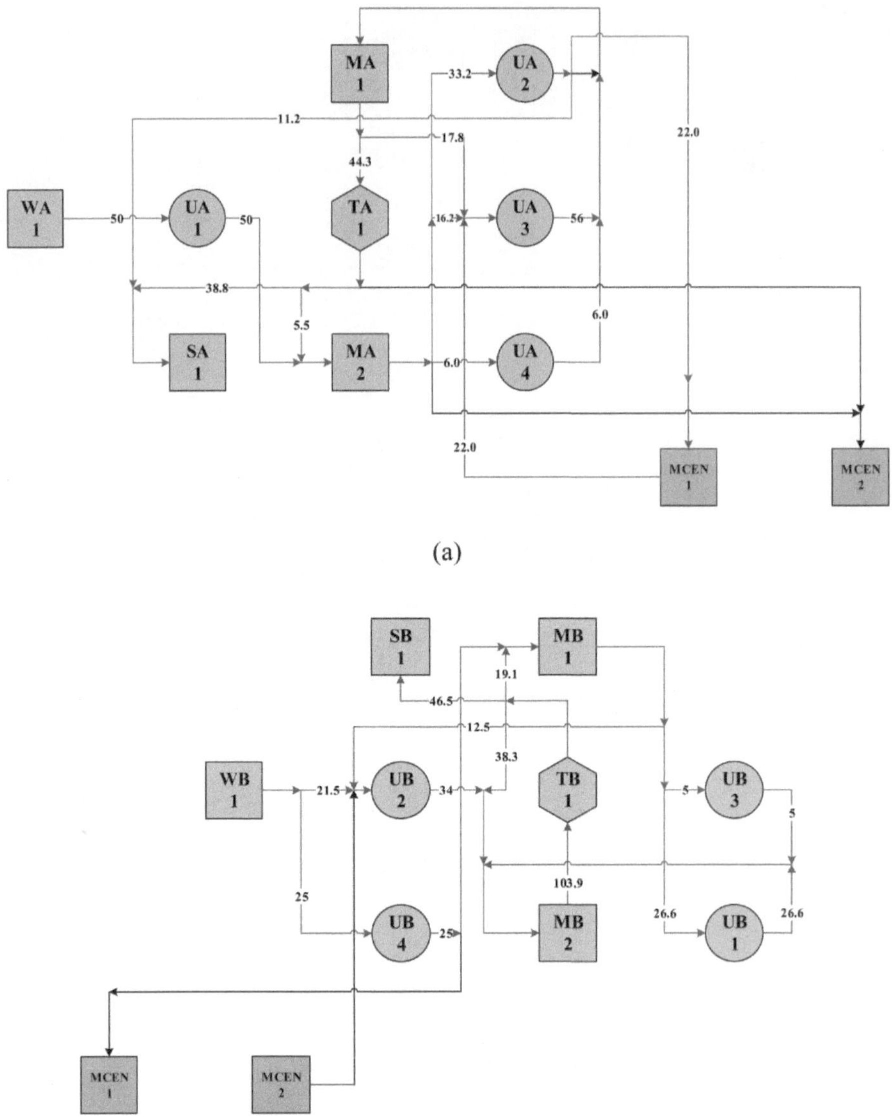

FIGURE A.6 Defective MPWNs of coalition $\{A, B\}$ using Strategy II. (a) Plant A runs alone after plant B shuts down; (b) plant B runs alone after plant A shuts down.

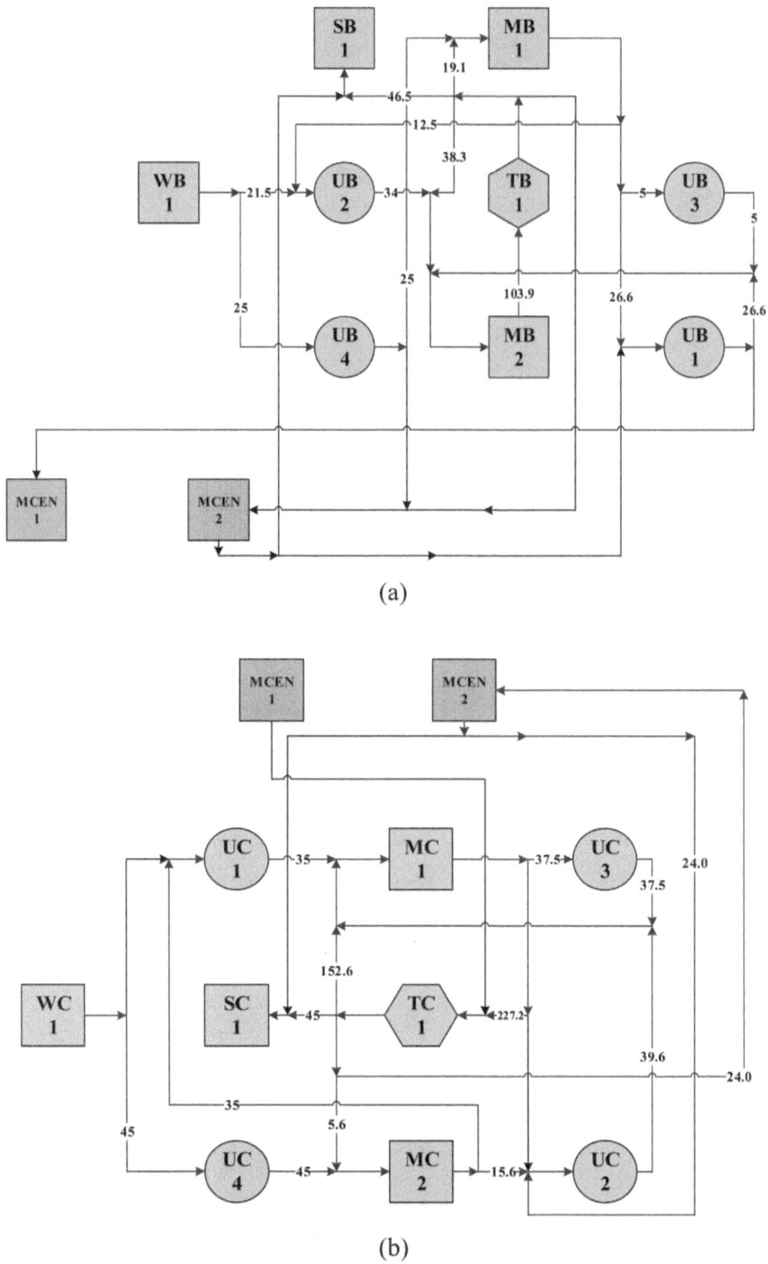

FIGURE A.7 Defective MPWNs of coalition {*B*, *C*} using Strategy II. (a) Plant B runs alone after plant C shuts down; (b) plant C runs alone after plant B shuts down.

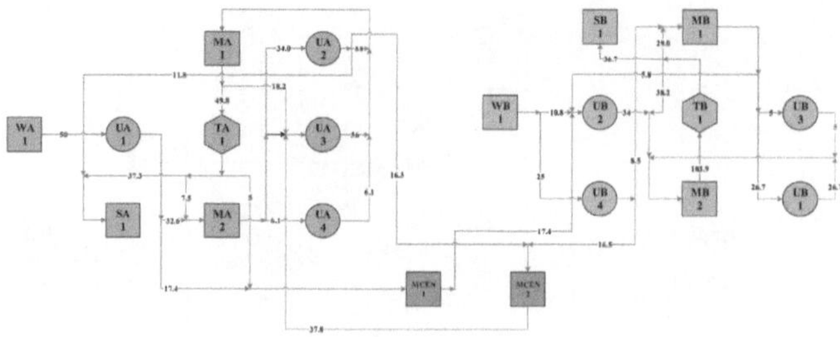

(a)

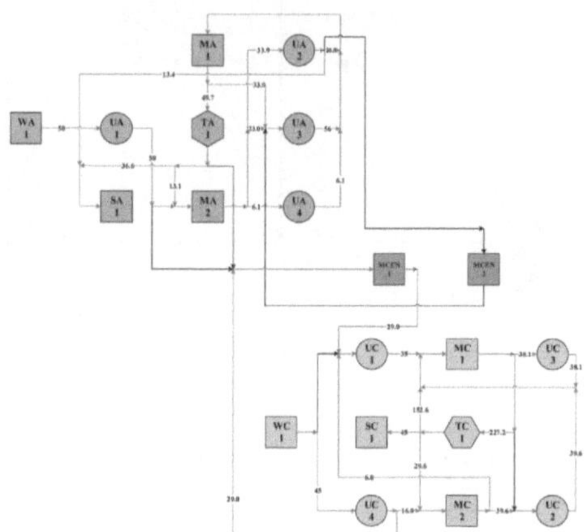

(b)

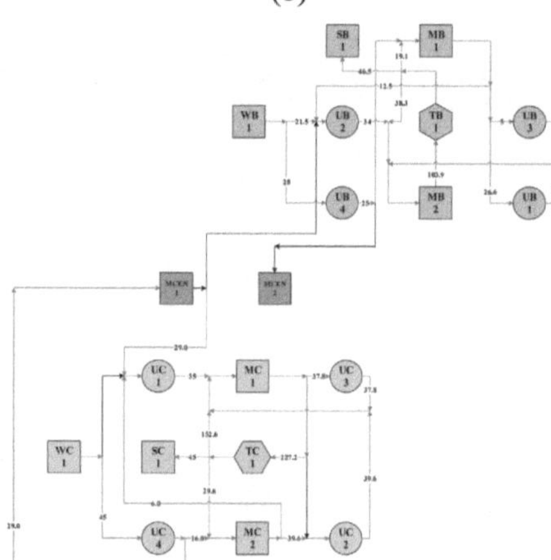

(c)

FIGURE A.8 Defective MPWNs of coalition {*A*, *B*, *C*} using Strategy II. (a) Plants A and B run together after plant C shuts down; (b) plants C and A run together after plant B shuts down; (c) plants B and C run together after plant A shuts down.

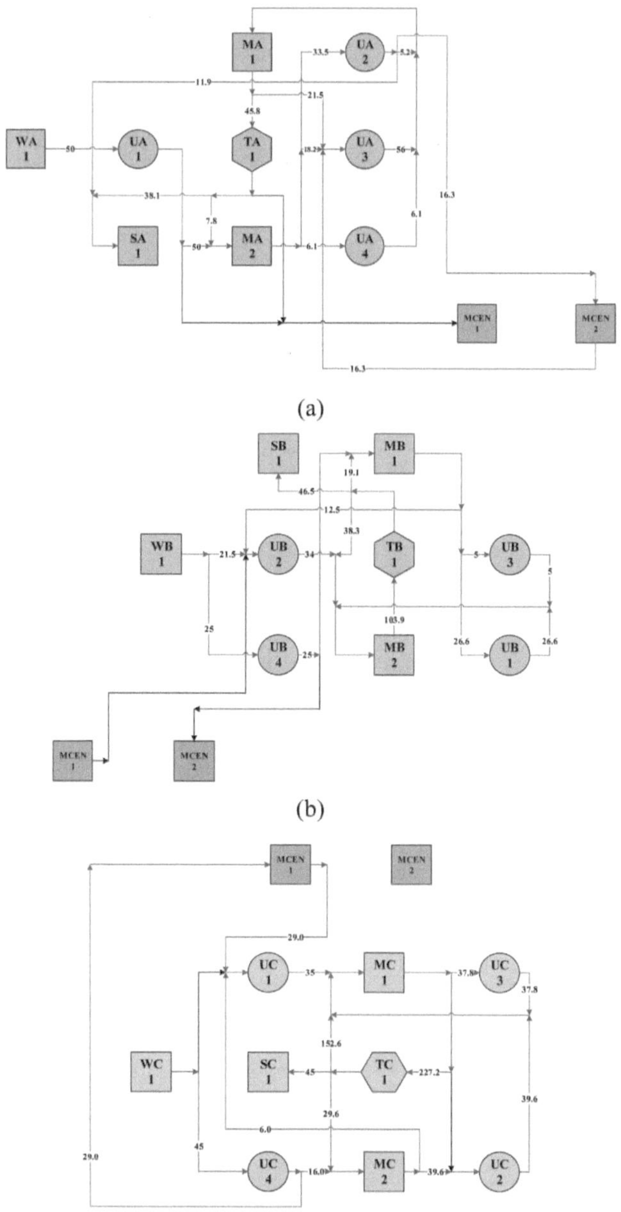

FIGURE A.9 Defective MPWNs of coalition {*A*, *B*, *C*} using Strategy II. (a) Plant A runs alone after plants B and C both shut down; (b) plant B runs alone after plants C and A both shut down; (c) plant C runs alone after plants A and B both shut down.

8 A Model-Based Method for Planning and Scheduling of Petroleum Supply Chain

The general goal of supply chain management is to create an efficient coordination framework to facilitate the suppliers, manufacturers, distributors, and retailers in the supply network to work closely together to acquire raw materials, to convert them into final products, and to deliver these products to retailers (Beamon, 1998; Tan, 2001; Chopra and Meindl, 2004). A petroleum supply chain can be roughly divided into four segments: (1) exploration, (2) transportation, (3) refining, and (4) distribution. Notice that the first segment of this chain may not exist in regions lacking oil resources. Crude oils are shipped directly from oversea sources by tankers to the oil terminals, which are then connected to refineries through a pipeline network. Logistic decisions at this level include supply planning, scheduling, and selection of the appropriate transportation modes. Crude oil is transformed into various products at the refineries. These refineries are often interconnected so as to achieve a high degree of flexibility for meeting fluctuating market demands. Products produced at the refineries are sent to the distribution centers, and then transported via pipelines, trucks, or rail cars to the customers. It should be noted that some of the aforementioned refining products are actually raw materials of a variety of other processes. For example, benzene can be alkylated to ethylbenzene and cumene, which are then used as the raw materials for the production of styrene and phenol, respectively. In other words, the petroleum supply chain may be further extended downstream. To keep the problem scope to a manageable size, the petrochemical processes discussed in this chapter are limited to those in a typical *conversion refinery* (Speight and Ozum, 2002).

A systematic approach is proposed in this chapter to create the production planning and scheduling schemes for the refinery segment of a petroleum supply chain. The planning and scheduling activities are in fact hierarchically linked. Both involve the allocation of resources over time to manufacture the required products. In the planning stage, the goal is to settle the higher-level decisions, such as the procurement amounts of crude oils and the inventory and production levels of various products, according to the given forecasts of market demands over a relatively long time horizon (say, months). The lower-level scheduling tasks, on the other hand, should be performed on the basis of shorter time intervals (say, weeks) to determine the specific timing in which every unit must be operated to meet the production goals set in the planning stage. However, if a sequential approach is taken to carry out the needed computation procedures, the detailed scheduling constraints are often ignored in the planning stage. As a result, there is really no guarantee that a feasible schedule can be obtained. It is thus desirable to address the planning and scheduling issues simultaneously.

This viewpoint has actually been adopted in several previous studies for different types of supply chains. Jackson and Grossmann (2003) adopted the Lagrangean decomposition (spatial and temporal) techniques for the solution of a non-linear programming (NLP) problem that models multisite production planning in a chemical company. Tang et al. (2001) proposed an integrated system to achieve the same goal for steel production planning and scheduling. Ho and Chang (2001) presented a computation framework to address related issues in the multistage manufacturing system by incorporating the concepts of material requirements planning (MRP) and just-in-time (JIT) production. Dogan and Grossmann (2006) constructed a multiperiod MILP model to integrate the planning and scheduling decisions for continuous multi-product plants that consist of a single processing unit. It should be pointed out that, although interesting results have been reported in the above papers, none of them were concerned specifically with the operation of conversion refineries.

The available planning and scheduling strategies have traditionally been developed *separately* for the petrochemical processes. A number of LP-based commercial software packages are available for generating production plans in the refineries, e.g., Refinery and Petrochemical Modeling System (RPMS) (Bonner and Moore, 1979) and Process Industry Modeling System (PIMS) (Bechtel, 1993). There are, however, relatively few generic scheduling tools on the market, and the existing ones do not allow rigorous representation of the plant particularities. Notice also that the issues concerning optimal planning and scheduling of the *continuous* multi-product petrochemical plants have not been studied extensively, as opposed to the large number of works devoted to batch manufacturing processes (Bonfill et al., 2004). A brief review of the related literature is given in the sequel: Pinto et al. (2000) and Pinto and Moro (2000) have studied the refinery operations in detail. The former work focused primarily on production scheduling for several specific areas in a refinery, whereas the latter developed a nonlinear production planning model for the same purpose. Gothe-Lundgren et al. (2002) studied the production planning and scheduling problem in an oil refinery company. Their scope concerned a production process with one distillation unit and two hydro-treatment units. Más and Pinto (2003) and Magalhães and Shah (2003) proposed a detailed MILP formulation for the optimal scheduling of an oil supply chain, which includes tankers, piers, storage tanks, substations, and refineries. The study of Jia and Ierapetritou (2004) focused on the short-term scheduling issues associated with refinery operations. Neiro and Pinto (2004) presented a comprehensive planning/scheduling framework for the front end of a petroleum supply chain. Li et al. (2005) integrated the CDU, FCC, and product blending models into a refinery planning model. Persson and Gothe-Lundgren (2005) proposed a mathematical programming model and the corresponding solution method for a shipment planning problem. Carvalho and Pinto (2006) presented an optimization model for planning the production infrastructure of offshore oilfields. Méndez et al. (2006) addressed the simultaneous optimization issues associated with off-line blending and scheduling problems in oil-refinery applications.

It should be pointed out that the scopes adopted in the aforementioned works are *not* comprehensive enough for realistic applications. This is due to the fact that a wide spectrum of important petrochemical products, e.g., LPG, gasoline, kerosene, ethylene, propylene, butadiene, benzene, toluene, and xylenes, etc., are produced in a conversion refinery and these previous studies only address the planning and/or scheduling issues concerned with a subset of these materials. To circumvent the drawbacks of aforementioned studies, a single integrated MILP model is presented in the present chapter to coordinate various planning and scheduling tasks for optimizing the supply-chain performance of a complete refining process. All critical decisions can be obtained from the optimal solution of the proposed model. On the corporate level, this model can be used to determine which crude to purchase and its quantity, which products to produce and their amounts, and which processing route to follow during each planning period over a given time horizon. On the plant level, the detailed operating conditions of each processing unit, e.g., temperatures, pressures, and throughputs, can be computed according to the higher-level decisions and the inherent process constraints and, in addition, the product transportation policies can be developed by considering the available storage spaces. More specifically, the appropriate production levels and distribution schedules for fuels, e.g., LPG, gasoline, kerosene, diesel, and jet fuel, and other intermediate petrochemical products, e.g., ethylene, propylene, butadiene, benzene, toluene, p-xylene, and o-xylene, can be determined at the shorter scheduling intervals.

8.1 PRODUCTION UNITS IN CONVERSION REFINERIES

The petroleum refining process can be regarded as a series of manufacturing steps by which the crude oil is converted into saleable products with the desired qualities and in quantities dictated by the market. A refinery is essentially a group of processing and storage facilities interconnected for the purpose of realizing this process. The refining units and their connection structure are selected to accommodate a wide variety of product distributions. The most versatile configuration is known as the *conversion refinery* (Speight and Ozum, 2002). A simplified flow diagram is shown in Figure 8.1. It should be noted that approximately two-thirds of the outputs generated from a modern conversion refinery are in the forms of unleaded gasoline, high jet fuel, liquefied gas (LPG), low-sulfur diesel fuel, lubricants, and petrochemical intermediates, i.e., ethylene, propylene, butadiene, benzene, toluene, and xylenes. The basic production units considered in this study are the atmospheric distillation units, the vacuum distillation units, the butadiene extraction units, the aromatics extraction units, the xylene fractionation units, and the parex units, naphtha crackers, cokers, fluidized catalytic cracking units, hydrotreaters, reforming units, the isomar units, the tatory units, the import and export facilities, and storage and transportation facilities (Watkins, 1979; Franck and Stadelhofer, 1988; Perry et al., 1997; Maples, 2000; Gary and Handwerk, 2001).

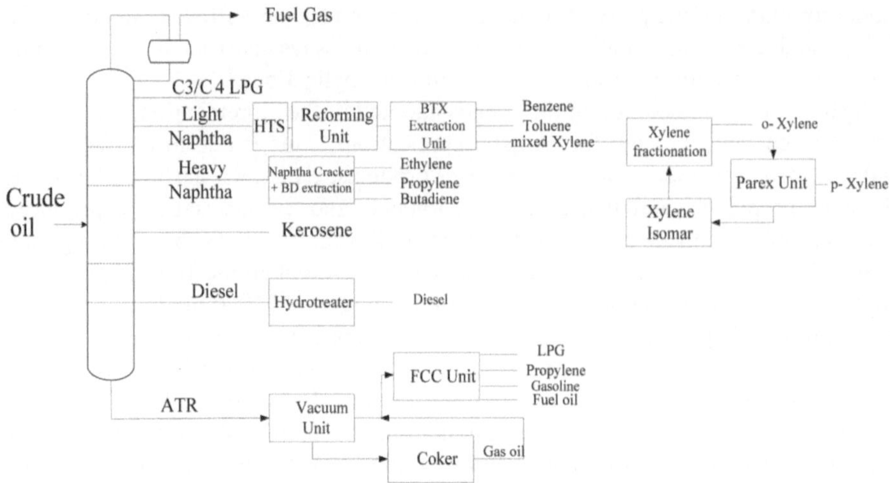

FIGURE 8.1 The simplified flow diagram of a typical conversion refinery. (Reproduced with permission from *Ind. Eng. Chem. Res.* 2008, 47: 1935–1954. Copyright 2008 American Chemical Society.)

8.2 BASIC UNIT MODELS

The processing units in the conversion refinery can be classified into three general types. They are referred to in this chapter as the reaction processes, the separation processes, and the storage processes. The naphtha crackers, the fluidized catalytic cracking units, the hydrotreaters, the cokers, the reforming units, the isomar units, and the tatory units are considered as the reaction processes. The separation processes are those in which only separators are present, i.e., the atmospheric distillation units, the vacuum distillation units, the butadiene extraction units, the aromatics extraction units, the xylene fractionation units, and the parex units. Detailed descriptions of these three basic models for planning purposes can be found in a paper published by Kuo and Chang (2008a). In a later study published by the same authors (2008b), both planning and scheduling issues were addressed simultaneously in an integrated mathematical program. Two different time scales were thus defined for these two tasks respectively. Specifically, two distinct sets of time intervals can be adopted, i.e.

$$\mathbf{TP} = \{tp \mid tp \text{ is the label of the } tp\text{th planning period}\} \tag{8.1}$$

$$\mathbf{TS}_{tp} = \{ts \mid ts \text{ is the label of the } ts\text{th scheduling interval in the } tp\text{th planning period}\} \tag{8.2}$$

The aforementioned unit models in Kuo and Chang (2008a) were modified accordingly for use in the integrated program. The specific mathematical formulations are presented below:

8.2.1 Reaction Processes

The naphtha crackers, cokers, fluidized catalytic cracking units, hydrotreaters, reforming units, the isomar units, and the tatory units are considered as the reaction processes. A sketch of the generalized process flow diagram is provided in Figure 8.2. It should be noted that a separation system is always included in the reaction process for the purpose of removing products and by-products from the unreacted raw materials. It is assumed that, after the catalytic reaction(s), the reactants can always be recovered and then recycled *completely*. To simplify the mathematical program, only the overall mass balances of the entire reaction process are considered in this study. The reaction yields of products and by-products are assumed to be dependent upon the feedstock compositions and operation modes. However, all of them are regarded as available model parameters. The generalized material-balance equation of the reaction processes can thus be written as

$$qp_{u,p,ts} = \sum_{s \in \mathbf{F}_u} qf_{u,s,ts} \left(\sum_{k \in \mathbf{K}_{u,s}} fi_{u,s,k,ts} YD_{u,s,k,p} \right) \tag{8.3}$$

$$\forall u \in \mathbf{UA}, \forall p \in \mathbf{P}_u, \forall ts \in \mathbf{TS}_{tp}, \forall tp \in \mathbf{TP}.$$

where **UA** is the union of all reaction sets, i.e., $\mathbf{UA} = \mathbf{U}_{nc} \cup \mathbf{U}_{cok} \cup \mathbf{U}_{fcc} \cup \mathbf{U}_{ht} \cup \mathbf{U}_{ref} \cup \mathbf{U}_{isomar} \cup \mathbf{U}_{tatory}$; $\mathbf{F}_u$ is the set of all allowable feeds of unit u; $\mathbf{P}_u$ is the set of all products (and by-products) of unit u; **TP** is the set of all planning periods; $\mathbf{K}_{u,s}$ is the set of all operation modes of unit u for processing feedstock s; $qf_{u,s,ts}$ and $qp_{u,p,ts}$ denote the process variables representing respectively the total amounts of consumed

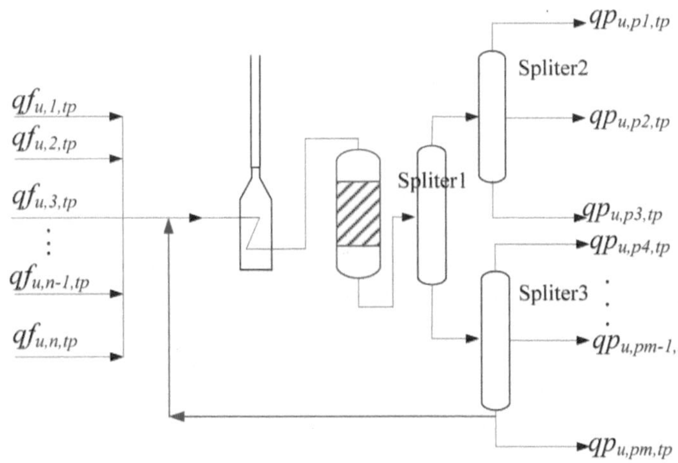

FIGURE 8.2 The generalized reaction process. (Reproduced with permission from *Ind. Eng. Chem. Res.* 2008, 47: 1935–1954. Copyright 2008 American Chemical Society.)

feedstock s and produced product p of unit u in scheduling interval ts; $fi_{u,s,k,ts}$ is the binary variable reflecting whether or not operation mode k of unit u is chosen to process feedstock s during scheduling interval ts; $YD_{u,s,k,p}$ is a given parameter representing the reaction yield of product p from feedstock s with operation mode k of unit u. To simplify the model formulation, let us next introduce a new variable

$$qfi_{u,s,k,ts} = fi_{u,s,k,ts} qf_{u,s,ts} \tag{8.4}$$

Thus, the generalized material balance in equation (8.3) can be written in a *linear* form as

$$qp_{u,p,ts} = \sum_{s \in \mathbf{F}_u} \sum_{k \in \mathbf{K}_{u,s}} qfi_{u,s,k,ts} YD_{u,s,k,p} \tag{8.5}$$

$$\forall u \in \mathbf{UA}, \forall p \in \mathbf{P}_u, \forall ts \in \mathbf{TS}_{tp}, \forall tp \in \mathbf{TP}.$$

It is also assumed in this study that, other than the tatory unit, only one input is allowed to be processed and only one operation mode can be adopted in every reaction unit during any scheduling interval. This feature is stipulated with the following constraints:

$$\sum_{s \in \mathbf{F}_u} \sum_{k \in \mathbf{K}_{u,s}} fi_{u,s,k,ts} = l_{u,ts} \tag{8.6}$$

$$\forall u \in \mathbf{UA}', \forall ts \in \mathbf{TS}_{tp}, \forall tp \in \mathbf{TP}.$$

where $\mathbf{UA}' = \mathbf{U}_{nc} \cup \mathbf{U}_{cok} \cup \mathbf{U}_{fcc} \cup \mathbf{U}_{ht} \cup \mathbf{U}_{ref} \cup \mathbf{U}_{isomar}$; $l_{u,ts} \in \{0,1\}$. Since the capacities of these processing units are finite, the upper and lower bounds of the throughputs in each scheduling interval must also be imposed, i.e.,

$$QFA^{L}_{u,s,k} fi_{u,s,k,ts} \le qfi_{u,s,k,ts} \le QFA^{U}_{u,s,k} fi_{u,s,k,ts} \tag{8.7}$$

$$\forall u \in \mathbf{UA}', \forall s \in \mathbf{F}_u, \forall k \in \mathbf{K}_{u,s}, \forall ts \in \mathbf{TS}_{tp}, \forall tp \in \mathbf{TP}.$$

where $QFA^{U}_{u,s,k}$ and $QFA^{L}_{u,s,k}$ denote respectively the maximum and minimum allowable throughputs of reaction unit u for feedstock s under operation mode k in every scheduling interval.

On the other hand, the decision to use one or more tatory units in a petroleum supply chain must be made on the basis of the market prices of its products (i.e., benzene and xylenes) and costs of its feeds (i.e., toluene and C_9-aromatics). It should also be noted that more than one feedstock is allowed to be processed in this unit. Consequently,

$$\sum_{k \in \mathbf{K}_{u,s}} fi_{u,s,k,ts} \le l_{u,ts} \tag{8.8}$$

$$\sum_{s \in \mathbf{F}_u} \sum_{k \in \mathbf{K}_{u,s}} fi_{u,s,k,ts} \geq l_{u,ts} \tag{8.9}$$

$$\forall u \in \mathbf{U}_{tatory}, \forall s \in \mathbf{F}_u, \forall ts \in \mathbf{TS}_{tp}, \forall tp \in \mathbf{TP}$$

Since every operation mode of any given unit can be used to process all its feeds, one can consider all the corresponding sets of operation modes are the same, i.e.

$$\mathbf{KT}_u = \mathbf{K}_{u,s} \tag{8.10}$$

$$\forall u \in \mathbf{U}_{tatory}, \forall s \in \mathbf{F}_u.$$

Also, since only one of the operation modes in $\mathbf{KT}_u$ can be activated, the following constraints must be incorporated in the mathematical model:

$$fi_{u,s,k,ts} = fi_{u,s',k,ts} \tag{8.11}$$

$$\forall u \in \mathbf{U}_{tatory}, \forall s, s' \in \mathbf{F}_u, \forall k \in \mathbf{KT}_u, \forall ts \in \mathbf{TS}_{tp}, \forall tp \in \mathbf{TP}$$

where $s \neq s'$. Furthermore, the flow ratio between any two feeds of a tatory unit is usually fixed under a selected operation mode, i.e.

$$RT_{u,s,k} qfi_{u,s,k,ts} = RT_{u,s',k} qfi_{u,s',k,ts} \tag{8.12}$$

$$\forall u \in \mathbf{U}_{tatory}, \forall s, s' \in \mathbf{F}_u, \forall k \in \mathbf{KT}_u, \forall ts \in \mathbf{TS}_{tp}, \forall tp \in \mathbf{TP}$$

where $RT_{u,s,k}$ and $RT_{u,s',k}$ are two constants. The throughput limits in this case should be written as

$$QFT_u^L l_{u,ts} \leq \sum_{s \in \mathbf{F}_u} \sum_{k \in \mathbf{K}_{u,s}} qfi_{u,s,k,ts} \leq QFT_u^U l_{u,ts} \tag{8.13}$$

$$u \in \mathbf{U}_{tatory}, \forall ts \in \mathbf{TS}_{tp}, \forall tp \in \mathbf{TP}$$

where QFT_u^U and QFT_u^L denote respectively the maximum and minimum allowable throughputs of tatory unit u. Finally, during any given planning period $tp \in \mathbf{TP}$, the following constraint is imposed to prohibit any alteration in the supply-chain configuration:

$$l_{u,ts} = l_{u,ts'} \tag{8.14}$$

$$\forall u \in \mathbf{UA}, \forall ts, ts' \in \mathbf{TS}_{tp}, \forall tp \in \mathbf{TP}.$$

where, $ts \neq ts'$.

8.2.2 Separation Processes

The separation processes in the petroleum supply chain are those in which only splitters are present, i.e., the atmospheric distillation units, the vacuum distillation units, the butadiene extraction units, the aromatics extraction units, the xylene fractionation units, and the parex units. On the basis of the flow diagram presented in Figure 8.3, the generalized material balances of the separation processes can be written as

$$qp_{u,p,ts} = \sum_{s \in \mathbf{F}_u} qf_{u,s,ts} X_{u,s,p} FC_{u,s,p} = \sum_{s \in \mathbf{F}_u} qf_{u,s,ts} RF_{u,s,p} \tag{8.15}$$

$$\forall u \in \mathbf{UB}, \forall p \in \mathbf{P}_u, \forall ts \in \mathbf{TS}_{tp}, \forall tp \in \mathbf{TP}.$$

where $\mathbf{UB} = \mathbf{U}_{atm} \cup \mathbf{U}_{vgo} \cup \mathbf{U}_{bd} \cup \mathbf{U}_{ext} \cup \mathbf{U}_{xylene} \cup \mathbf{U}_{parex}$; $X_{u,s,p}$ and $FC_{u,s,p}$ are the design parameters denoting respectively the volume fraction and recovery ratio of product p in feedstock s of unit u; $RF_{u,s,p}$ $(= X_{u,s,p} FC_{u,s,p})$ is referred to in this chapter as the *recovery efficiency* of product p in feedstock s of unit u. Notice that the definitions of $qf_{u,s,ts}$ and $qp_{u,p,ts}$ have already been defined before.

Since only a single operation mode is implemented in each separation process and mixed feeds are allowed, the throughput limits can be expressed as

$$QFB_u^L l_{u,ts} \le \sum_{s \in \mathbf{F}_u} qf_{u,s,ts} \le QFB_u^U l_{u,ts} \tag{8.16}$$

$$\forall u \in \mathbf{UB}, \forall ts \in \mathbf{TS}_{tp}, \forall tp \in \mathbf{TP}$$

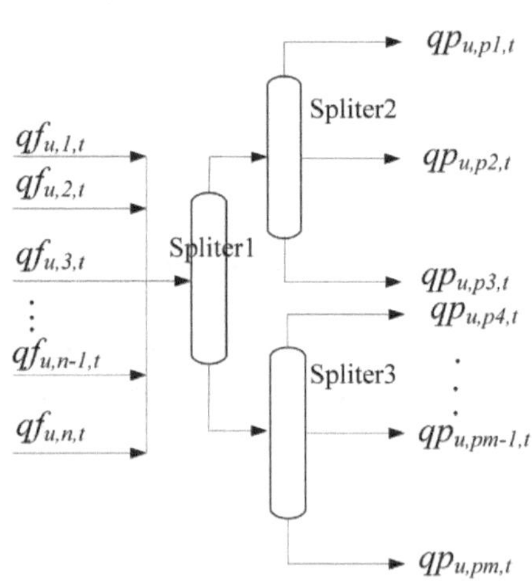

FIGURE 8.3 The generalized separation process. (Reproduced with permission from *Ind. Eng. Chem. Res.* 2008, 47: 1935–1954. Copyright 2008 American Chemical Society.)

where $l_{u,ts} \in \{0,1\}$; QFB_u^U and QFB_u^L denote respectively the maximum and minimum allowable throughputs of separation unit u in a scheduling interval. Finally, since it is also desirable to maintain a fixed process configuration *within a planning period*, the following logic constraints are also imposed:

$$l_{u,ts} = l_{u,ts'} \tag{8.17}$$

$$\forall u \in \mathbf{UA}, \forall ts, ts' \in \mathbf{TS}_{tp}, \forall tp \in \mathbf{TP}$$

where $ts \neq ts'$.

8.2.3 Storage Processes

As mentioned before, the planning and scheduling tasks are usually carried out over a fixed time horizon on two different time scales according to the forecasts of feedstock supplies and market demands. The material-balance model of a generalized storage process (see Figure 8.4) is formulated on the basis of the finer scheduling intervals, i.e.

$$vin_{u,m,ts} = vin_{u,m,ts-1} + qti_{u,m,ts} - qto_{u,m,ts} \tag{8.18}$$

$$\forall u \in \mathbf{U}, \forall m \in \mathbf{M}_u, \forall ts \in \mathbf{TS}_{tp}, \forall tp \in \mathbf{TP}$$

where **U** is the set of all processing units in which storage tanks may be present, i.e., $\mathbf{U} = \mathbf{UA} \cup \mathbf{UB} \cup \mathbf{UT} \cup \mathbf{UD}$, and **UT** and **UD** denote respectively the sets of all import/export facilities and product distribution terminals; $\mathbf{M}_u$ is the set of all process materials in unit u, i.e., $\mathbf{M}_u = \mathbf{F}_u \cup \mathbf{P}_u$; $vin_{u,m,ts}$ represents the inventory of material m in unit u at the end of scheduling interval ts; $qti_{u,m,ts}$ and $qto_{u,m,ts}$ denote respectively the *total* amounts of material m delivered to and withdrawn from unit u during interval ts. Notice that $vin_{u,m,ts}$ is the same as the inventory level at the end of the previous planning period, i.e.

$$vin_{u,m,0} = vin'_{u,m,ts_f} \tag{8.19}$$

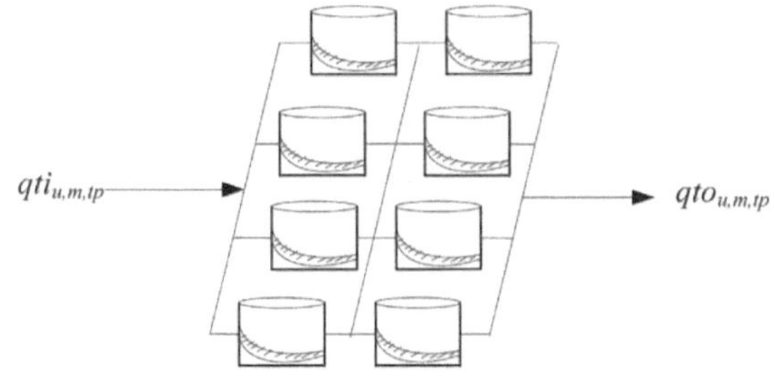

FIGURE 8.4 The generalized storage process. (Reproduced with permission from *Ind. Eng. Chem. Res.* 2008, 47: 1935–1954. Copyright 2008 American Chemical Society.)

where, $ts'_f = \max_{ts' \in \mathbf{TS}_{tp-1}} ts'$. Since in practice the storage capability cannot be unlimited, it is also necessary to impose the following inequality constraints:

$$INV^L_{u,m} \le vin_{u,m,ts} \le INV^U_{u,m} \tag{8.20}$$

$$\forall u \in \mathbf{U}, \forall m \in \mathbf{M}_u, \forall ts \in \mathbf{TS}_{tp}, \forall tp \in \mathbf{TP}$$

where $INV^U_{u,m}$ and $INV^L_{u,m}$ represent respectively the upper and lower bounds of the inventory of material m in unit u.

8.3 SUPPLY CHAIN STRUCTURE

To facilitate the construction of a comprehensive supply-chain structure, let us assume that one buffer-tank system is available for every feedstock and one for every product in each processing unit. More specifically, the storage facilities are installed in the reaction and separation processes according to the frameworks presented in Figures 8.5 and 8.6 respectively. It should be emphasized that this assumption is by no means restrictive. If a particular buffer system is not present in a practical application, it is only necessary to set both the upper and lower bounds of the corresponding inventory in the mathematical model to be zeros. On the other hand, there are certainly various storage tanks in the product distribution terminals and also in the import/export terminals for raw materials, intermediates, and final products. Usually, their capacities are larger than those in the reaction and separation processes. The general framework of the corresponding storage and transportation processes can be found in Figure 8.7.

Notice that the symbols "M" and "D" in all three general structures mentioned above represent *mixers* and *distributors* respectively. The inputs of mixers in a

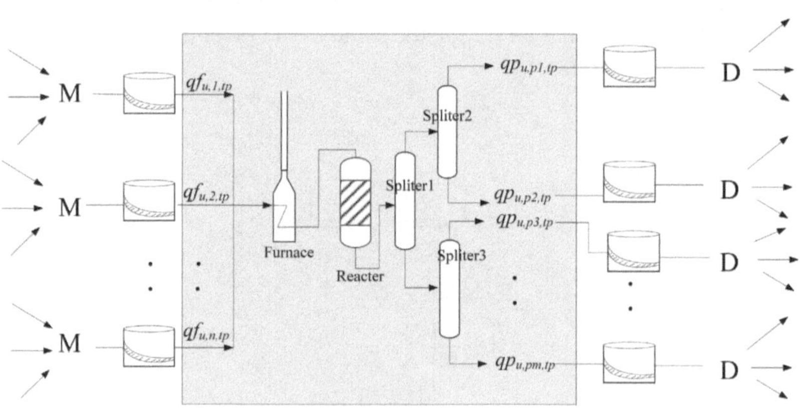

FIGURE 8.5 The flow structure around a reaction process. (Reproduced with permission from *Ind. Eng. Chem. Res.* 2008, 47: 1935–1954. Copyright 2008 American Chemical Society.)

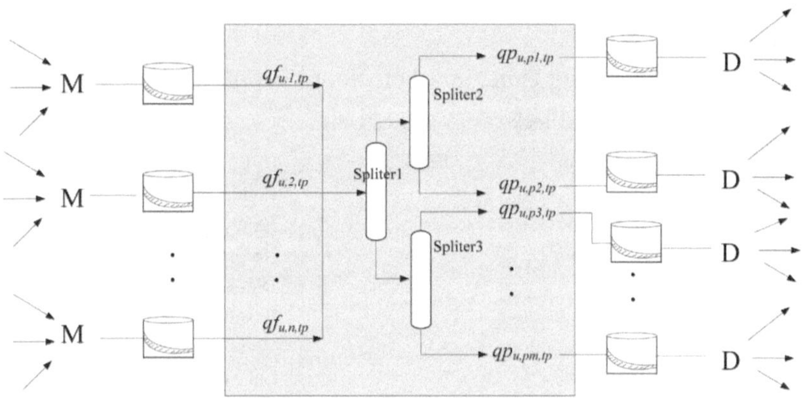

FIGURE 8.6 The flow structure around a separation process. (Reproduced with permission from *Ind. Eng. Chem. Res.* 2008, 47: 1935–1954. Copyright 2008 American Chemical Society.)

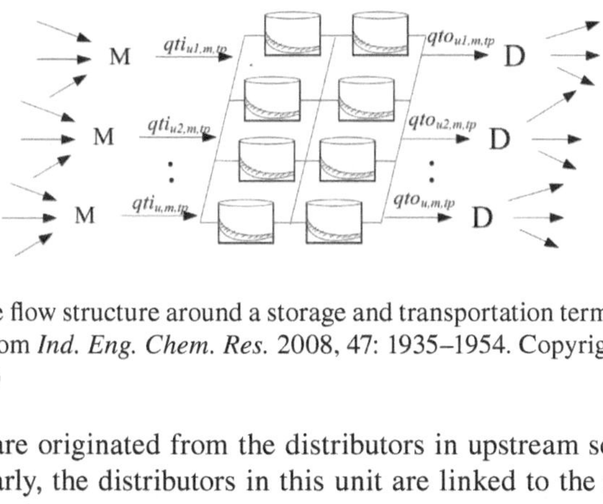

FIGURE 8.7 The flow structure around a storage and transportation terminal. (Reproduced with permission from *Ind. Eng. Chem. Res.* 2008, 47: 1935–1954. Copyright 2008 American Chemical Society.)

production unit are originated from the distributors in upstream source units of its feeds and, similarly, the distributors in this unit are linked to the mixers in downstream sink units of its products. It should be realized that the mixers and distributors may or may not be present physically in the supply chain. They are adopted conceptually here to facilitate accurate formulation of the mathematical programming model. The detailed connections in the general configuration of a petroleum supply chain are listed in Table 8.1. For the sake of conciseness, every unit is represented with a two- or three-letter code and these codes are defined in the first column of the table. In addition, a hyphen indicates that the corresponding sink unit is not modeled in the present chapter. Finally, the feeds of parex and isomar units are denoted as *parexf* and *isomarf.* The former is mainly a mixture of o-xylene, p-xylene, ethylbenzene, and m-xylene, and the latter consists of o-xylene, ethylbenzene, and m-xylene.

To construct an integrated planning and scheduling model, the basic unit models presented in Section 8.2 must be augmented with a complete mathematical description of the aforementioned supply chain structure. These formulations are outlined below:

TABLE 8.1
Connections of Processing Units in a Petroleum Supply Chain

Processing Unit	Feedstock	Source Unit	Product	Sink Unit
Atmospheric distillation (AD)	Crude oil	IEF	C_3/C_4 LPG	PDT
			Light naphtha	HT
			Heavy naphtha	NC
			Kerosene	PDT
			Diesel	HTD
			Residual oil	VD
Vacuum distillation (VD)	Residual oil	AD	Vacuum gas oil	FCC
			Lube oil	PDT
			Asphalt	PDT
			Crude oil residua	CK
Coking (CK)	Crude oil residua	VD	C_3/C_4 LPG	PDT
			Heavy naphtha	NC
			Fuel oil	PDT
			Gas oil	PDT
			Coke	PDT
Fluidized catalytic cracking (FCC)	Vacuum gas oil	VD, CK	C_3/C_4 LPG	PDT
			Propylene	IEF, PDT
			Gasoline	PDT
			Light cycle oil	PDT
Naphtha cracking (NC)	Heavy naphtha	IEF, AD, CK	C_3/C_4 LPG	PDT
			Ethylene	IEF, PDT
			Propylene	IEF, PDT
			Crude C_4	BD
			Pyrolysis gasoline	AET
Butadiene extraction (BD)	Crude C_4	NC	BBR	PDT
			Butadiene	IEF, PDT
Hydroytreating (HT)	Light naphtha	IEF, AD	Light naphtha	RF
(HTD)	Diesel	AD	Diesel	PDT
Reforming (RF)	Light naphtha	IEF, HT	C_3/C_4 LPG	PDT
			C_5 gasoline	PDT
			Reformate	AET
			C_9-aromatic	TT
Aromatic extraction (AET)	Reformate	RF	C_6-gasoline	PDT
	Pyrolysis gasoline	Naphtha cracker	Benzene	PDT, IEF
			Toluene	TT, PDT, IEF
			Mixed xylenes	XF, PDT, IEF
Xylene fractionation (XF)	Mixed xylenes	AET, TT, IM, IEF	o-Xylene	PDT, IEF
			C_{9+}gasoline	PDT
			Parexf	PR

(*Continued*)

TABLE 8.1 (CONTINUED)
Connections of Processing Units in a Petroleum Supply Chain

Processing Unit	Feedstock	Source Unit	Product	Sink Unit
Tatory (TT)	Toluene	AET, PDT	Benzene	PDT, IEF
	C_9	RF	Mixed xylenes	XF, PDT, IEF
			C_{10}-gasoline	PDT
Parex (PR)	Parexf	XF	p-Xylene	PDT, IEF
			Isomarf	IM
Xylene isomar (IM)	Isomarf	PR	Mixed xylenes	XF, PDT, IEF
Import and export facility (IEF)	Benzene	ET, TT, OS	Benzene	OC, PDT
	Butadiene	BD, OS	Butadiene	OC, PDT
	Crude oil	OS	Crude oil	AD
	Ethylene	NC, OS	Ethylene	OC, PDT
	Gasoline	FCC	Gasoline	OC
	Heavy naphtha	OS	Heavy naphtha	NC
	Light naphtha	OS	Light naphtha	RF
	Mixed xylenes	OS, ET, TT, IM, OS	Mixed xylenes	OC, XF, PDT
	o-Xylene	XF, OS	o-Xylene	OC, PDT
	Propylene	NC, FCC, OS	Propylene	OC, PDT
	p-Xylene	PR, OS	p-Xylene	OC, PDT
	Toluene	ET, OS	Toluene	OC, PDT
Product distribution terminal (PDT)	Asphalt	VD	Asphalt	DC
	Benzene	ET, TT, IEF	Benzene	DC
	BBR	BD	BBR	DC
	Butadiene	BD, IEF	Butadiene	DC
	C_3/C_4 LPG	AD, CK, FCC	C_3/C_4 LPG	DC
	Coke	CK	Coke	DC
	Diesel	HT	Diesel	DC
	Ethylene	NC, IEF	Ethylene	DC
	Fuel oil	CK	Fuel oil	DC
	Gas oil	CK	Gas oil	DC
	Gasoline	AET, FCC, RF, TT, XF	Gasoline	DC
	Kerosene	AD	Kerosene	DC
	Light cycle oil	FCC	Light cycle oil	DC
	Lube oil	VD	Lube oil	DC
	Mixed xylenes	ET, TT, IM, IEF	Mixed xylene	DC
	o-Xylene	XF, IEF	o-Xylene	DC
	p-Xylene	PR, IEF	p-Xylene	DC
	Propylene	NC, FCC, IEF	Propylene	DC
	Toluene	ET, IEF	Toluene	DC, TT

8.3.1 Mixer and Distributor Connections

The material balance around a mixer can be written as

$$qti_{u,s,ts} = \sum_{u' \in \mathbf{UI}_{u,s}} q_{u',u,s,ts} \tag{8.21}$$

$$\forall u \in \mathbf{U}, \forall s \in \mathbf{F}_u, \forall ts \in \mathbf{TS}_{tp}, \forall tp \in \mathbf{TP}$$

where $\mathbf{UI}_{u,s}$ denotes the set of all source units (or suppliers) of feedstock s received by unit u; $qti_{u,s,ts}$ denotes the accumulated amount of feedstock s sent to the buffer tank in unit u during interval ts; $q_{u',u,s,ts}$ represents the accumulated amount of feedstock s transported from unit u' to unit u during interval ts. The material balance around a distributor can be formulated with a similar equation, i.e.

$$qto_{u,p,ts} = \sum_{u' \in \mathbf{UO}_{u,p}} q_{u,u',p,ts} \tag{8.22}$$

$$\forall u \in \mathbf{U}, \forall p \in \mathbf{P}_u, \forall ts \in \mathbf{TS}_{tp}, \forall tp \in \mathbf{TP}$$

where $\mathbf{UO}_{u,p}$ denotes the set of all sink units (or customers) of the product p generated in unit u; $qto_{u,p,ts}$ denotes the accumulated amount of product p withdrawn from the buffer tank in unit u during interval ts; $q_{u,u',p,ts}$ represents the accumulated amount of product p transported from unit u to unit u' during interval ts.

Notice that the mixers and distributors may be present in all processes of the supply chain. In the import/export facility or product distribution terminal, every storage tank is equipped with both mixers and distributors (see Figure 8.7). However, in the cases of reaction and separation processes, the mixers are attached to the feedstock buffer tanks, while the distributors are dedicated to the product buffer tanks (see Figures 8.5 and 8.6). In other words, the outputs of the former tanks can be regarded as the inputs of reaction and separation processes, i.e.

$$qto_{u,s,ts} = \sum_{k \in \mathbf{K}_{u,s}} qfi_{u,s,k,ts} \tag{8.23}$$

$$\forall u \in \mathbf{UA}, \forall s \in \mathbf{F}_u, \forall ts \in \mathbf{TS}_{tp}, \forall tp \in \mathbf{TP}$$

$$qto_{u,s,ts} = qf_{u,s,ts} \tag{8.24}$$

$$\forall u \in \mathbf{UB}, \forall s \in \mathbf{F}_u, \forall ts \in \mathbf{TS}_{tp}, \forall tp \in \mathbf{TP}$$

Similarly, the inputs of the product tanks should be considered to be the same as the outputs of these processes, i.e.

$$qti_{u,p,ts} = qp_{u,p,ts} \tag{8.25}$$

$$\forall u \in \mathbf{U}', \forall p \in \mathbf{P}_u, \forall ts \in \mathbf{TS}_{tp}, \forall tp \in \mathbf{TP}$$

where $\mathbf{U}' = \mathbf{UA} \cup \mathbf{UB}$.

8.3.2 Transportation Capacity

It is obvious that the process materials must be transported from one unit to another with pumps and pipelines. The transportation method for importing the raw materials and intermediates should be by tanker. The products and byproducts could be delivered to domestic customers either via pipelines or by trucks. Since the corresponding transportation capacities should always be limited in practice, the following inequalities are included in our model:

$$Q^L_{v,v',m} \le q_{v,v',m,ts} \le Q^U_{v,v',m} \tag{8.26}$$

$$\forall v, v' \in \mathbf{USC}, \forall m \in \mathbf{M}, \forall ts \in \mathbf{TS}_{tp}, \forall tp \in \mathbf{TP}$$

where $\mathbf{M} = \bigcup_{u \in \mathbf{U}} \mathbf{M}_u$; $\mathbf{USC} = \mathbf{U} \cup \mathbf{S} \cup \mathbf{C}$; **S** is the set of all suppliers of the raw materials and/or intermediates; **C** is the set of all customers of the final products and/or byproducts; $Q^U_{v,v',m}$ and $Q^L_{v,v',m}$ denote respectively the upper and lower limits of the transportation capacity for material m from unit v to unit v'; $q_{v,v',m,ts}$ represents the accumulated amount of delivered material m from unit v to unit v' during scheduling interval ts.

8.3.3 Input Constraints

It is assumed that the crude oil is purchased from the international market and its quality, quantity, and shipment dates can be reasonably predicted over a relatively small number of planning periods. These forecasts are characterized with the following equality constraints in this study:

$$\sum_{v' \in \mathbf{UT}} q_{v,v',m,ts} = QCS_{v,m,ts} i^{Cr}_{v,m,ts} + QCC_{v,m,ts} \tag{8.27}$$

$$\forall v \in \mathbf{S}_{crude}, \forall m \in \mathbf{P}_v, \forall ts \in \mathbf{TS}^{v,m}_{tp}, \forall tp \in \mathbf{TP}$$

where $\mathbf{S}_{crude}$ denotes the set of overseas suppliers of crude oils; $\mathbf{P}_v$ represents the set of different types of crudes provided by supplier v; $\mathbf{TS}^{v,m}_{tp}$ denotes the set of scheduling intervals (within the planning period tp) in which shipment of crude m from supplier v can be scheduled (i.e., $\mathbf{TS}^{v,m}_{tp} \subset \mathbf{TS}_{tp}$); $QCS_{v,m,ts}$ represents the amount of crude

m that can be obtained from supplier v during the scheduling interval ts. $QCC_{v,m,ts}$ denotes a fixed amount of crude m guaranteed in a long-term contract by supplier v during the scheduling interval ts. If $ts \in \mathbf{TS}_{tp}^{v,m}$, the binary variable $i_{v,m,ts}^{Cr} \in \{0,1\}$ denotes whether or not the crude is selected. However, the constraint $i_{v,m,ts}^{Cr} = 0$ should be imposed if $ts \notin \mathbf{TS}_{tp}^{v,m}$.

Let us further assume that three intermediates of the petroleum supply chain, i.e., heavy naphtha, light naphtha, and mixed xylenes, can be imported for use in the naphtha crackers, reforming units and xylene fractionation units respectively. Therefore, additional equality constraints are imposed upon the total imported quantity of each of these intermediates:

$$\sum_{v' \in \mathbf{UT}} q_{v,v',m,ts} = QHN_{v,m,ts} i_{v,m,ts}^{Hn} + QHNC_{v,m,s} \tag{8.28}$$

$$\forall v \in \mathbf{S}_{hvynap}, \forall m \in \mathbf{P}_v, \forall ts \in \mathbf{TS}_{tp}^{v,m}, \forall tp \in \mathbf{TP}$$

$$\sum_{v' \in \mathbf{UT}} q_{v,v',m,ts} = QLN_{v,m,ts} i_{v,m,ts}^{Ln} + QLNC_{v,m,s} \tag{8.29}$$

$$\forall v \in \mathbf{S}_{litnap}, \forall m \in \mathbf{P}_v, \forall ts \in \mathbf{TS}_{tp}^{v,m}, \forall tp \in \mathbf{TP}$$

$$\sum_{v' \in \mathbf{UT}} q_{v,v',m,ts} = QXI_{v,m,ts} i_{v,m,ts}^{Mx} + QXIC_{v,m,s} \tag{8.30}$$

$$\forall v \in \mathbf{S}_{mxyl}, \forall m \in \mathbf{P}_v, \forall ts \in \mathbf{TS}_{tp}^{v,m}, \forall tp \in \mathbf{TP}$$

where $\mathbf{S}_{hvynap}$, $\mathbf{S}_{litnap}$, and $\mathbf{S}_{mxyl}$ denote the sets of overseas suppliers of heavy naphtha, light naphtha, and mixed xylene respectively; $\mathbf{P}_v$ represents the set of different types of heavy naphtha, light naphtha, or mixed xylene provided by supplier v; $\mathbf{TS}_{tp}^{v,m}$ denotes the set of scheduling intervals (within the planning period tp) in which shipment of process material m from supplier v can be scheduled (i.e., $\mathbf{TS}_{tp}^{v,m} \subset \mathbf{TS}_{tp}$); $QHN_{v,m,ts}$, $QLN_{v,m,ts}$, and $QXI_{v,m,ts}$ represent respectively the amounts of heavy naphtha, light naphtha, and mixed xylene m that can be purchased from supplier v during the scheduling interval ts. $QHNC_{v,m,s}$, $QLNC_{v,m,s}$, and $QXI_{v,m,ts}$ denote respectively the fixed amounts of heavy naphtha, light naphtha, and mixed xylene (m) guaranteed in a long-term contract by supplier v during the scheduling interval ts. If $ts \in \mathbf{TS}_{tp}^{v,m}$, the binary variable $i_{v,m,ts}^{Hn}$, $i_{v,m,ts}^{Ln}$, or $i_{v,m,ts}^{Mx}$ denotes whether or not intermediate m is chosen. If, on the other hand, $ts \notin \mathbf{TS}_{tp}^{v,m}$, then the constraints $i_{v,m,ts}^{Hn} = 0$, $i_{v,m,ts}^{Ln} = 0$, and $i_{v,m,ts}^{Mx} = 0$ should be imposed.

It is assumed that the major petrochemical products of the supply chain, i.e., ethylene, propylene, butadiene, benzene, toluene, and xylene, can also be purchased from oversea suppliers to fulfill contract obligations to the customers. A set of equality constraints is thus included in the model, i.e.

$$\sum_{v' \in \mathbf{UT}} q_{v,v',m,ts} = QPET_{v,m,ts} i^{Pet}_{v,m,ts} + QPETC_{v,m,ts} \tag{8.31}$$

$$\forall v \in \mathbf{S}_{pet}, \forall m \in \mathbf{P}_v, \forall ts \in \mathbf{TS}^{v,m}_{tp}, \forall tp \in \mathbf{TP}$$

where $\mathbf{S}_{pet}$ denotes the set of overseas suppliers of the above-mentioned petrochemical products; $\mathbf{P}_v$ represents the set of these products provided by supplier v; $\mathbf{TS}^{v,m}_{tp}$ denotes the set of scheduling intervals within the planning period tp in which shipment of product m from supplier v can be scheduled (i.e., $\mathbf{TS}^{v,m}_{tp} \subset \mathbf{TS}_{tp}$); $QPET_{v,m,ts}$ represents the amount of product m that can be purchased from supplier v during the scheduling interval ts. $QPETC_{v,m,ts}$ represents a fixed amount of product m guaranteed in a long-term contract by supplier v during the scheduling interval ts. If $ts \in \mathbf{TS}^{v,m}_{tp}$, the binary variable $i^{Pet}_{v,m,ts}$ denotes whether or not the petrochemical product m is actually purchased. Again, the constraint $i^{Pet}_{v,m,ts} = 0$ should be imposed if $ts \notin \mathbf{TS}^{v,m}_{tp}$. Notice finally that the set of all possible inputs of the petroleum supply chain can be expressed as $\mathbf{S} = \mathbf{S}_{crude} \cup \mathbf{S}_{hvynap} \cup \mathbf{S}_{litnap} \cup \mathbf{S}_{mxyl} \cup \mathbf{S}_{pet}$.

8.3.4 Output Constraints

In this chapter, the customer demands for the petrochemical products (i.e., ethylene, propylene, butadiene, benzene, toluene, and xylenes) are assumed to be predictable at least in the short term. This is due to the fact that these products are consumed mainly in the downstream production processes. It is also assumed that the total amounts actually shipped to these customers during a particular planning period are allowed to vary in several given ranges, i.e.

$$\sum_{l \in \mathbf{L}_{v',m,tp}} QPL^{L}_{v',m,tp,l} j_{v',m,tp,l} \le \tilde{q}_{v',m,tp} \le \sum_{l \in \mathbf{L}_{v',m,tp}} QPL^{U}_{v',m,tp,l} j_{v',m,tp,l} \tag{8.32}$$

$$\forall v' \in \mathbf{C}_{pet}, \forall m \in \mathbf{F}_{v'}, \forall tp \in \mathbf{TP}$$

$$\sum_{l \in \mathbf{L}_{v',m,tp}} j_{v',m,tp,l} = 1 \tag{8.33}$$

$$\forall v' \in \mathbf{C}_{pet}, \forall m \in \mathbf{F}_{v'}, \forall tp \in \mathbf{TP}$$

where $\tilde{q}_{v',m,tp} = \sum_{ts \in \mathbf{TS}_{tp}} \sum_{v \in \mathbf{UD}} q_{v,v',m,ts}$; $QPL^{U}_{v',m,tp,l}$ and $QPL^{L}_{v',m,tp,l}$ represent respectively the upper and lower bound of the lth range of the amount of material m delivered to customer v' during planning period tp. The binary variable $j_{v',m,tp,l}$ is used to reflect if the amount of material m sent to customer v' during period tp is in range l ($j_{v',m,tp,l}=1$) or not ($j_{v',m,tp,l}=0$).

The aforementioned petrochemical products should also be allowed to be sold on the overseas market so that the overall performance of the supply chain may be further enhanced. This practice is justifiable if the selling prices and/or demands for some of the products are high enough to compensate for the losses due to the need to cut down inventories of the other products and to sell them at lower prices and/or higher transportation costs. The quantities delivered to overseas customers should be subject to simple inequality constraints, i.e.

$$QOS^L_{v',m,tp} \le \sum_{ts\in \mathbf{TS}_{tp}} \sum_{v\in \mathbf{UD}} q_{v,v',m,ts} \le QOS^U_{v',m,tp} \tag{8.34}$$

$$\forall v' \in \mathbf{C}_{ovs}, \forall m \in \mathbf{F}_{v'}, \forall tp \in \mathbf{TP}$$

where $\mathbf{C}_{ovs}$ denotes the sets of overseas customers for the petrochemical products; $QOS^U_{v',m,tp}$ and $QOS^L_{v',m,tp}$ represent respectively the acceptable maximum and minimum amounts of product m transported to customer v' in period tp.

Finally, it is assumed in this work that the remaining products and by-products of the supply chain, i.e. liquefied petroleum gas, mixed xylenes, gasoline, kerosene, and diesel, can always be sold to domestic consumers. The corresponding constraints are

$$QXO^L_{v',m,tp} \le \sum_{ts\in \mathbf{TS}_{tp}} \sum_{v\in \mathbf{UD}} q_{v,v',m,ts} \le QXO^U_{v',m,tp} \tag{8.35}$$

$$\forall v' \in \mathbf{C}_{dt}, \forall m \in \mathbf{F}_{v'}, \forall tp \in \mathbf{TP}$$

where $\mathbf{C}_{dt}$ denotes the sets of these domestic customers; $QXO^U_{v',m,tp}$ and $QXO^L_{v',m,tp}$ represent respectively the acceptable maximum and minimum amounts of product m delivered to customer v' in period tp. Notice finally that the set of all possible outputs of the petroleum supply chain can be expressed as $\mathbf{C}=\mathbf{C}_{pet}\cup\mathbf{C}_{dt}\cup\mathbf{C}_{ovs}$ in the present model.

8.4 OBJECTIVE FUNCTION

The objective function of the present optimization problem is the total profit realized over a specified planning horizon, i.e.

$$\text{Total Profit} = \sum_{tp\in \mathbf{TP}} pf_{tp} = \sum_{tp\in \mathbf{TP}} \left(rs_{tp} - cr_{tp} - co_{tp} - ct_{tp} - ci_{tp}\right) \tag{8.36}$$

where pf_{tp}, rs_{tp}, cr_{tp}, co_{tp}, ct_{tp}, and ci_{tp} denote respectively the net profit, the total revenue secured from various product sales, the total cost of raw materials, the total operating cost, the total transportation cost, and the total inventory cost in planning period tp. The total sale revenue can be expressed as

$$rs_{tp} = \sum_{v' \in \mathbf{C}_{pet}} \sum_{m \in \mathbf{F}_{v'}} \left(\sum_{l \in \mathbf{L}_{v',m,tp}} SP_{v',m,tp,l}\, qj_{v',m,tp,l} \right) + \sum_{v' \in \mathbf{C}_{ovs}} \sum_{m \in \mathbf{F}_{v'}} \left[SPO_{v',m,tp} \left(\sum_{v \in \mathbf{UD}} \sum_{ts \in \mathbf{TS}_{tp}} q_{v,v',m,ts} \right) \right] + \sum_{v' \in \mathbf{C}_{dt}} \sum_{m \in \mathbf{F}_{v'}} \left[SPD_{v',m,tp} \left(\sum_{v \in \mathbf{UD}} \sum_{ts \in \mathbf{TS}_{tp}} q_{v,v',m,ts} \right) \right] \quad (8.37)$$

It should be noted that a new variable $qj_{v',m,tp.l}$ is introduced here to covert the revenue model into linear form. Specifically, the total amount of product m sent to customer v' in planning period tp, i.e., $\tilde{q}_{v',m,tp}$, can be related to this variable with the following two constraints:

$$\tilde{q}_{v',m,tp} = \sum_{l \in \mathbf{L}_{v',m,tp}} qj_{v',m,tp.l} \quad (8.38)$$

$$0 \leq qj_{v',m,tp.l} \leq j_{v',m,tp.l} U \quad (8.39)$$

$$\forall v' \in \mathbf{C}_{pet}, \forall m \in \mathbf{F}_{v'}, \forall tp \in \mathbf{TP}, l \in \mathbf{L}_{v',m,tp}$$

where U is a sufficiently large constant. $SP_{v',m,tp,l}$ represents the selling price of product m sold to customer v' during period tp in level range l; $SPO_{v',m,tp}$ denotes the selling price of intermediate or final product m sold to overseas customer v' during period tp; $SPD_{v',m,tp}$ denotes the selling price of intermediate or final product m sold to domestic customer v' during period tp. The general relationship between the delivered amount of a product and its selling price is sketched in Figure 8.8.

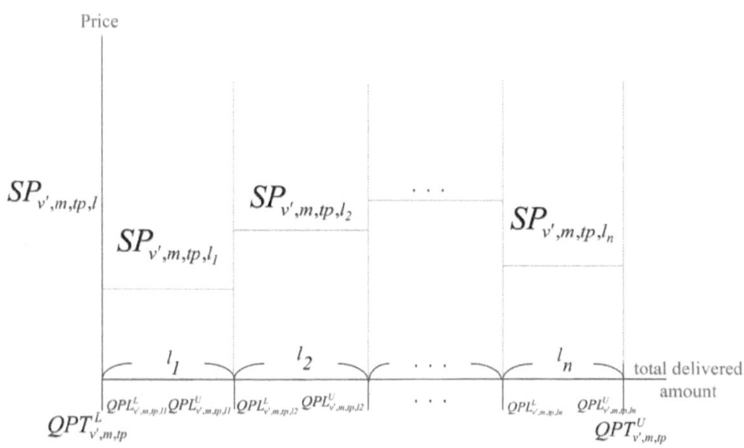

FIGURE 8.8 The general relationship between the amount of delivered product and its selling price. (Reproduced with permission from *Ind. Eng. Chem. Res.* 2008, 47: 1935–1954. Copyright 2008 American Chemical Society.)

The second term on the right-hand side of equation (8.36) is the sum of the purchasing costs of the crudes, the imported intermediates (i.e., heavy naphtha, light naphtha, and mixed xylenes), and also possibly the final products provided by domestic suppliers. In particular, this term can be written as

$$\begin{aligned} cr_{tp} = & \sum_{v \in \mathbf{S}_{cr}} \sum_{m \in \mathbf{P}_v} \left[CRO_{v,m,tp} \sum_{ts \in \mathbf{TS}_{tp}} \left(QCS_{v,m,ts} i^{Cr}_{v,m,ts} \right) + CROC_{v,m,tp} \sum_{ts \in \mathbf{TS}_{tp}} QCC_{v,m,ts} \right] \\ & + \sum_{v \in \mathbf{S}_{hvynap}} \sum_{m \in \mathbf{P}_v} \left[CHN_{v,m,tp} \sum_{ts \in \mathbf{TS}_{tp}} \left(QHN_{v,m,ts} i^{Hn}_{v,m,ts} \right) + CHNC_{v,m,tp} \sum_{ts \in \mathbf{TS}_{tp}} QHNC_{v,m,ts} \right] \\ & + \sum_{v \in \mathbf{S}_{litnap}} \sum_{m \in \mathbf{P}_v} \left[CLN_{v,m,tp} \sum_{ts \in \mathbf{TS}_{tp}} \left(QLN_{v,m,ts} i^{Ln}_{v,m,ts} \right) + CLNC_{v,m,tp} \sum_{ts \in \mathbf{TS}_{tp}} QLNC_{v,m,ts} \right] \\ & + \sum_{v \in \mathbf{S}_{mxyl}} \sum_{m \in \mathbf{P}_v} \left[CMX_{v,m,tp} \sum_{ts \in \mathbf{TS}_{tp}} \left(QXI_{v,m,ts} i^{Mx}_{v,m,ts} \right) + CMXC_{v,m,tp} \sum_{ts \in \mathbf{TS}_{tp}} QXIC_{v,m,ts} \right] \\ & + \sum_{v \in \mathbf{S}_{pet}} \sum_{m \in \mathbf{P}_v} \left[CPET_{v,m,tp} \sum_{ts \in \mathbf{TS}_{tp}} \left(QPET_{v,m,ts} i^{Pet}_{v,m,ts} \right) + CPETC_{v,m,tp} \sum_{ts \in \mathbf{TS}_{tp}} QPETC_{v,m,ts} \right] \end{aligned} \quad (8.40)$$

where $CRO_{v,m,tp}$, $CHN_{v,m,tp}$, $CLN_{v,m,tp}$, $CMX_{v,m,tp}$, and $CPET_{v,m,tp}$ denote respectively the unit costs for purchasing crude oils, heavy naphtha, light naphtha, mixed xylenes, and final products; $CROC_{v,m,tp}$, $CHNC_{v,m,tp}$, $CLNC_{v,m,tp}$, $CMXC_{v,m,tp}$, and $CPETC_{v,m,tp}$ represent the unit costs specified in long-term contracts for crude oils, heavy naphtha, light naphtha, mixed xylenes, and final products respectively. The total operating cost can be considered as the sum of the operating costs for running all reaction and separation processes, i.e.

$$\begin{aligned} co_{tp} = & \sum_{ts \in \mathbf{TS}_{tp}} \sum_{u \in \mathbf{UA}} \left[l_{u,ts} CXA_{u,ts} + \sum_{s \in \mathbf{F}_u} \sum_{k \in \mathbf{K}_{u,s}} \left(qfi_{u,s,k,ts} CVA_{u,s,k,ts} \right) \right] \\ & + \sum_{ts \in \mathbf{TS}_{tp}} \sum_{u \in \mathbf{UB}} \left[l_{u,ts} CXB_{u,ts} + \sum_{s \in \mathbf{F}_u} \left(qf_{u,s,ts} CVB_{u,s,ts} \right) \right] \end{aligned} \quad (8.41)$$

where $CXA_{u,ts}$ and $CXB_{u,ts}$ denote the fixed operating costs for unit u in **UA** and **UB** during scheduling interval ts respectively; $CVA_{u,s,k,ts}$ represents the variable operating cost of reaction unit u for processing feedstock s under operation mode k during scheduling interval ts; $CVB_{u,s,ts}$ is the variable operating cost of separation unit u for processing feedstock s during scheduling interval ts.

The overall transportation cost can be expressed as

$$ct_{tp} = \sum_{v \in \mathbf{USC}} \sum_{v' \in \mathbf{USC}} \sum_{m \in \mathbf{M}} \sum_{ts \in \mathbf{TS}_{tp}} q_{v,v',m,ts} CTP_{v,v',m,ts} \tag{8.42}$$

where $CTP_{v,v',m,ts}$ represents the unit transportation cost for moving material m from unit v to unit v' during scheduling interval ts. Finally, the unprocessed inventories remaining in the supply chain at the end of each scheduling interval can be penalized by incorporating additional costs, i.e.

$$ci_{tp} = \sum_{u \in \mathbf{U}} \sum_{m \in \mathbf{M}_u} \sum_{ts \in \mathbf{TS}_{tp}} vin_{u,m,ts} CIN_{u,m,ts} \tag{8.43}$$

where $CIN_{u,m,ts}$ represents the inventory cost per unit of process material m in unit u during scheduling interval ts.

8.5 CASE STUDIES

The petroleum supply network considered in the following case studies is sketched in Figure 8.9. It is assumed that the crude oils and intermediates, i.e., heavy naphtha,

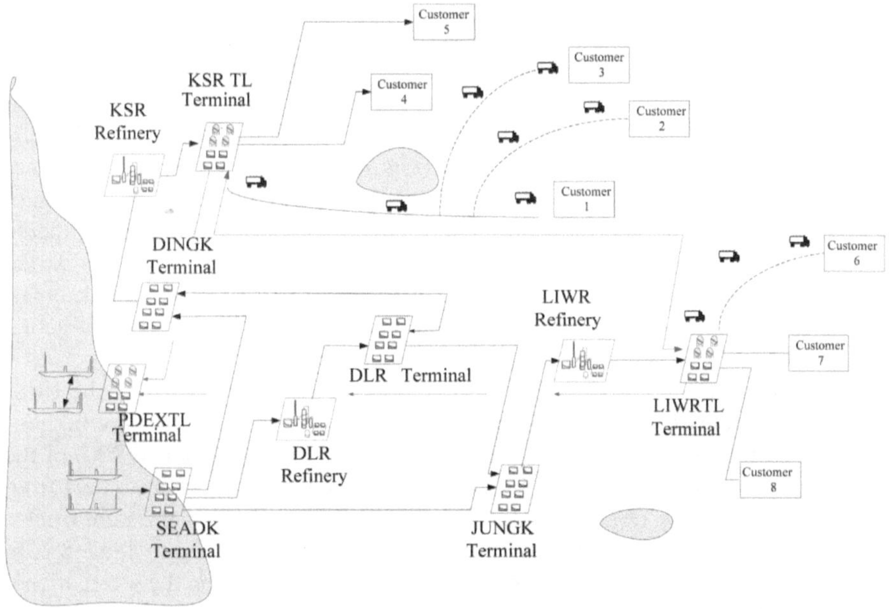

FIGURE 8.9 The petroleum supply network in case studies. (Reproduced with permission from *Ind. Eng. Chem. Res.* 2008, 47: 1935–1954. Copyright 2008 American Chemical Society.)

light naphtha, and mixed xylenes, are shipped from foreign suppliers by oil tankers via terminal SEADK, while the products and by-products of the supply chain, i.e., ethylene, propylene, liquefied petroleum gas, butadiene, benzene, toluene, xylene, gasoline, kerosene, and diesel, are transported via terminal PDEXTL to and from the overseas customers and suppliers respectively. This terminal is equipped with refrigerated and pressurized tanks as well as vaporization facilities for storing and transporting ethylene and liquefied petroleum gas. There are three refineries (KSR, DLK, and LIWR) in this example. The crude oils are transferred from terminal SEADK to refinery DLR directly, but sent to refineries KSR and LIWR through terminals DINGK and JUNGK respectively (see Figure 8.9).

It is assumed that the network configuration of this supply chain is *fixed* and it consists of three (3) atmospheric distillation units, three (3) diesel hydro-treaters, three (3) vacuum distillation units, three (3) naphtha crackers, three (3) butadiene extraction units, three (3) naphtha hydro-treaters and reforming units, three (3) cokers, three (3) aromatic extraction units, two (2) fluidized catalytic cracking units, two (2) xylene fractionation units, two (2) tatory units, two (2) parex units, and two (2) xylene isomar units. The actual locations of these units are given in the detailed process flow diagrams shown in Figures 8.10, 8.11, and 8.12. The first complete set of production units, i.e., an atmospheric distillation unit, a naphtha cracker, a butadiene extraction unit, a fluidized catalytic cracking unit, a vacuum distillation unit, a diesel hydro-treater, a coker, a naphtha hydro-treater, a reforming unit, an aromatic extraction unit, a xylene fractionation unit, a tatory unit, a parex unit, and a xylene isomar unit, are situated in the KSR refinery (Figure 8.10). The second atmospheric distillation unit, diesel hydro-treater, naphtha hydro-treater, reforming unit, naphtha cracker, butadiene extraction unit, fluidized catalytic cracking unit, vacuum distillation unit, and coker are located in the DLR refinery (Figure 8.11), while all other production units are in the LIWR refinery (Figure 8.12). In the case studies, these units are identified according to the codes given in Table 8.1. Since each reaction or separation process can be carried out in alternative units, these units are numbered according to their locations. Finally, it is assumed that the first aromatic extraction unit contains two processing trains designed for two distinct feedstock types, i.e., reformer gasoline and pyrolysis gasoline, and the remaining aromatic extraction units are only capable of processing the reformer gasoline.

The selling prices of the petrochemical products in set $\mathbf{C}_{pet}$ are given in three different demand ranges. The price in the target range of each product is the highest, while those in the lower and higher delivery ranges are set at 80% and 90% of the target level respectively. All these products are shipped to the local customers from the product distribution terminals KSRTL and LIWRTL via pipelines and/or trucks. The allowed delivery ranges in every planning period are presented in Table 8.2. In addition, it is assumed that ten different crude oils are available on the international market, and there are five providers for each species in every scheduling interval (i.e., 1 week). It is further assumed that no long-term supply contracts of crude oils or intermedia oils are in effect with any of these providers. In addition, the initial inventories of crude oils, the intermediates, and final products are assumed to be zero, and all transportation capacities are assumed to be limitless in order to simplify

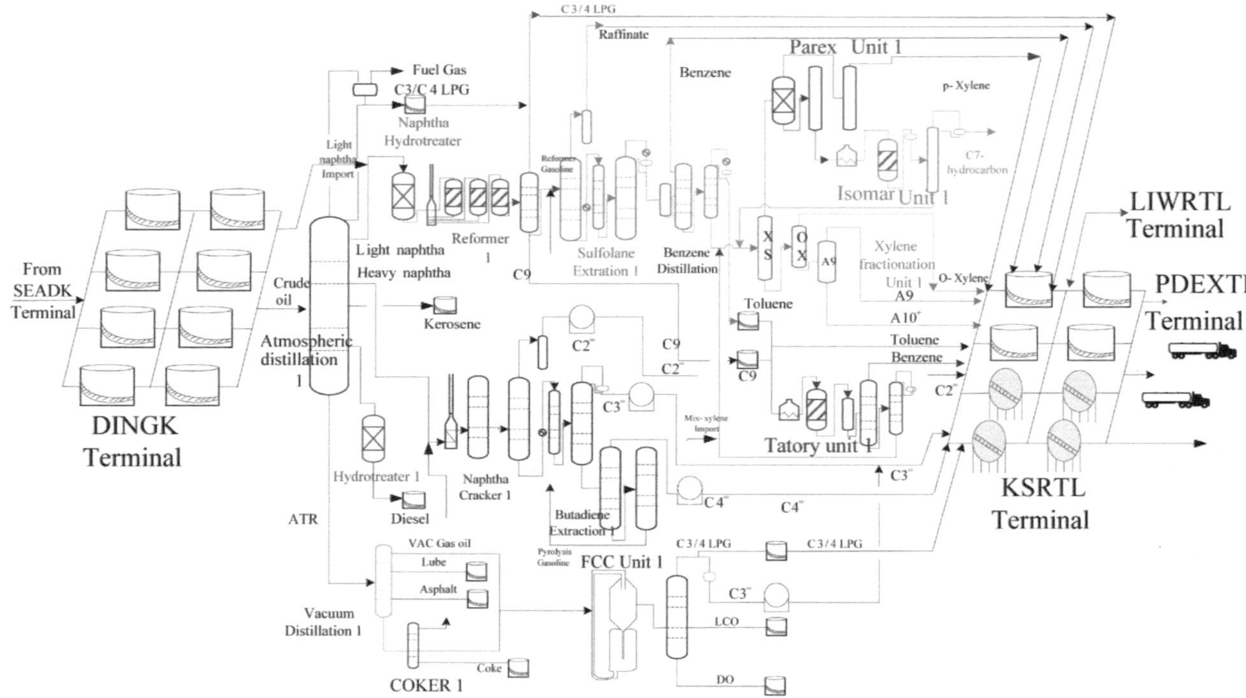

FIGURE 8.10 The KSR Refinery. (Reproduced with permission from *Ind. Eng. Chem. Res.* 2008, 47: 1935–1954. Copyright 2008 American Chemical Society.)

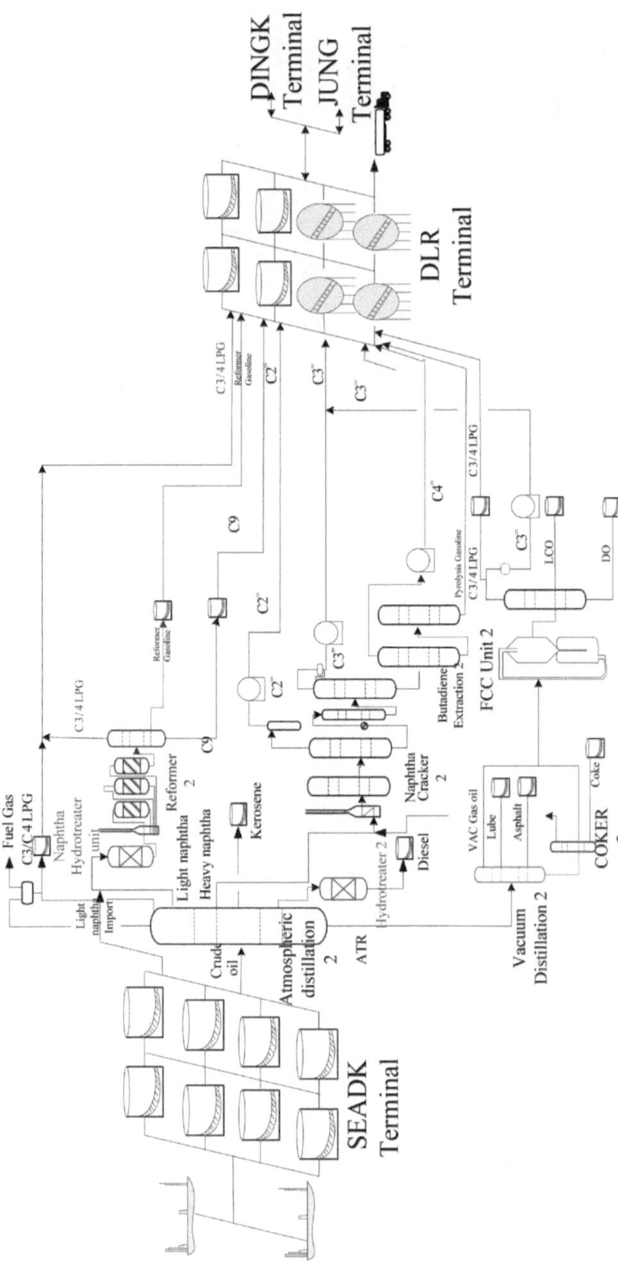

FIGURE 8.11 The DLR Refinery. (Reproduced with permission from *Ind. Eng. Chem. Res.* 2008, 47: 1935–1954. Copyright 2008 American Chemical Society.)

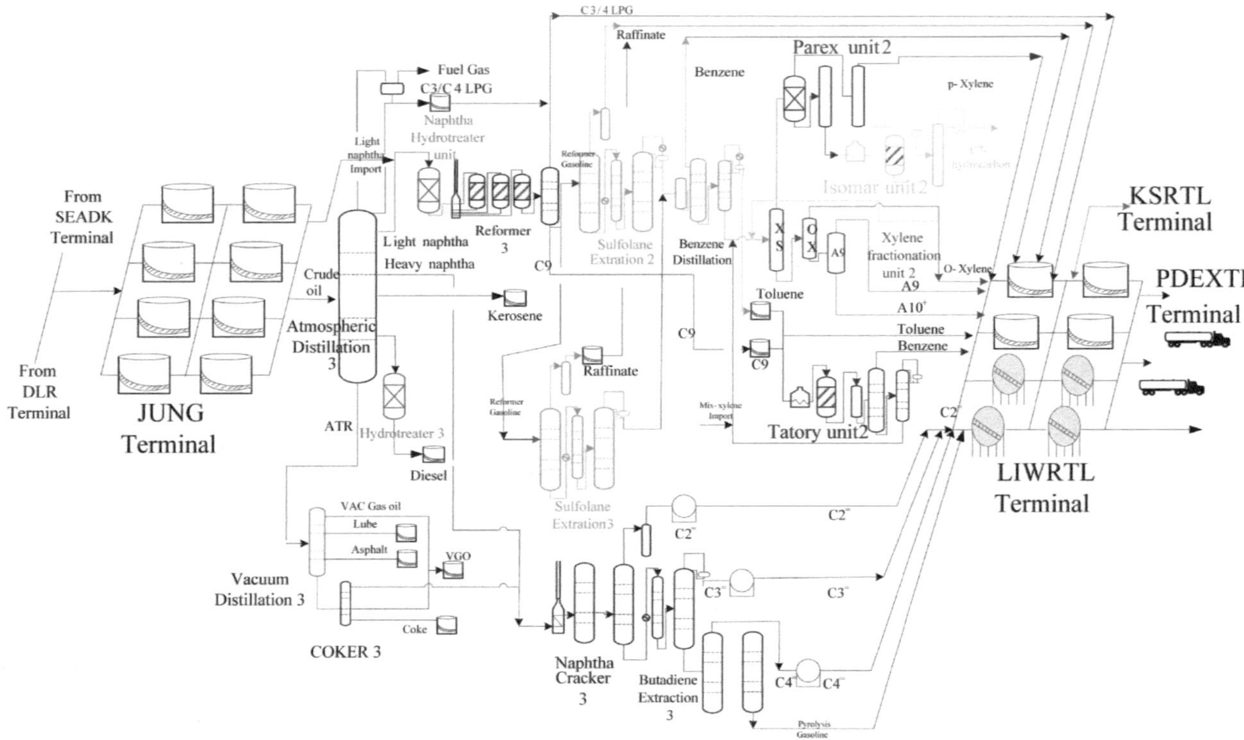

FIGURE 8.12 The LIWR Refinery. (Reproduced with permission from *Ind. Eng. Chem. Res.* 2008, 47: 1935–1954. Copyright 2008 American Chemical Society.)

TABLE 8.2
Allowable Delivery Ranges of the Main Products per Month

Product			tp1			tp2			tp3		
(unit)	Customer	Bounds	Range 1	Range 2	Range 3	Range 1	Range 2	Range 3	Range 1	Range 2	Range 3
Ethylene	Kuco2	Lower	17425	19475	21525	17000	19000	21000	12325	13775	15225
(en)		Upper	19475	21525	23575	19000	21000	23000	13775	15225	16675
(kg)	LIWRco3	Lower	172040	192280	212520	333285	372495	411705	245650	274550	303450
		Upper	192280	212520	232760	372495	411705	450915	274550	303450	332350
Propylene	Kuco3	Lower	481100	537700	594300	129200	144400	159600	21250	23750	26250
(pn)		Upper	537700	594300	650900	144400	159600	174800	23750	26250	28750
(m^3)	Kuco5	Lower	144500	161500	178500	17000	19000	21000	773840	864880	955920
		Upper	161500	178500	195500	19000	21000	23000	864880	955920	1046960
	LIWRco2	Lower	8330	9310	10290	34000	38000	42000	51170	57190	63210
		Upper	9310	10290	11270	38000	42000	46000	57190	63210	69230
Butadiene	Kuco3	Lower	850	950	1050	1275	1425	1575	3230	3610	3990
(bd) (m^3)		Upper	950	1050	1150	1425	1575	1725	3610	3990	4370
Benzene	Kuco2	Lower	17850	19950	22050	17850	19950	22050	18700	20900	23100
(bz)		Upper	19950	22050	24150	19950	22050	24150	20900	23100	25300
(m^3)	Kuco5	Lower	53380	59660	65940	17000	19000	21000	57715	64505	71295
		Upper	59660	65940	72220	19000	21000	23000	64505	71295	78085
Toluene	Kuco1	Lower	2550	2850	3150	2550	2850	3150	9690	10830	11970
(tl)		Upper	2850	3150	3450	2850	3150	3450	10830	11970	13110
(m^3)	LIWRco2	Lower	5100	5700	6300	5100	5700	6300	5270	5890	6510
		Upper	5700	6300	6900	5700	6300	6900	5890	6510	7130

(Continued)

TABLE 8.2 (CONTINUED)
Allowable Delivery Ranges of the Main Products per Month

Product			tp1			tp2			tp3		
(unit)	Customer	Bounds	Range 1	Range 2	Range 3	Range 1	Range 2	Range 3	Range 1	Range 2	Range 3
Mixed	Kuco1	Lower	11900	13300	14700	17000	19000	21000	15300	17100	18900
xylenes		Upper	13300	14700	16100	19000	21000	23000	17100	18900	20700
(mx)	Kuco5	Lower	25500	28500	31500	28560	31920	35280	35955	40185	44415
(m^3)		Upper	28500	31500	34500	31920	35280	38640	40185	44415	48645
	LIWRco2	Lower	2754	3078	3402	2040	2280	2520	64260	71820	79380
		Upper	3078	3402	3726	2280	2520	2760	71820	79380	86940
p-Xylene	Kuco1	Lower	3400	3800	4200	3400	3800	4200	3400	3800	4200
(px)		Upper	3800	4200	4600	3800	4200	4600	3800	4200	4600
(m^3)	Kuco3	Lower	37400	41800	46200	34000	38000	42000	68000	76000	84000
		Upper	41800	46200	50600	38000	42000	46000	76000	84000	92000
o-Xylene	Kuco1	Lower	10710	11970	13230	8500	9500	10500	17850	19950	22050
(ox)		Upper	11970	13230	14490	9500	10500	11500	19950	22050	24150
(m^3)	Kuco3	Lower	21760	24320	26880	5100	5700	6300	21250	23750	26250
		Upper	24320	26880	29440	5700	6300	6900	23750	26250	28750
	LIWRco1	Lower	18530	20710	22890	9350	10450	11550	47175	52725	58275
		Upper	20710	22890	25070	10450	11550	12650	52725	58275	63825

the model formulation. For the sake of brevity, the remaining model parameters are presented elsewhere (Kuo and Chang, 2008b). These parameters include the upper and lower throughput limits of all production units, the purchasing costs and supply rates of crude oils in each scheduling period, the purchasing costs of final products if obtained from domestic suppliers, the purchasing costs and supply amounts of light naphtha, heavy naphtha, and mixed xylenes if bought from overseas suppliers, the selling prices of the products and by-products, the largest available quantities of the raw materials, intermediates, and final products that can be purchased from the international and domestic markets, the maximum amounts of exportable products and by-products, the performance indices of the reaction and separation processes, i.e., the product yields of the former processes and the recovery efficiencies of the latter, the feed compositions of separation units, the upper limits of all inventories and their costs, and all transportation costs.

There are 22,449 variables (3518 binary variables and 18,931 real variables) and 20,335 constraints in the corresponding MILP model. The base case was solved with module CPLEX of the commercial software GAMS in 0.438 seconds (CPU time) on a personal computer with Pentium IV 3.0 CPU and 1024 KB RAM. The most appropriate types and quantities of crude oils, intermediates, and products to be purchased in each week are presented in Tables 8.3 and 8.4. It can be seen that the total amounts of purchased enthylene, light naphtha, and heavy naphtha in three months are 69,000 kg, 80,000 m^3, and 11,000 m^3 respectively. The optimal throughputs of all production units in 12 scheduling intervals can be identified from the solution of the MILP model, and these data are summarized in Table 8.5. Let us use the production schedule of unit FCC1 as an example to illustrate these results. Notice that this FCC unit is located in refinery KSR (see Figure 8.10). It can be observed from Table 8.5 that FCC1 has been chosen to operate in all 12 scheduling intervals, and the total throughputs in the three planning periods are found to be 6,593,506 m^3, 6,530,645 m^3, and 6,600,000 m^3 respectively. Notice also that all aromatic extraction units (i.e., AET1, AET2, and AET3) and all xylene fractionation units (i.e., XF1 and XF2) are required to be operated at full capacities in all 12 weeks. Thus, it is clear that these two groups of units are the bottlenecks of the supply chain. It should be noted that the units (and also the network configurations of the supply chain) chosen during the three planning periods are *not* the same. The best delivery schedules for the products and by-products can be found in Table 8.6. In particular, the qualities and quantities of products sent to every customer and the corresponding delivery time and terminal are shown in Figure 8.13, e.g., the amounts of propylene sent to customer Kuoco3 in week 2 via terminal LIWR and KSR are 284,980 m^3 and 317,154 m^3 respectively. It is interesting to note that the suggested delivery amounts almost always fall within the second range (with the highest selling price) so as to maximize profit. The total profit over the planning horizon in this base case is predicted to be 15,525,720,311 US dollars.

Several scenarios have been studied with the proposed model, and the results of five of them are presented below. In the first case, we assumed that the first reforming unit (RF-1) in the first planning period, the second atmospheric distillation unit (AD-2) in the second period, and the first aromatic extraction unit (AET-1) in the

TABLE 8.3
Optimal Amounts of Crude Oils to Be Purchased in the Base Case (10^3 m^3/wk)

	Month	tp1				tp2				tp3			
	Week												
Material	Supplier	ts1	ts2	ts3	ts4	ts5	ts6	ts7	ts8	ts9	ts10	ts11	ts12
cro 2	Sup1		500	650		400							
	Sup3	800	200			250							
	Sup4			300	600								
cro 3	Sup1	900			500					100	250		400
	Sup2							700	700	450			
	Sup3						300						
	Sup4		900										
	Sup5					750	500			750		1600	400
cro 4	Sup1	400		850						500		100	150
	Sup2	400		400				550	400	750	700		200
	Sup3		900		600	350				350			600
	Sup4		800			400							300
	Sup5	750			300			500			550	800	400
cro 5	Sup1		250	350						300	750	350	400
	Sup2	400	400	400		550		500	400	750			200
	Sup3					450			400				
	Sup4	500			300								
	Sup5						700			650	1650	600	600
cro 7	Sup1	100	250										
	Sup2					450							
	Sup3					650			600				
	Sup4	500										500	
	Sup5							500		650			

(Continued)

TABLE 8.3 (CONTINUED)
Optimal Amounts of Crude Oils to Be Purchased in the Base Case (10^3 m^3/wk)

	Month	tp1				tp2				tp3			
Material	Week / Supplier	ts1	ts2	ts3	ts4	ts5	ts6	ts7	ts8	ts9	ts10	ts11	ts12
cro 8	Sup1	200	250	350							350	350	400
	Sup2	700		200		650	500	200	1600		550	300	200
	Sup3						500	400					
	Sup4	400	1800	400	300								
	Sup5						600	500	400	550	350	800	700
cro 9	Sup1	200	250	350								350	
	Sup2	400	700										
	Sup3							400					
	Sup4	500		600									
	Sup5					550			300				
cro 10	Sup1	500	250	350								350	
	Sup2	600	400	400								500	
	Sup3								300				
	Sup4	600	700	200	1800								
Total(month)		28750				16800				22500			
Sum		68050											

TABLE 8.4
Optimal Amounts of Intermediate Oils and Final Products to Be Purchased in the Base Case

Material	Month / Week / Supplier	tp1 ts4	tp2 ts5	ts6	ts7	ts8	tp3 ts9	ts10	ts11	ts12
en	Sup2	21								
(10^3Kg)	Sup4		42							
	Sup5	6								
	Total (month)	27		42						
ln1	Sup1						6.5	6.5		15.5
(10^3m^3)	Total (month)							18.5		
ln2	Sup1		6			0.5		5.5		
(10^3m^3)	Sup2		11			0.5		6.5	5.5	
	Sup3				16					
	Sup4		1					1.5	6	
	Sup5							1.5		
	Total (month)			35				26.5		
hn3	Sup1	1					1.5	1.5		
(10^3m^3)	Sup2						1.5	1.5		
	Sup4						1.5	1.5		
	Sup5							1.5		
	Total (month)	1						10.5		

third period are not operable due to scheduled maintenance. The resulting optimal throughputs of all production units in the 12 scheduling intervals are shown in Table 8.7. Notice that, due to the availabilities of other units (i.e., the second reforming unit RF-2 and the first atmospheric distillation unit AD-1), the required schedules can be maintained in the first two planning periods. However, as a result of AET-1 outage, the production capacities of the supply chain have to be reduced in the third period and, consequently, it becomes necessary to import final products, e.g., benzene and xylenes, to satisfy customer demands. The optimal levels of purchased crude oils, intermediate oils, and final products in this scenario are presented in Table 8.8. It was found that the total profit is about 2.49 thousand million US dollars less than that achieved with the complete set of production units in the base case.

In the second scenario, it was assumed that no suitable petrochemical products (i.e., ethylene, propylene, butadiene, benzene, toluene, and xylenes) could be purchased in all planning periods. Thus, the quantities of raw materials consumed in

TABLE 8.5
Optimal Throughputs of All Production Units in the Base Case

Week / Unit	1st Month				2nd Month				3rd Month			
	1	2	3	4	5	6	7	8	9	10	11	12
AD1												
AD2	5858900	5517500	4676230	3512006	5585090	2462527	1727159	4688248	2877576	2029846	4458590	3033280
AD3	2040410	2713190	1958563			502382	2626181	282479	3039260	2973517	2248038	2860440
VD1	1700000	1700000	1692549	1416827	1700000	1563805	1700000	1677900	1700000	1700000	1700000	1700000
VD2	1800000	1800000	1676694					450000	450000	450000	450000	450000
VD3	524777	702180		432000	1188007		523632	432000	951679	467491	1305928	956823
CK1	375000	375000	375000	375000	375000	375000	375000	375000	375000	375000	375000	375000
CK2	322142	316445	125000			143099		125000	125000		125000	125000
CK3	450000	450000	450000	192671	383642		274815	144710	450000	414186	450000	440823
FCC1	1650000	1650000	1650000	1643506	1650000	1580645	1650000	1650000	1650000	1650000	1650000	1650000
FCC2	565276	1625000	225000				225000		828845	292654	793051	631263
NC1	359112	375794	350000	350000	413696		325000	350929	350000		350000	350000
NC2												
NC3			354773						300000	300000		351566
BD1	11904							23266				
BD2	11904							30000	14010			28471
BD3		11904	11822	23205	27428		25468	32000	32000	32000	32000	32000
HTD1				350000					350000			
HTD2	1133427	1076661	927546	274136	1109124	273697	360627	573889	275323	416286	878531	596876
HTD3	424786	558329	364359			350000	492114	350000	623179	584383	455284	563152
RF1	857483	406333	305000	411515	387878	355555	315255	376835	364664	348205	363958	387878
RF2		802224	682249		457208		405578	310000	479862	310000	633935	421627

(Continued)

TABLE 8.5 (CONTINUED)
Optimal Throughputs of All Production Units in the Base Case

Week / Unit	1st Month				2nd Month				3rd Month			
	1	2	3	4	5	6	7	8	9	10	11	12
RF3	295000											
AET1	240000	241255	264642	381048	240000	240000	206460	240000	361963	255407	264173	329573
AET2	220000	212357	220000	220000	220000	220000	220000	220000	220000	220000	220000	220000
AET3	220000	220000	220000	220000	220000	220000	220000	220000	220000	220000	220000	220000
XF1	110000	110000	110000	110000	110000	110000	110000	110000	110000	110000	110000	110000
XF2	100000	100000	100000	100000	100000	100000	100000	100000	100000	100000	100000	100000
TT1	60534	108000	87740	41151	88038	60029	79455	108000	108000	108000	108000	86582
TT2	54000	93450	79209		65117	35000	51600	52157	52157	52157	52157	51739
PR1	70682	72739	63131	68473	74249	75605	63010	71604	79604	71604	71604	71158
PR2	55639	55815	48710	53781	62160	59170	55674	45735	65735	55735	55735	55689
IM1	68392	67242	55684	81423	62161	74279	68165	67764	67764	67764	67764	68065
IM2												

TABLE 8.6
The Suggested Amounts of Products and By-Products to Be Delivered to the Customers in Each Planning Period in the Base Case

Product (unit)	Month	tp1		tp2		tp3	
	Customer	Amount	Level	Amount	Level	Amount	Level
Ethylene	Kuco2	21525	2	21000	2	15225	2
(en) (kg)	LIWRco3	212520	2	411705	2	303450	2
Propylene (pn) (m^3)	Kuco3	650900	3	174800	3	28750	3
	Kuco5	195500	3	23000	3	1046960	3
	LIWRco2	11270	3	46000	3	69230	3
Butadiene (bd) (m^3)	Kuco3	1050	2	1575	2	3990	2
Benzene	Kuco2	22050	2	22050	2	23100	2
(bz) (m^3)	Kuco5	65940	2	21000	2	71295	2
Toluene	Kuco1	3150	2	3150	2	11970	2
(tl) (m^3)	LIWRco2	6300	2	6300	2	6510	2
Mixed	Kuco1	14700	2	21000	2	18900	2
xylenes	Kuco5	31500	2	35280	2	44415	2
(mx) (m^3)	LIWRco2	3402	2	2520	2	79380	2
p-Xylene	Kuco1	4200	2	4200	2	4200	2
(px)(m^3)	Kuco3	46200	2	42000	2	84000	2
o-Xylene	Kuco1	13230	2	10500	2	22050	2
(ox) (m^3)	Kuco3	26880	2	6300	2	26250	2
	LIWRco1	22890	2	11550	2	58275	2

this situation must be larger than those used in the base case. The optimal supply rates of the purchased materials can be identified from the MILP solution (see Table 8.8). It can be observed that the total amounts of purchased crude oils and naphtha (intermediates) are more than those required in the base case by 13,300,000 m^3 and 12,000 m^3 respectively. The optimal throughputs of all production units in this case are shown in Table 8.9. When compared with Table 8.5, it can be observed that the throughputs of production units here are almost all larger than those in the base case. Profit in this case is reduced to 15,520,219,285 US dollars.

The impacts of demand variations are examined in the third case study. Let us assume that all ethylene orders from customer LIWRco3 are cancelled unexpectedly just before the first week due to its equipment problems. As a result, the total demand for ethylene drops to 94% of the original level. This reduction in product demand in turn forces some production units in the supply chain to lower their processing rates and, consequently, the total amounts of purchased crude oils and ethylene must be decreased accordingly. Since the production rates of other intermedia oils, i.e., light naphtha, are also reduced as a result, it becomes necessary to make up by increasing their purchase levels (see Table 8.8). Notice that the total amount of purchased crude oils is indeed less than that in the base case by 1,100,000 m^3 (1.6%) and no

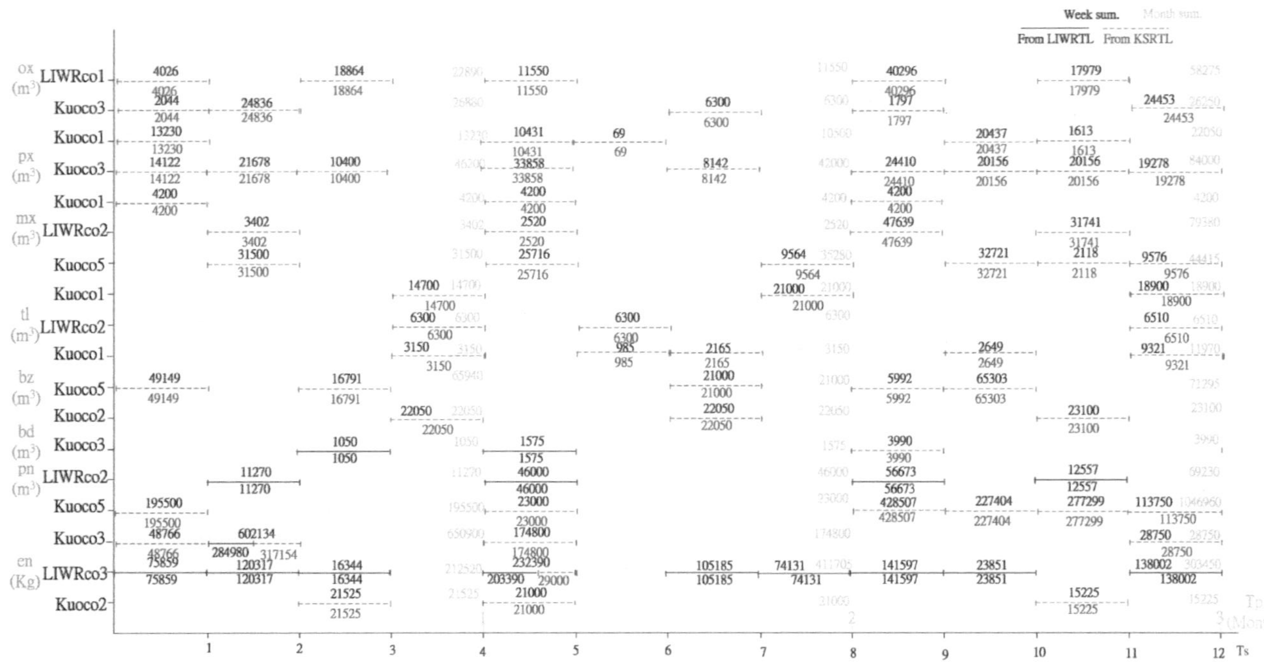

FIGURE 8.13 Suggested amounts of products and by-products to be delivered in base case. (Reproduced with permission from *Ind. Eng. Chem. Res.* 2008, 47: 1935–1954. Copyright 2008 American Chemical Society.)

TABLE 8.7
Optimal Throughputs of All Production Units in Scenario 1

Unit \ Week	1st Month				2nd Month				3rd Month			
	1	2	3	4	5	6	7	8	9	10	11	12
AD1					394820	1582110		2427666		255000	255000	
AD2	5864200	6483812	5994170	4795230	Scheduled maintenance				3033280	3033280	3033280	2162327
AD3	2049170	322817	1281024		2887088	3455750	3109809	3496130	1826742	1977902	3012106	2098571
VD1	1700000	1700000	1700000	1654321	1700000	1700000	1656785	1700000	1700000	1700000	1700000	1700000
VD2	1800000	1800000	1576855	450000				518804				
VD3	552133		432000	432000		914114		854339	880043	1067148	1600000	580479
CK1	375000	375000	375000	375000	375000	375000	375000	375000	375000	375000	375000	375000
CK2	255095	233035	150773		125000		125000	125000			125000	
CK3	450000	450000	450000	410370		366449		284071	441335	416486	450000	287645
FCC1	1650000	1650000	1650000	1650000	1650000	1650000	1650000	1650000	1650000	1650000	1650000	1650000
FCC2	399321	1546306	225000	239938		225000			803922	567257	562341	225000
NC1	429617	468967	430535	350000	325000		325000				363069	
NC2								340000			350000	
NC3						300000		300000		333177		
BD1	4674	7230						30000				
BD2	11904						24645	30000			8621	
BD3	11904			13613	25469	31800	32000	32000		19991	32000	13969
HTD1				350000		357903		438980		350000		
HTD2	1134307	1250000	1099640	521058					631877	336192	596877	432222
HTD3	426416		350000		686283	660679	606355	654629	410256	402405	646904	411421
RF1	Scheduled maintenance				399324	305000	376131	323333				
RF2	826394	984615	884519	663441		483210		487390	421627	315000	398586	315000

(Continued)

TABLE 8.7 (CONTINUED)
Optimal Throughputs of All Production Units in Scenario 1

Week / Unit	1st Month				2nd Month				3rd Month			
	1	2	3	4	5	6	7	8	9	10	11	12
RF3	308542		295000						295000	401627	411627	305000
AET1	240000	240000	246903		367100	255408	257100	308003		Scheduled maintenance		
AET2	220000	220000	220000		220000	220000	220000	220000				
AET3	220000	220000	220000		220000	220000	220000	220000				
XF1	110000	110000	110000	25000	98572	110000	110000	110000	75180			
XF2	100000	100000	100000	20023	100000	100000	100000	100000	38834			
TT1	52826	99800	108000		40558	108000	25181	108000				
TT2	61708	68763	120000		35000	93758	35000	114561				
PR1	69526	69968	60896		70576	63792	76145	66017	58088			
PR2	56786	57561	52874	30000	60286	48676	62786	49596	30000			
IM1	68392	67649	42239	14551	67202	54947	81386	42314	42546			
IM2			12000	12000				12000	12000			

TABLE 8.8
Optimal Amounts of Crude Oils, Intermediates and Final Products to Be Purchased for Different Cases (10^3 m^3 or 10^3 kg)

	Material \ Period		tp1	tp2	tp3	Total
Base case	Crude oil		28750	16800	22500	68050
	en		27	42		69
	ln1				18.5	18.5
	ln2			35	26.5	61.5
	hn3		1		10.5	11.5
Case 1	Crude oil		28050	27550	19700	75300
	en		21	18.5	1	40.5
	ln1		1	33.5		34.5
	ln3		51	19.5	48	118.5
	hn1			10		10
	hn3		2	11		13
	bz			11.5		11.5
	mx		3	25	74	102
	ox				9	9
Case 2	Crude oil		27800	31850	21700	81350
	ln2			27.5	24.5	52
	ln3		18		20	38
	hn2		3	10.5		13.5
Case 3	Crude oil		26950	18000	22000	66950
	ln1		13	50		63
	ln2			16.5	5.5	22
	ln3		35.5		45.5	81
	hn2				27.5	27.5
	hn3			4	18	22
Case 4	Crude oil		24850	20100	21500	66450
	ln2			25	3	28
	ln3			4		4
	hn2		24	10	39.5	73.5
	hn3		11	22		33
	mx				34	34
Case 5	Crude oil	Purchased	21450	13200	15000	49650
		Contract	5000	5000	5000	15000
		Month sum	26450	18200	20000	64650
	ln1	Purchased		15.5		15.5
		Contract	20	20	20	60
		Month sum	20	35.5	20	75.5
	en		27	81		108
	ln2			25.5		25.5
	ln3		18.5	18.5		37
	hn2		13.5			13.5
	hn3			27.5		27.5

TABLE 8.9
Optimal Throughputs of All Production Units in Scenario 2

Week / Unit	1st Month				2nd Month				3rd Month			
	1	2	3	4	5	6	7	8	9	10	11	12
AD1					343040							
AD2	44846190	5026326	1996173	5264200	4150480	3129724	4476410	2068954	2947872	2029846	33702866	3937016
AD3	1916810	2442633	4764147			279800	329541	2390846	2026251	2266230	2963887	2266230
VD1	1700000	1700000	1700000	1700000	1700000	1700000	1700000	1700000	1700000	1700000	1700000	1700000
VD2	450000	511381	450000	450000				590988	899050		450000	450000
VD3	1312106	1600000	1307571	588253	612642	931222	776899			552635	1338791	1085224
CK1	375000	375000	375000	375000	375000	375000	375000	375000	375000	375000	375000	375000
CK2	163402	266722	134255					241193			162429	125000
CK3	450000	450000	450000	450000	267384	339205	296834		448680	279118	450000	450000
FCC1	1650000	1650000	1650000	1650000	1650000	1650000	1650000	1650000	1650000	1650000	1650000	1650000
FCC2	247863	1625000	535172			225000			613916	388745	857056	682605
NC1	350000	359681		379217	350000	350000	350000		357723		357439	350000
NC2												
NC3		300000	300000					338809		304238		304238
BD1	11300							30000				
BD2	11904							30000	14010			28471
BD3		11904	13543	25142	23205	23205	29119	32000	23717	32000	23948	32000
HTD1				350000		431811			350000			
HTD2	951505	991057	412084	624707	931668	593023	913718	381277	301586	416286	723284	778289
HTD3	424786	558329	364359					439986	450221	459857	584017	459557
RF1	305000	344755	704465	377222	333889	377222	364242	348205	387879	343210	400105	365484
RF2	686228	688633	324661	346157	310000	306848	330960	310949	315000	315000	517595	556616

(Continued)

TABLE 8.9 (CONTINUED)
Optimal Throughputs of All Production Units in Scenario 2

Week / Unit	1st Month				2nd Month				3rd Month			
	1	2	3	4	5	6	7	8	9	10	11	12
RF3												
AET1	240000	240000	241888	390038	264642	264642	254642	272575	266069	256190	266016	320833
AET2	220000	220000	220000	220000	220000	220000	220000	220000	220000	220000	220000	220000
AET3	220000	220000	220000	220000	220000	220000	220000	220000	220000	220000	220000	220000
XF1	110000	110000	110000	110000	110000	110000	110000	110000	110000	110000	110000	110000
XF2	110000	110000	110000	110000	110000	110000	110000	110000	110000	110000	110000	110000
TT1	63250	63250	62863	78720	58146	10800	43674	46952	108000	107362	108000	86582
TT2	51283	51283	51276	84412	57368	118132	35000	52157	52158	52145	52158	52158
PR1	70673	70673	65665	65315	78566	64353	70296	69476	79605	71592	71605	71605
PR2	55639	55639	54109	50489	66414	48202	63362	46683	65735	55735	55735	55735
IM1	68393	68393	68398	67804	70699		83280	68487	67764	67764	67764	67764
IM2						51830						

TABLE 8.10
The Shipment Schedules of Crude Oil and Light Naphtha in the Last Case Study

	Month	tp1				tp2				tp3			
Material	Week / **Supplier**	**ts1**	**ts2**	**ts3**	**ts4**	**ts5**	**ts6**	**ts7**	**ts8**	**ts9**	**ts10**	**ts11**	**ts12**
cro 1 (10^3 m^3)	Sup1	2500		2500		2500		2500		2500		2500	
	Total (month)		5000				5000				5000		
ln1	Sup1		20				20				20		
(10^3 m^3)	Total (month)		20				20				20		

ethylene needs to be purchased. In addition, about 86,000 m^3 extra light naphtha (about 107.5% more) must be bought from the suppliers. The corresponding total profit is about 444.99 million US dollars less than that of the base case.

The fourth scenario is concerned with export reductions. Let us assume that the maximum exportable amounts of benzene and toluene are both less than their original predictions in the base case by 800,000 m^3 (i.e., 80% lower) in every planning period. As a result of these changes, it is not beneficial to operate the supply chain at a high throughput level. In this case, the total amount of purchased crude oils was found to be less than that in the base case by 1,600,000 m^3 (2.35%). Consequently, the total amounts of purchased naphthas and mixed xylenes must be increased to 47,000 m^3 and 34,000 m^3 respectively during the three-month period (see Table 8.8). The profit in this case is lowered to 13,855,925,468 US dollars.

In order to secure enough raw materials to meet the demand targets, it is a common practice for a refinery to sign long-term contacts with its upstream suppliers. In this last case study, it is assumed that the suppliers of crude oil and light naphtha are required to ship fixed amounts of their products according to predetermined schedules (see Table 8.10). Under this condition, the amounts of raw materials purchased from other sources must be reduced. It can be found in the optimization results that the total amount of purchased crude oils over 3 months in this case is 64,650,000 m^3 (in which 15,000,000 m^3 is obtained via the long-term contracts). Since this quantity is at approximately 95% of the base-case level, the amounts of purchased intermediate oils must be increased. For examples, the extra amounts of purchased ethylene, light naphtha (including the long-term supplies), and heavy naphtha were found to be higher than their original levels by 39,000 kg (56.5%), 58,000 m^3 (72.5%), and 29,500 m^3 (256.5%) (see Table 8.8). In other words, the flexibility in selecting raw materials during the planning/scheduling stage is to some degree lost due to the need to maintain supply stability. The total profit in this case is 15,188,162,035 USD, which is lower than that in the base case by 337 million USD.

8.6 CONCLUDING REMARKS

An integrated mixed-integer linear program has been presented in this chapter to coordinate various planning and scheduling decisions for optimizing the performance of a comprehensive petroleum supply network in typical conversion refineries. Several realistic scenarios can be efficiently examined accordingly. It is clear that the proposed approach can be used not only to generate the proper procurement and delivery plans on the basis of given supply and demand rates, but also simultaneously to select the optimal schedules for producing various petrochemical products over the specified planning horizon. This capability is believed to be superior to that achieved by the sequential procedures.

REFERENCES

Beamon, B.M. 1998. Supply chain design and analysis: models and methods. *International Journal on Production Economics* 55: 281–294.

Bechtel 1993. *PIMS (Process Industry Modeling System). User's Manual.* Version 6.0. Houston, TX: Bechtel Corp.

Bonfill, A., Bagajewicz, M.J., Espune, A., Puigjaner, L. 2004. Risk management in the scheduling of batch plants under uncertain market demand. *Industrial and Engineering Chemistry Research* 43: 741–750.

Bonner and Moore Management Science. 1979. *RPMS (Refinery and Petrochemical Modeling System): A System Description.* Houston, NY: Bonner and Moore Management Science.

Carvalho, M.C.A., Pinto, J.M. 2006. An MILP model and solution technique for the planning of infrastructure in offshore oilfields. *Journal of Petroleum Science and Engineering.* 51: 97–110.

Chopra, S., Meindl, P. 2004. *Supply Chain Management – Strategy, Planning and Operation.* 2nd ed. Upper Saddle River, NJ: Pearson Education, Inc.

Dogan, M.E., Grossmann, I.E. 2006. Design of multi-echelon supply chain networks under demand uncertainty. *Industrial and Engineering Chemistry Research* 45: 299–315.

Franck, H.G., Stadelhofer, J.W. 1988. *Industrial Aromatic Chemistry: Raw Materials, Processes, Products.* Berlin: Spring-Verlag.

Gary, J.H., Handwerk, G.E. 2001. *Petroleum Refining Technology and Economics.* 4th ed. New York: Marcel Dekker.

Göthe-Lundgren, M., Lundren, J.T., Persson, J.A. 2002. An optimization model for refinery production scheduling. *International Journal of Production Economics* 78: 255–270.

Ho, J.C., Chang, Y.L. 2001. An integrated MRP and JIT framework. *Computers and Industrial Engineering* 41: 173–185.

Jackson, J., Grossmann, I.E. 2003. Temporal decomposition scheme for nonlinear multisite production planning and distribution models. *Industrial and Engineering Chemical Research* 42: 3045–3055.

Jia, Z., Ierapetritou, M. 2004. Efficient short-term scheduling of refinery operations based on a continuous time formulation. *Computers and Chemical Engineering* 28: 1001–1019.

Kuo, T.H., Chang, C.T. 2008a. Optimal planning strategy for the supply chains of light aromatic compounds in petrochemical industries. *Computers and Chemical Engineering* 32(6): 1147–1166.

Kuo, T.H., Chang, C.T. 2008b. Application of a mathematic programming model for integrated planning and scheduling of petroleum supply networks. *Industrial and Engineering Chemistry Research* 47(6): 1935–1954.

Li, W.W., Hui, C.W., Li, A. 2005. Integrating CDU, FCC and product blending models into refinery planning. *Computers and Chemical Engineering* 29: 2010–2028.

Magalhães, M.V. and Shah, N. 2003. Crude oil scheduling. In I.E. Grossmann and C.M. McDonald, editors. *Proceedings of* Fourth International Conference *on* Foundations *of* Computer-Aided Process Operations, 323–326. Coral Springs: CAChE.

Maples, R.E. 2000. *Petroleum Refinery Process Economics.* Tulsa, OK: PennWell Corp.

Más, R., Pinto, J.M. 2003. A mixed-integer optimization strategy for oil supply in distribution complexes. *Optimization and Engineering* 4: 23–64.

Méndez, C.A., Grossmann, I.E., Harjunkoski, I., Kaboré, P. 2006. A simultaneous optimization approach for off-line blending and scheduling of oil-refinery operations. *Computers and Chemical Engineering* 30: 614–634.

Neiro, S.M.S., Pinto, J.M. 2004. A general modeling framework for the operational planning of petroleum supply chain. *Computers and Chemical Engineering* 28: 871–896.

Persson, J.A., Göthe-Lundgren, M. 2005. Shipment planning at oil refineries using column generation and valid inequalities. *European Journal of Operational Research* 163: 631–652.

Perry, R.H., Green, D.W., Maloney, J.O. 1997. *Perry's Chemical Engineering's Handbook.* 7th ed. New York: McGraw-Hill.

Pinto, J.M., Moro, L.F.L. 2000. A planning model for petroleum refineries. *Brazilian Journal of Chemical Engineering* 17(4–7): 575–585.

Pinto, J.M., Joly, M., Moro, L.F.L. 2000. Planning and scheduling models for refinery operations. *Computers and Chemical Engineering* 24: 2259–2276.

Speight, J.G., Ozum, B. 2002. *Petroleum Refining Processes.* New York: Marcel Dekker.

Tan, K.C. 2001. A framework of supply chain management literature. *European Journal of Purchasing and Supply Management* 7: 39–48.

Tang, L., Liu, J., Rong, A., Yang, Z.. 2001. A review of planning and scheduling systems and methods for integrated steel production. *European Journal of Operation Research* 133: 1–20.

Watkins, R.N. 1979. *Petroleum Refining Distillation.* 2nd ed. Houston: Gulf Publishing Co.

9 A Decentralized Petroleum Supply-Chain Management Model for Maximum Overall Profit and Reasonable Benefit Allocation

Traditional petrochemical industries are facing steep challenges today due to the irreversible trend of continuing integration of the global economy. How to manage the petroleum supply chain has become an important issue for oil and petrochemical companies in recent years. The supply chain management (SCM) tools presented in this chapter are aimed to allocate benefits fairly among the participating members while maximizing the overall profit. In particular, game theory is adopted to facilitate the development of long-term and stable cooperative relations among the supply chain members. Nash (1951) proposed a nonlinear model for analyzing the Nash equilibrium in a multiplayer non-zero sum non-cooperative game. In 1953, he also showed that there is a unique solution to the two-person bargaining problem (Nash, 1953). Since this is the first time a game-related issue was described in mathematical language, the work is regarded as a milestone. Roth (1979) proposed the generalized Nash bargaining equation considering the asymmetry in the status of the members and extended the two-person game to the multiplayer game. The currently popular trend in SCM is for buyers and sellers to form alliances in order to increase negotiation efficiency and improve overall profitability. In addition, to tackle the problems in constructing cooperative coalitions, the widely used benefit distribution methods, i.e., core solution and Shapley value, are also discussed in the present chapter. As an illustrative example, the petroleum supply network discussed in the previous chapter is divided into several processes owned by different companies. A reasonable benefit allocation plan can be determined by computing the Shapley value of all members (companies) and then checking the stability of the alliance with core. Finally, the Nash bargaining equation was used to analyze the asymmetry of the negotiation powers among members.

9.1 PROFIT ALLOCATION METHODS AMONG SUPPLY CHAIN MEMBERS: A SIMPLE EXAMPLE

Let us assume that, as an example, three members in a supply chain agree to cooperate and form a coalition so as to gain a larger overall profit, and also to allocate the additional profit amongst them fairly. For illustration convenience, let us consider this case with arbitrarily selected parameters as follows:

- Before forming a coalition

  It is assumed that the profit of actor i (denoted as $Profit_i$) working independently can be calculated as follows

$$Profit_i = \omega_i^{out} F_i^{out} - \omega_i^{in} F_i^{in} - OC_i F_i^{in} \tag{9.1}$$

where i = 1,2,3; ω_i^{in} and ω_i^{out} respectively denote the unit market prices of feedstock purchased and product sold by actor i before forming coalition ($/m^3); F_i^{in} and F_i^{out} respectively denote the volumetric flowrates of feedstock to and product from actor i before forming coalition (m^3/day); OC_i is the operating cost of actor i per unit volume of feedstock ($/m^3). Let us further assume that the operating costs of the three actors in this example are 0.03, 0.05, and 0.045, respectively, and the corresponding market prices and input and output flowrates of all actors are listed in the sketch given in Figure 9.1. The resulting profits can be determined by substituting the above parameters into equation (9.1), and their values can be found to be 119,940, 259,900, and 599,910, respectively. Therefore, under the condition that all three actors work independently, their total profit is 979,750 $/day.

It can be observed from Figure 9.1 that, since actor 1 obviously wishes to raise its product price and actor 2 to lower its feedstock cost, there is incentive for the two parties to cooperate and form a coalition. Similarly, there is also incentive for actor 2 and actor 3 to cooperate.

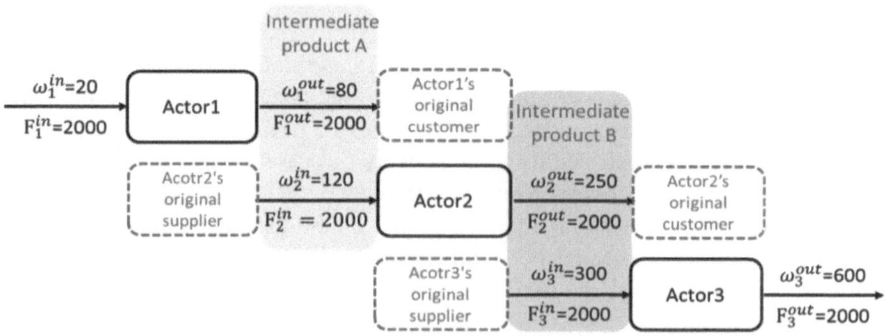

FIGURE 9.1 Three actors in a fictitious market without forming coalition.

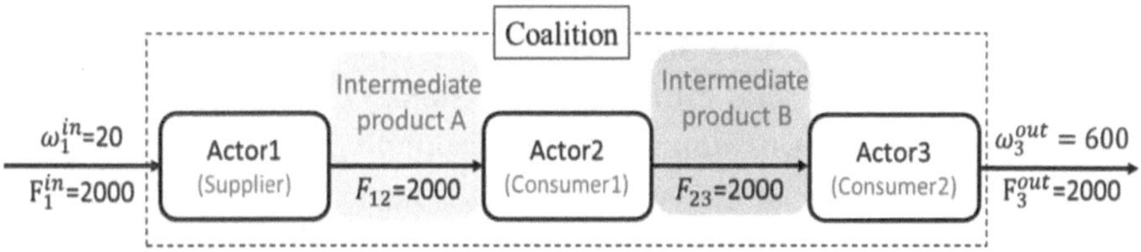

FIGURE 9.2 A three-actor supply chain in a fictitious market.

- After forming a coalition
  As shown in Figure 9.2, the prices of intermediates A and B can be ignored after the above actors join the same coalition. The goal of running the supply chain becomes maximization of the overall profit of coalition (*TP*) which can be expressed as

$$TP = \omega_3^{out} * F_3^{out} - \omega_1^{in} * F_1^{in} - \sum_{i=1}^{3} OC_i * F_i^{in} \tag{9.2}$$

  By substituting the aforementioned parameter values into equation (9.2), it can be found that the total profit of the coalition is 1,259,750 \$/day, which represents a 28.6% increase from the total profit before forming a coalition. The main question in this situation is obviously how to allocate the extra profit fairly.

9.1.1 Benefit Allocation Plan

The benefit allocation plan for any given supply chain can be produced on the basis of the corresponding Shapley values. A detailed computation procedure of these values can be found in Section 5.1.2. A concise version is summarized below:

1. Determine the total profits of coalitions formed by all possible combinations of actors.
2. Determine the marginal benefits of each actor according to the results of step 1 and equations (5.6) and (5.7) in Section 5.1.2 for all possible precedence orders of this actor joining the coalition.
3. Determine the Shapley value of each actor by taking the arithmetic average of the marginal benefits obtained in step 2 according to equation (5.8) in Section 5.1.2.

By applying step 1, the total profits of all possible coalitions can be determined, and they are summarized in Table 9.1.

Notice that, since there are no intermediate transfers between actors 1 and 3, the total profit of coalition {1,3} in the above table is essentially the sum of profits obtained by the two actors working independently.

TABLE 9.1
Total Profits of All Possible Coalitions in Simple Example

Coalition	Total Profit ($/day)
{1}	119,940
{2}	259,900
{3}	599,910
{1,2}	459,840
{1,3}	719,850
{2,3}	959,810
{1,2,3}	1,259,750

TABLE 9.2
Marginal Benefits for Actors Joining a Coalition under All Possible Precedence Orders

	Marginal Benefit ($/day)		
Precedence Order	**Actor 1**	**Actor 2**	**Actor 3**
1,2,3	119,940	339,900	799,910
1,3,2	119,940	539,900	599,910
2,1,3	199,940	259,900	799,910
2,3,1	299,940	259,900	699,910
3,1,2	119,940	539,900	599,910
3,2,1	299,940	359,900	599,910

By applying step 2, the marginal benefits of each actor associated with all possible precedence orders can be obtained, and they are given in Table 9.2.

In step 3, the Shapley value of each actor is determined by taking the arithmetic average of the marginal benefits listed in the corresponding column of Table 9.2. The Shapley values of the three actors in this example, along with the corresponding fractions of the total profit of coalition {1,2,3}, are presented in Table 9.3.

It can be observed from the Shapley values in Table 9.3 that all actors gain extra profits by joining the coalition {1,2,3}. Thus, this benefit allocation scheme should at least be acceptable to every member in the supply chain if profit making is the only criterion for forming a long-lasting coalition.

9.1.2 Coalition Stability

The additional requirements implied by a "core" are needed to ensure coalition stability. The definition of core has already been given in Section 5.1.1 in Chapter 5.

TABLE 9.3
Shapley Values and the Corresponding Fractions in Total Profit

	Actor 1	Actor 2	Actor 3	Total Profit
Shapley value ($/day)	193,273	383,233	683,243	1,259,750
Fraction	0.15	0.30	0.54	1.00

Specifically, the above-mentioned additional requirements include the individual rationality described in equation (5.1), the group rationality in equation (5.2), the coalitional rationality in equation (5.3), and the no subsidy principle in equation (5.4). In the simple example discussed previously in Section 9.1.1, the following constraints can be formulated according to equations (5.1)–(5.3).

$$x_{\{1,2,3\},1} \geq \nu\left(\{1\}\right) \tag{9.3}$$

$$x_{\{1,2,3\},2} \geq \nu\left(\{2\}\right) \tag{9.4}$$

$$x_{\{1,2,3\},3} \geq \nu\left(\{3\}\right) \tag{9.5}$$

$$x_{\{1,2,3\},1} + x_{\{1,2,3\},2} \geq \nu\left(\{1,2\}\right) \tag{9.6}$$

$$x_{\{1,2,3\},1} + x_{\{1,2,3\},3} \geq \nu\left(\{1,3\}\right) \tag{9.7}$$

$$x_{\{1,2,3\},2} + x_{\{1,2,3\},3} \geq \nu\left(\{2,3\}\right) \tag{9.8}$$

$$x_{\{1,2,3\},1} + x_{\{1,2,3\},2} + x_{\{1,2,3\},3} = \nu\left(\{1,2,3\}\right) \tag{9.9}$$

Substituting equation (9.9) into equations (9.3)–(9.8) yields

$$x_{\{1,2,3\},2} + x_{\{1,2,3\},3} \leq \nu\left(\{1,2,3\}\right) - \nu\left(\{1\}\right) \tag{9.10}$$

$$x_{\{1,2,3\},1} + x_{\{1,2,3\},3} \leq \nu\left(\{1,2,3\}\right) - \nu\left(\{2\}\right) \tag{9.11}$$

$$x_{\{1,2,3\},1} + x_{\{1,2,3\},2} \leq \nu\left(\{1,2,3\}\right) - \nu\left(\{3\}\right) \tag{9.12}$$

$$x_{\{1,2,3\},3} \leq \nu\left(\{1,2,3\}\right) - \nu\left(\{1,2\}\right) \tag{9.13}$$

$$x_{\{1,2,3\},2} \leq \nu\left(\{1,2,3\}\right) - \nu\left(\{1,3\}\right) \tag{9.14}$$

$$x_{\{1,2,3\},1} \leq \nu\left(\{1,2,3\}\right) - \nu\left(\{2,3\}\right) \tag{9.15}$$

TABLE 9.4
Boundaries of Core for Coalition {1,2,3} in the Simple Example

	Actor 1		Actor 2		Actor 3	
Boundaries	$x_{\{1,2,3\},1}$ **($/day)**	**Fraction**	$x_{\{1,2,3\},2}$ **($/day)**	**Fraction**	$x_{\{1,2,3\},3}$ **($/day)**	**Fraction**
Upper	299,940	0.24	539,900	0.43	659,840	0.52
Lower	119,940	0.10	259,900	0.21	459,840	0.37

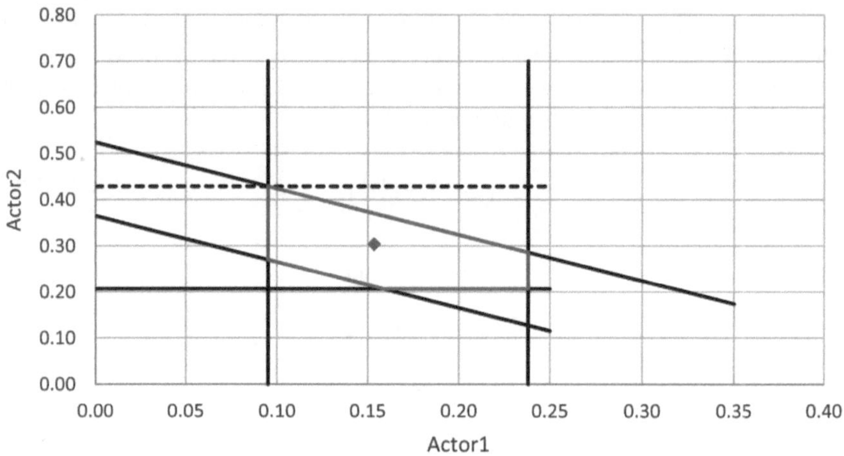

FIGURE 9.3 Core of coalition {1,2,3} in the simple example.

Notice that equations (9.13)–(9.15) can also be deduced from equation (5.4). One can next substitute the total profits of all coalitions (listed in Table 9.1) into equations (9.3), (9.4), (9.6), (9.12), (9.14), and (9.15) to determine the upper and lower bounds of $x_{\{1,2,3\},1}$, $x_{\{1,2,3\},2}$, and $x_{\{1,2,3\},1} + x_{\{1,2,3\},2}$ (or $x_{\{1,2,3\},3}$), and then divide these bounds by $v(\{1,2,3\})$ to determine the corresponding allocated fractions of total profit. The results of the above calculations are summarized in Table 9.4, and the upper and lower limits of allocated profit fractions are also plotted in Figure 9.3. Notice that the intersection of the ranges between these limits, i.e., the core, is marked with red line segments in this figure. Finally, the allocated profit fractions determined on the basis of the Shapley values (see Table 9.3) can also be used to locate a point (in red) in Figure 9.3. Based on the observation that this point is situated within the marked region, it can be concluded that all requirements of core are satisfied in the present example and therefore coalition {1,2,3} is stable.

9.1.3 Negotiation Power

By considering different strengths of negotiation power of multiple players in a cooperative game, Roth (1979) proposed to resolve the bargaining problem among the players with the following approach

$$\max \prod_{i=1}^{n} \left(AP_i - d_i\right)^{\gamma_i} \tag{9.16}$$

where n is the total number of players in the cooperative game; Ap_i ($> d_i$) is the profit allocated to player i (or actor i) after cooperation; d_i denotes the status-quo point of player i and, in this application, it is chosen to be the maximum profit obtained by player i working independently; γ_i denotes the negotiation-power indicator of player i. It should be noted that the negotiation-power indicator can be viewed abstractly as the influence level of a player in negotiation. Specifically, equation (9.16) implies that the profit allocated to player i should be proportional to γ_i. To facilitate clearer illustration, let us take the natural log of equation (9.16) and impose additional constraints to the resulting maximization problem, i.e.

$$\max \sum_{i=1}^{n} \gamma_i \ln\left(AP_i - d_i\right) \tag{9.17}$$

subject to

$$d_i = Profit_i \tag{9.18}$$

$$TP = \sum_{i=1}^{n} AP_i \tag{9.19}$$

$$\sum_{i=1}^{n} \gamma_i = 1 \tag{9.20}$$

where $Profit_i$ is the profit of actor i working independently which has already been defined previously in equation (9.1); TP is the total profit obtained after cooperation which can also be calculated according to equation (9.2). By substituting equations (9.18) and (9.19) into the objective function in equation (9.17), one can obtain

$$\sum_{i=1}^{n} \gamma_i \ln\left(AP_i - Profit_i\right) = \sum_{i=1}^{n-1} \gamma_i \ln\left(AP_i - Profit_i\right) + \gamma_n \ln\left(TP - \sum_{i=1}^{n-1} AP_i - Profit_n\right) \tag{9.21}$$

Differentiating the right side of equation (9.21) with respect to AP_i ($i = 1,2, \ldots, n-1$) and setting the result to be zero should end up with

$$\frac{\gamma_1}{AP_1 - Profit_1} = \frac{\gamma_2}{AP_2 - Profit_2} = \cdots = \frac{\gamma_{n-1}}{AP_{n-1} - Profit_{n-1}} = \frac{\gamma_n}{AP_n - Profit_n} \tag{9.22}$$

By making use of equations (9.19) and (9.20), the above relations can be transformed to

$$\gamma_i = \frac{AP_i - Profit_i}{TP - \sum_i Profit_i} \tag{9.23}$$

TABLE 9.5
Negotiation-Power Indicators of All Actors in the Simple Example

	Actor 1	Actor 2	Actor 3
AP_i (\$/day)	193,273	383,233	683,243
$Profit_i$ (\$/day)	119,940	259,900	599,910
γ_i	0.26	0.44	0.3

where $i = 1,2, \ldots, n$. Notice that, when compared with the scenario that all actors work independently, the denominator in the right side of the above equation represents the total extra profit gained by forming an n-actor coalition and the numerator is the extra profit received by actor i in this coalition. Notice also that, if a benefit allocation plan is stipulated according to the Shapley values, equation (9.23) offers a means to quantify the corresponding negotiation power of actor i and this index can then be adopted to compare with the actor's intuitive perception so as to check if the given plan is consistent. Table 9.5 shows the negotiation-power indicators of all actors in the aforementioned simple example.

Since $AP_i > Profit_i$ ($i = 1,2,3$) in this table, it is clear that all actors are willing to cooperate in a coalition. For actor 1 or 3, the cooperation partner is actor 2 in either case. Thus, actor 2 is indispensable in forming the three-actor coalition. The contribution of actor 1 to the profit enhancement of the entire coalition can be attributed to its revenue increase, and that of actor 2 to cost reduction. On the other hand, since both the revenue and cost of actor 2 are improved after cooperation, the impact on its overall profit should also be the greatest among the three actors. Finally, notice from Figure 9.1 that, since the input-output price difference of intermediate B is larger than that of intermediate A, the effects of the former on improvements of the overall profit and also the negotiation space are also larger. As a result, the negotiation strength of actor 3 should be greater than that of actor 1 (see Table 9.5).

9.2 PETROLEUM SUPPLY CHAIN: A REALISTIC EXAMPLE

A detailed description of the petroleum supply chain has already been given in Chapter 8. For illustration convenience, let us consider only the KSR refinery presented in Figure 8.10 as a realistic example. Let us further assume that this refinery process can be divided into four sections and owned separately by different companies. A sketch of this structure is shown in Figure 9.4. As shown in this structure, the four actors in the supply chain are denoted as S, $C1$, $C2$, and $C3$ respectively. Among a large variety of products, actor S produces naphtha and diesel to be consumed by actor $C1$ and actor $C3$ respectively. Also, one of the products of actor $C1$, i.e., mixed xylene, is delivered to $C2$. A simplified block flow diagram (BFD) of the supply chain is also presented in Figure 9.5 to facilitate clearer illustration. In this figure,

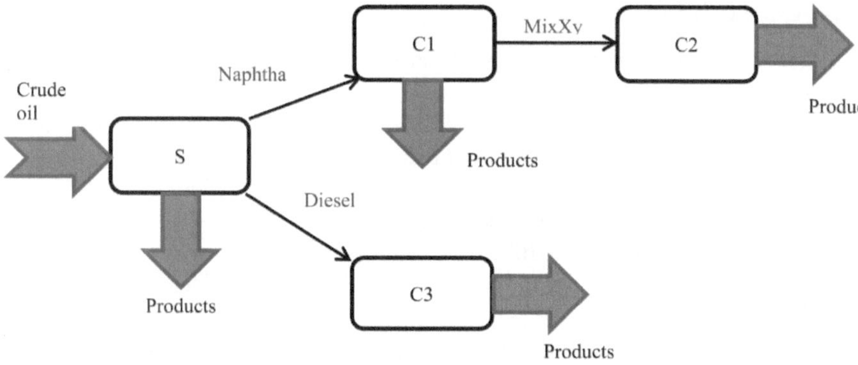

FIGURE 9.4 Petroleum supply chain structure.

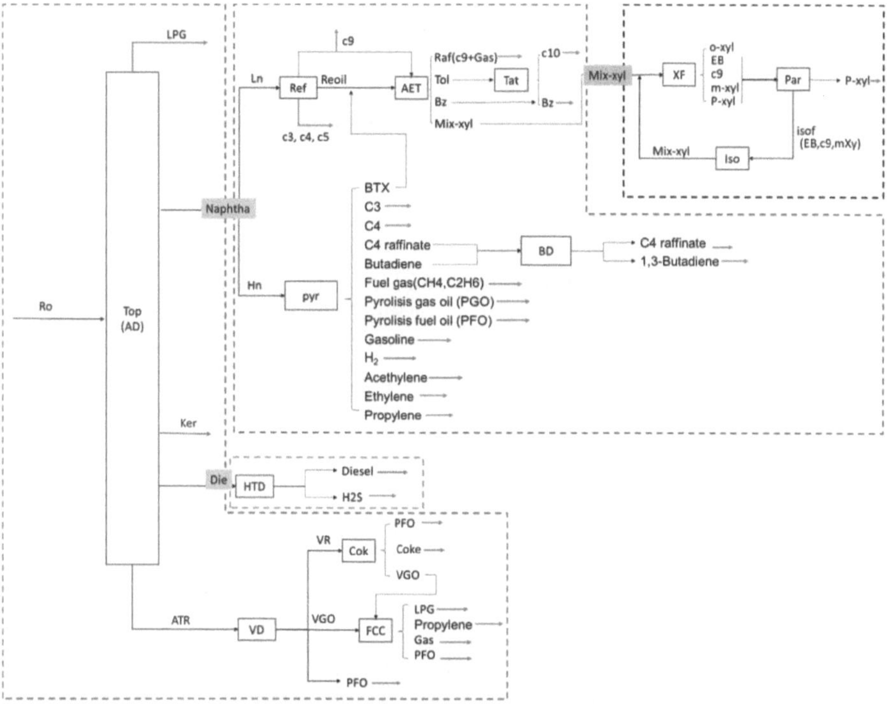

FIGURE 9.5 Simplified BFD for the petroleum supply chain.

actors *S*, *C*1, *C*2, and *C*3 are enclosed with red, blue, black, and green dashed lines, respectively, and the processing units are denoted by black solid-line rectangles. Each black arrow is used to represent a material flow to be delivered to a downstream unit. The green arrows are products to be sold to the domestic market, while the blue ones represent products to be consumed by domestic and overseas markets.

9.3 CASE STUDIES

To simplify the subsequent discussions, let us assume that

- The oil import price and the prices of the final products in the domestic and overseas markets remain constant at least for a short period of time.
- The intermediates (i.e., naphtha, diesel, and mixed xylenes) can only be purchased or sold in the overseas markets before the coalition is formed and, due to exchange-rate differences and tariffs, their import prices are always larger than or equal to the corresponding export prices.

In the following case studies concerning the profit allocation plan in a supply chain, the effects of different intermediate export prices before forming the coalition and the corresponding import and export price differences will be investigated on the basis of the results of extensive optimization runs.

9.3.1 CASE 1

In this case, the import and export price differences of naphtha, diesel, and mixed xylenes are fixed to be 0, 3000, and 2160 NTD/m^3, respectively. The export price of naphtha varies from 7500 to 16,500 NTD/m^3 at 1500-NTD/m^3 intervals in seven (7) scenarios, while those of diesel and mixed xylenes are kept unchanged at 3000 and 15,840 NTD/m^3 respectively.

Based on the mathematical models presented in Chapter 8 and the aforementioned prices and price differences, the maximum profits of all possible coalitions in seven different scenarios can be determined and the results are presented in Table 9.6. It can be observed that, as the naphtha price increases, the profit of actor S increases but the profit of actor $C1$ decreases. Since the prices of diesel and mixed xylenes are kept constant, the profits of actors $C2$ and $C3$ remain unchanged.

On the other hand, based upon the calculation procedure described in Section 9.1.1 and the data listed in Table 9.6, one can determine the Shapley values in every scenario. These values are given in Table 9.7. By comparing the profit of each actor working independently in every scenario (see Table 9.6) and the corresponding Shapley value in Table 9.7, one can see that the former is always smaller than the latter. Thus, the profit allocation plan may be acceptable to all actors.

Since, according to Section 9.1.2, the next item to be checked should be coalition stability, the boundaries of core are thus determined and a subset of them is presented in Table 9.8. Notice that the upper bounds of actor S in all scenarios are slightly smaller than the corresponding lower bounds. This observation implies that a feasible core does not exist in the present case and, thus, the coalition {S, $C1$, $C2$, $C3$} is unstable.

9.3.2 CASE 2

In the second case, the import and export price differences of naphtha, diesel, and mixed xylenes are fixed to be 0, 4500, and 2160 NTD/m^3, respectively. The export

TABLE 9.6
Total Profits of All Possible Coalitions of Petroleum Supply Chain in Case 1

	Profit (10^8NTD/Mon.)						
Coalition	Scenario 1	Scenario 2	Scenario 3	Scenario 4	Scenario 5	Scenario 6	Scenario 7
{*S*}	8.91	9.54	10.17	10.83	11.46	12.09	12.75
{*C*1}	4.05	3.42	2.77	2.13	1.49	0.85	0.21
{*C*2}	3.18	3.18	3.18	3.18	3.18	3.18	3.18
{*C*3}	3.03	3.03	3.03	3.03	3.03	3.03	3.03
{*S*, *C*1}	12.96	12.96	12.96	12.96	12.96	12.96	12.96
{*S*, *C*2}	12.09	12.72	13.35	14.01	14.64	15.30	15.93
{*S*, *C*3}	12.96	13.59	14.25	14.88	15.51	16.17	16.80
{*C*1, *C*2}	7.68	7.02	6.39	5.76	5.10	4.47	3.84
{*C*1, *C*3}	7.08	6.45	5.79	5.16	4.50	3.87	3.24
{*C*2, *C*3}	6.21	6.21	6.21	6.21	6.21	6.21	6.21
{*S*, *C*1, *C*2}	16.56	16.56	16.56	16.56	16.56	16.56	16.56
{*S*, *C*1, *C*3}	17.01	17.01	17.01	17.01	17.01	17.01	17.01
{*S*, *C*2, *C*3}	16.14	16.80	17.43	18.06	18.72	19.35	19.98
{*C*1, *C*2, *C*3}	11.79	11.13	10.50	9.87	9.21	8.58	7.95
{*S*, *C*1, *C*2, *C*3}	20.64	20.64	20.64	20.64	20.64	20.64	20.64

TABLE 9.7
Shapley Values of All Actors of Petroleum Supply Chain in Case 1

	Shapley Values (10^8NTD/mon)			
Scenario No.	Actor *S*	Actor *C*1	Actor *C*2	Actor *C*3
1	9.15	4.35	3.48	3.63
2	9.78	3.72	3.48	3.63
3	10.44	3.09	3.48	3.63
4	11.07	2.44	3.48	3.63
5	11.70	1.80	3.48	3.63
6	12.36	1.16	3.48	3.63
7	12.99	0.52	3.48	3.63

price of naphtha varies from 7500 to 16,500 NTD/m^3 at 1500-NTD/m^3 intervals in seven (7) scenarios, while those of diesel and mixed xylenes are kept unchanged at 3000 and 15,840 NTD/m^3 respectively. Notice that, in the above data, only the price difference of diesel is not the same as that in case 1.

Based on the mathematical models presented in Chapter 8 and the aforementioned prices and price differences, the maximum profits of all possible coalitions in seven different scenarios can be determined, and the results are presented in Table 9.9. Notice that only the bold-faced values are different from those in Table 9.6. It can also be observed that, as the naphtha price increases, the profit of actor *S* increases but the profit of actor *C*1 decreases. Since the prices of diesel and mixed xylenes are kept constant, the profits of actors *C*2 and *C*3 remain unchanged.

On the other hand, based upon the calculation procedure described in Section 9.1.1 and the data listed in Table 9.9, one can determine the Shapley values in every scenario (see Table 9.10). By comparing the profit of each actor working independently in every scenario (see Table 9.9) and the corresponding Shapley value in Table 9.10, one can see that the former is always smaller than the latter. Thus, the corresponding profit allocation plan may be acceptable to all actors.

The next item to be checked is again the coalition stability according to the description in Section 9.1.2. Since the feasible region of core is present in every scenario in this case, the corresponding Shapley values are plotted against the backdrop of core in Figures 9.6–9.12, respectively. From part (c) in each of these figures, it can be observed that the point corresponding to the Shapley values is located outside of the core. This observation also implies that the coalition {*S*, *C*1, *C*2, *C*3} is unstable in all scenarios of case 2.

9.3.3 Case 3

In the third case, the import and export price differences of naphtha, diesel, and mixed xylenes are fixed to be 3000, 3000, and 2160 NTD/m^3, respectively. The export price of naphtha varies from 7500 to 13,500 NTD/m^3 at 1500-NTD/m^3 intervals in

TABLE 9.8
Partial Boundaries of Core for Coalition {S, $C1$, $C2$, $C3$} in Case 1

Scenario No.	Boundaries	$x_{\{S,C1,C2,C3\},S}$ (10^8NTD/mon)	$x_{\{S,C1,C2,C3\},C1}$ (10^8NTD/mon)	$x_{\{S,C1,C2,C3\},C2}$ (10^8NTD/mon)	$x_{\{S,C1,C2,C3\},C3}$ (10^8NTD/mon)
1	Upper	8.85	4.47	3.63	4.08
	Lower	8.91	4.05	3.18	3.03
2	Upper	9.48	3.84	3.63	4.08
	Lower	9.54	3.42	3.18	3.03
3	Upper	10.14	3.21	3.63	4.08
	Lower	10.17	2.77	3.18	3.03
4	Upper	10.77	2.56	3.63	4.08
	Lower	10.83	2.13	3.18	3.03
5	Upper	11.40	1.92	3.63	4.08
	Lower	11.46	1.49	3.18	3.03
6	Upper	12.06	1.28	3.63	4.08
	Lower	12.09	0.852	3.18	3.03
7	Upper	12.69	0.639	3.63	4.08
	Lower	12.75	0.212	3.18	3.03

TABLE 9.9
Total Profits of All Possible Coalitions of Petroleum Supply Chain in Case 2

	Profit (10^8NTD/mon)						
Coalition	**Scenario 1**	**Scenario 2**	**Scenario 3**	**Scenario 4**	**Scenario 5**	**Scenario 6**	**Scenario 7**
{*S*}	8.91	9.54	10.17	10.83	11.46	12.09	12.75
{*C*1}	4.05	3.42	2.77	2.13	1.49	0.85	0.21
{*C*2}	3.18	3.18	3.18	3.18	3.18	3.18	3.18
{*C*3}	**2.50**	**2.50**	**2.50**	**2.50**	**2.50**	**2.50**	**2.50**
{*S*, *C*1}	12.96	12.96	12.96	12.96	12.96	12.96	12.96
{*S*, *C*2}	12.09	12.72	13.35	14.01	14.64	15.30	15.93
{*S*, *C*3}	12.96	13.59	14.25	14.88	15.51	16.17	16.80
{*C*1, *C*2}	7.68	7.02	6.39	5.76	5.10	4.47	3.84
{*C*1, *C*3}	**6.54**	**5.71**	**5.28**	**4.62**	**3.99**	**3.36**	**2.71**
{*C*2, *C*3}	**5.70**	**5.70**	**5.70**	**5.70**	**5.70**	**5.70**	**5.70**
{*S*, *C*1, *C*2}	16.56	16.56	16.56	16.56	16.56	16.56	16.56
{*S*, *C*1, *C*3}	17.01	17.01	17.01	17.01	17.01	17.01	17.01
{*S*, *C*2, *C*3}	16.14	16.80	17.43	18.06	18.72	19.35	19.98
{*C*1, *C*2, *C*3}	**11.07**	**10.44**	**9.78**	**9.15**	**8.52**	**7.86**	**7.23**
{*S*, *C*1, *C*2, *C*3}	20.64	20.64	20.64	20.64	20.64	20.64	20.64

TABLE 9.10
Shapley Values of All Actors of Petroleum Supply Chain in Case 2

	Shapley values (10^8NTD/mon)			
Scenario No.	**Actor *S***	**Actor *C*1**	**Actor *C*2**	**Actor *C*3**
1	9.45	4.35	3.48	3.36
2	10.08	3.69	3.48	3.36
3	10.74	3.06	3.48	3.36
4	11.37	2.42	3.48	3.36
5	12.00	1.78	3.48	3.36
6	12.66	1.14	3.48	3.36
7	13.29	0.501	3.48	3.36

five (5) scenarios, while those of diesel and mixed xylenes are kept unchanged at 3000 and 15,840 NTD/m^3 respectively. Notice that, in the above data, mainly the price difference of naphtha is different from that in case 1 while only five (instead of seven) scenarios are considered.

Based on the mathematical models presented in Chapter 8 and the aforementioned prices and price differences, the maximum profits of all possible coalitions in five different scenarios can be determined and the results are presented in Table 9.11. Notice that only the bold-faced values are different from those in Table 9.6. It can also be observed that, as the naphtha price increases, the profit of actor *S* increases but the profit of actor *C*1 decreases. Since the prices of diesel and mixed xylenes are kept constant, the profits of actors *C*2 and *C*3 remain unchanged.

On the other hand, based upon the calculation procedure described in Section 9.1.1 and the data listed in Table 9.11, one can determine the Shapley values in every scenario (see Table 9.12). By comparing the profit of each actor working independently in every scenario (see Table 9.11) and the corresponding Shapley value in Table 9.12, one can see that the former is always smaller than the latter. Thus, the corresponding profit allocation plan may be acceptable to all actors.

The next item to be checked is obviously coalition stability as described in Section 9.1.2. Since a nonempty core is present in every scenario in this case, the corresponding Shapley values are plotted against the backdrop of core in Figures 9.13–9.17, respectively. From these figures, it can be observed that the point corresponding to the Shapley values is always located inside of the core. This observation also implies that the coalition {*S*, *C*1, *C*2, *C*3} is stable in all scenarios of case 3.

In the last step, let's calculate the negotiation-power indicators of all actors by substituting the data in Tables 9.11 and 9.12 into equation (9.23). It can be found that the resulting actor powers remain unchanged. i.e., $\gamma_S = 0.32$, $\gamma_{C1} = 0.34$, $\gamma_{C2} = 0.11$, and $\gamma_{C3} = 0.22$, in all scenarios. Since *C*1 cooperates with both *S* and *C*2 and, thus, gains extra savings in its feedstock and revenue in its product, the participation of *C*1 in the coalition should cause the greatest improvement in the overall profit. On

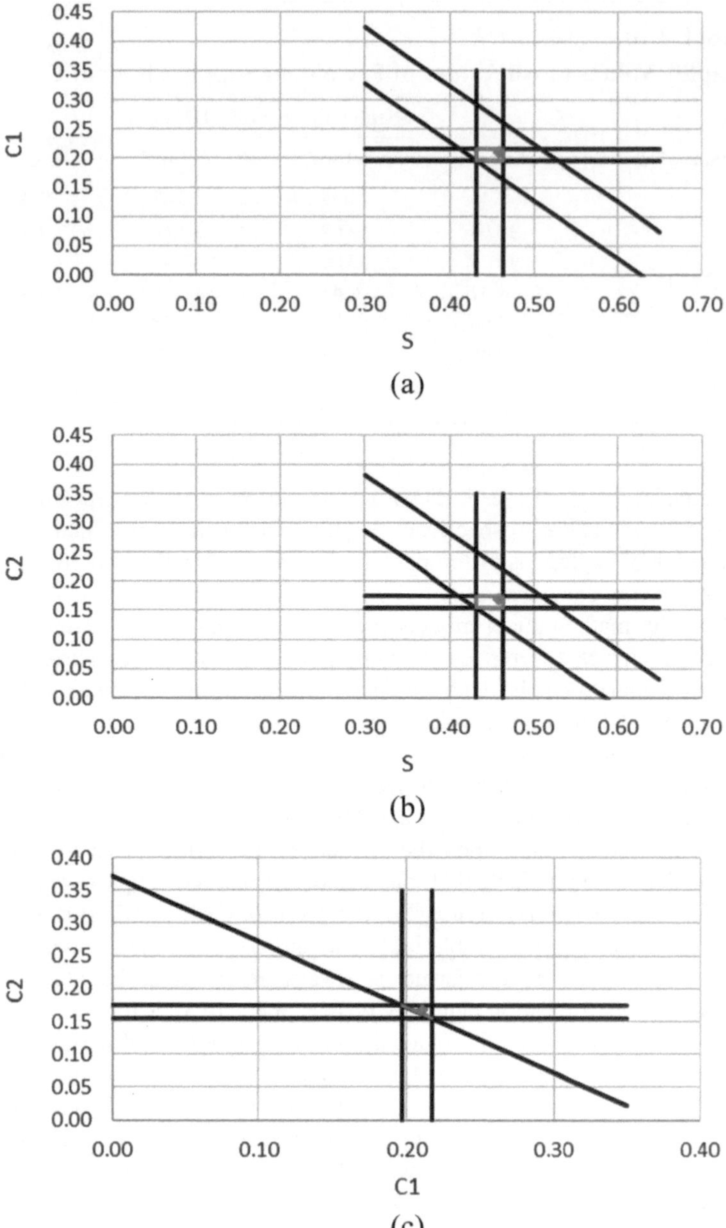

FIGURE 9.6 Core and Shapley values in scenario 1 of case 2: (a) Projection to *S*-*C*1 plane; (b) Projection to *S*-*C*2 plane; (c) Projection to *C*1-*C*2 plane.

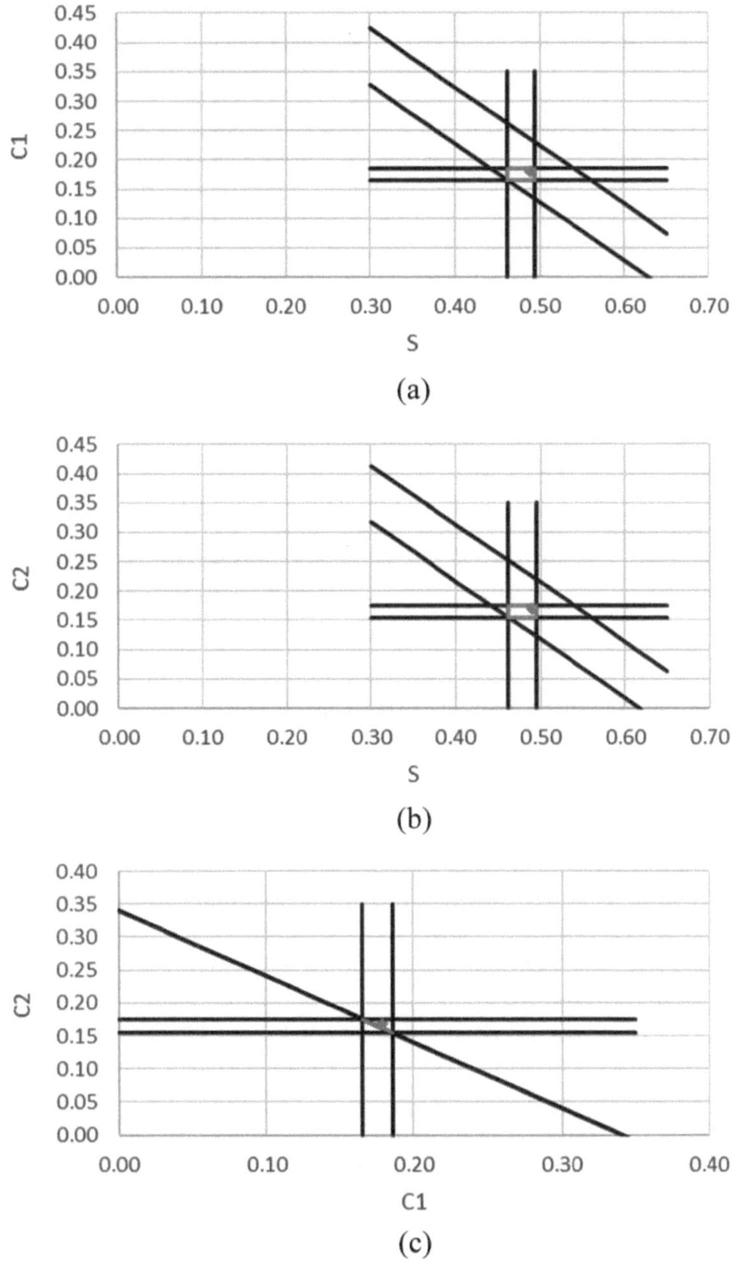

FIGURE 9.7 Core and Shapley values in scenario 2 of case 2: (a) Projection to S-$C1$ plane; (b) Projection to S-$C2$ plane; (c) Projection to $C1$-$C2$ plane.

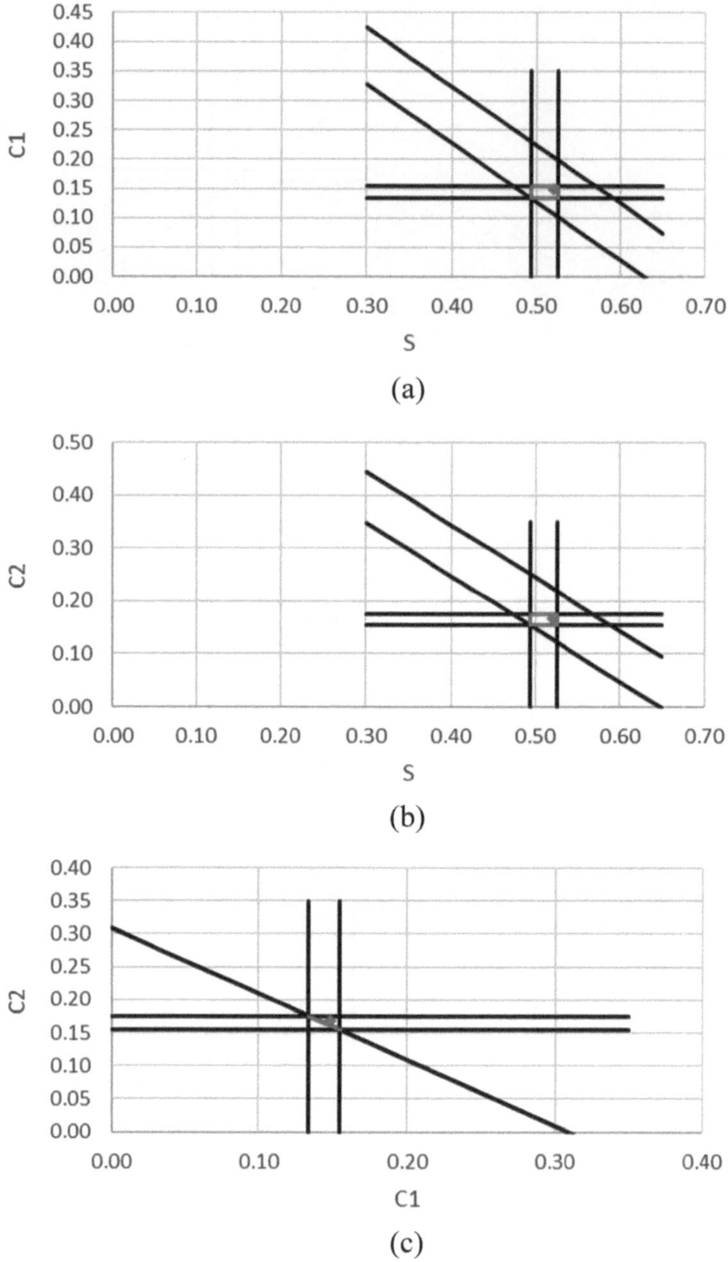

FIGURE 9.8 Core and Shapley values in scenario 3 of case 2: (a) Projection to *S*-*C*1 plane; (b) Projection to *S*-*C*2 plane; (c) Projection to *C*1-*C*2 plane.

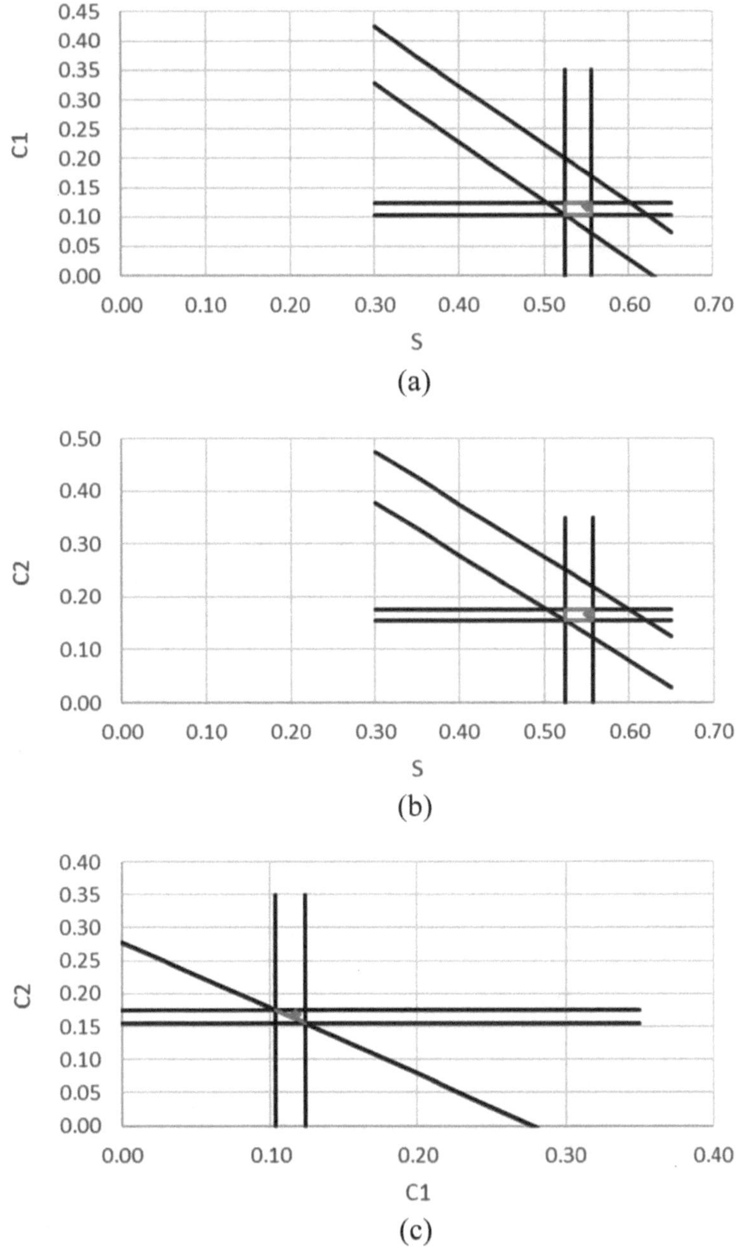

FIGURE 9.9 Core and Shapley values in scenario 4 of case 2: (a) Projection to *S*-*C*1 plane; (b) Projection to *S*-*C*2 plane; (c) Projection to *C*1-*C*2 plane.

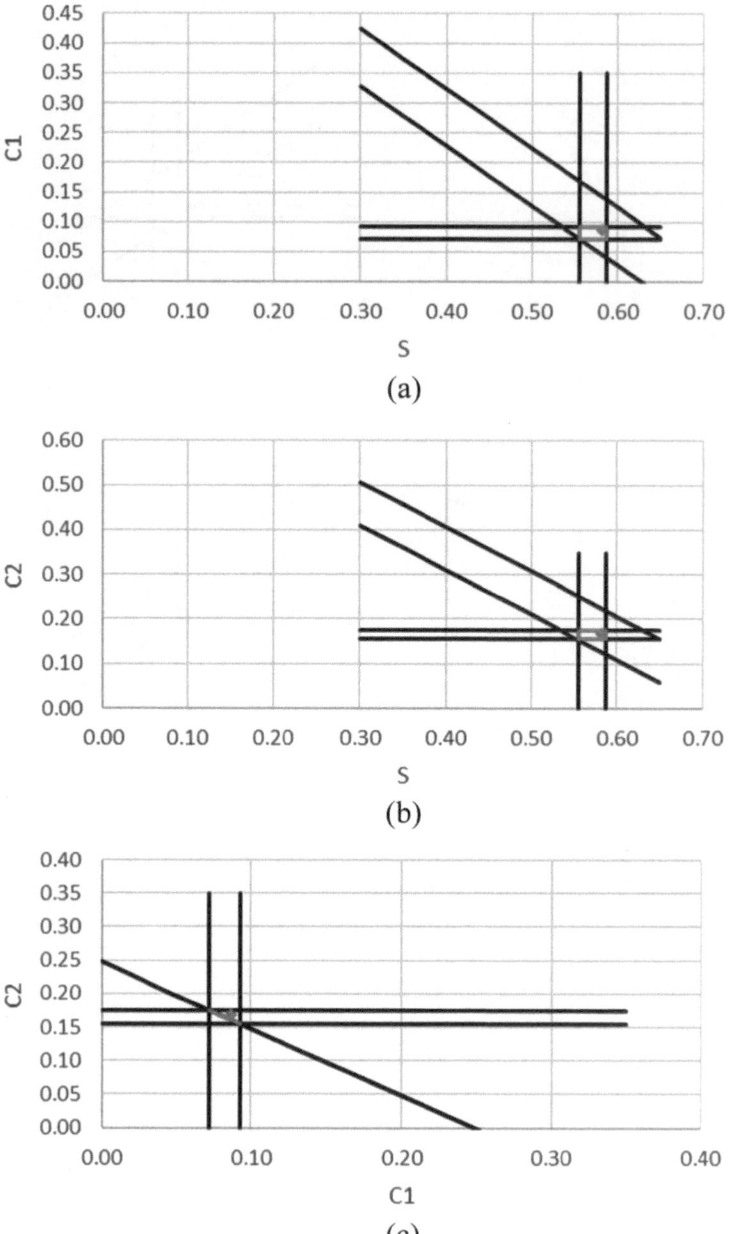

FIGURE 9.10 Core and Shapley values in scenario 5 of case 2: (a) Projection to *S*-*C*1 plane; (b) Projection to *S*-*C*2 plane; (c) Projection to *C*1-*C*2 plane.

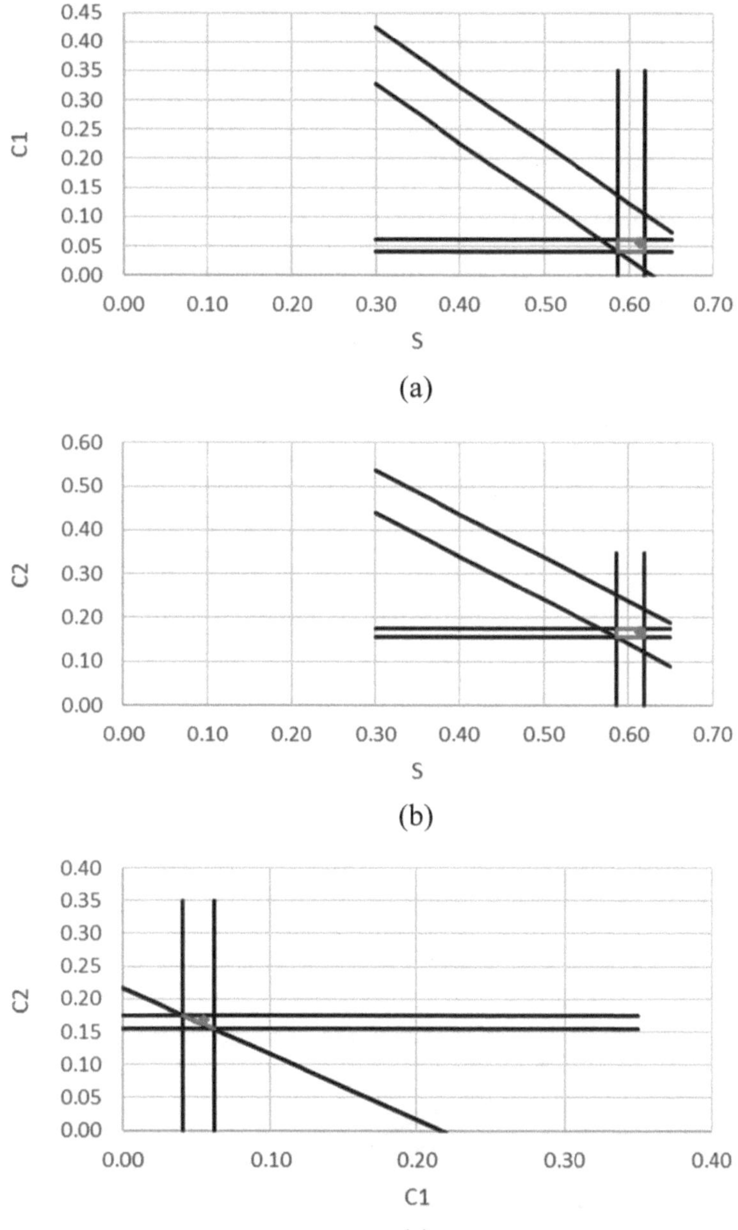

FIGURE 9.11 Core and Shapley values in scenario 6 of case 2: (a) Projection to *S*-*C*1 plane; (b) Projection to *S*-*C*2 plane; (c) Projection to *C*1-*C*2 plane.

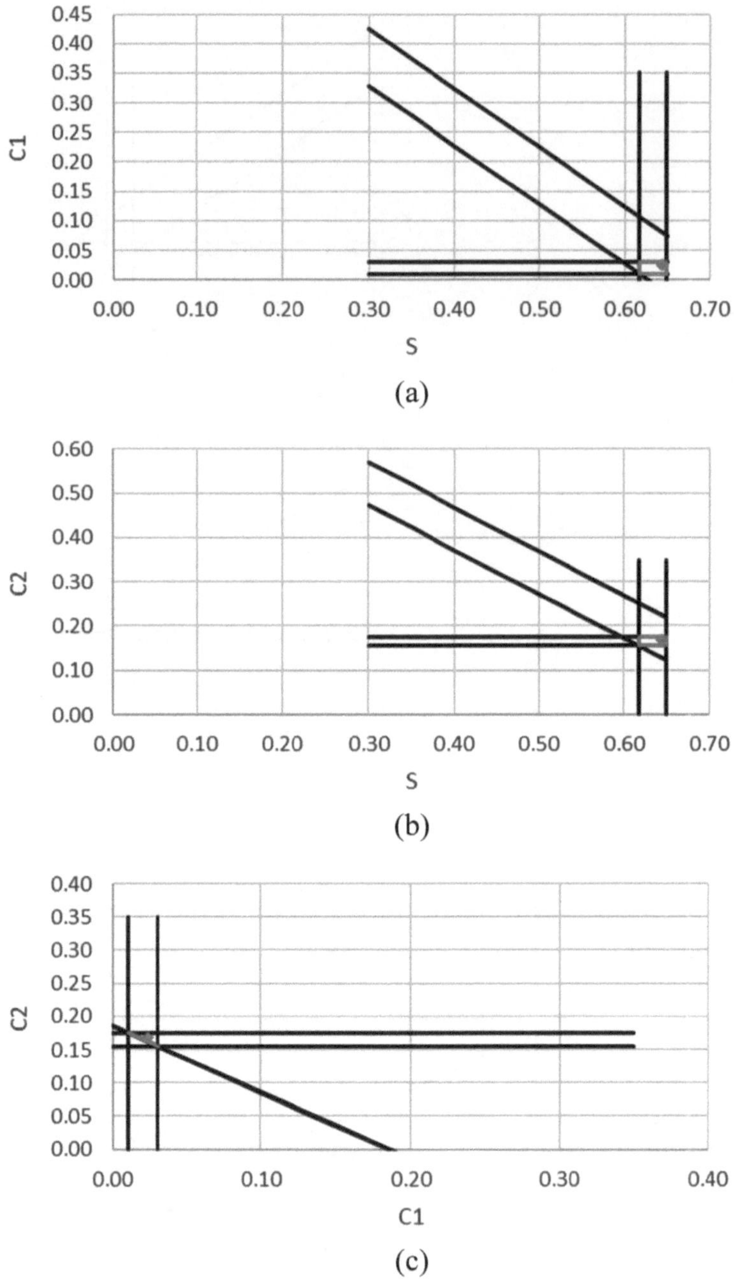

FIGURE 9.12 Core and Shapley values in scenario 7 of case 2: (a) Projection to *S*-*C*1 plane; (b) Projection to *S*-*C*2 plane; (c) Projection to *C*1-*C*2 plane.

TABLE 9.11
Total Profits of All Possible Coalitions of Petroleum Supply Chain in Case 3

	Profit (10^8NTD/mon)				
Coalition	**Scenario 1**	**Scenario 2**	**Scenario 3**	**Scenario 4**	**Scenario 5**
{*S*}	8.91	9.54	10.17	10.83	11.46
{*C*1}	**2.77**	**2.13**	**1.49**	**0.852**	**0.212**
{*C*2}	3.18	3.18	3.18	3.18	3.18
{*C*3}	3.03	3.03	3.03	3.03	3.03
{*S*, *C*1}	12.96	12.96	12.96	12.96	12.96
{*S*, *C*2}	12.09	12.72	13.35	14.01	14.64
{*S*, *C*3}	12.96	13.59	14.25	14.88	15.51
{*C*1, *C*2}	**6.39**	**5.76**	**5.10**	**4.47**	**3.84**
{*C*1, *C*3}	**5.79**	**5.16**	**4.50**	**3.87**	**3.24**
{*C*2, *C*3}	6.21	6.21	6.21	6.21	6.21
{*S*, *C*1, *C*2}	16.56	16.56	16.56	16.56	16.56
{*S*, *C*1, *C*3}	17.01	17.01	17.01	17.01	17.01
{*S*, *C*2, *C*3}	16.14	16.80	17.43	18.06	18.72
{*C*1, *C*2, *C*3}	**10.50**	**9.87**	**9.21**	**8.58**	**7.95**
{*S*, *C*1, *C*2, *C*3}	20.64	20.64	20.64	20.64	20.64

TABLE 9.12
Shapley Values of All Actors of Petroleum Supply Chain in Case 3

	Shapley Values (10^8NTD/mon)			
Scenario No.	**Actor *S***	**Actor C1**	**Actor C2**	**Actor C3**
1	9.78	3.72	3.48	3.63
2	10.44	3.09	3.48	3.63
3	11.07	2.44	3.48	3.63
4	11.70	1.80	3.48	3.63
5	12.33	1.16	3.48	3.63

the other hand, since actor *S* is the provider of intermediates, i.e., naphta and diesel, to actors *C*1 and *C*3, its impact on the overall profit is also great. Finally, the negotiation power of *C*3 should be higher than that of *C*2 since the import and export price difference of diesel is larger than that of mixed xylenes. Therefore, based on the above discussion, the precedence order of negotiation-power indicators clearly should be: $\gamma_{C1} > \gamma_S > \gamma_{C3} > \gamma_{C2}$ and the corresponding profit allocation plan is reasonable and fair.

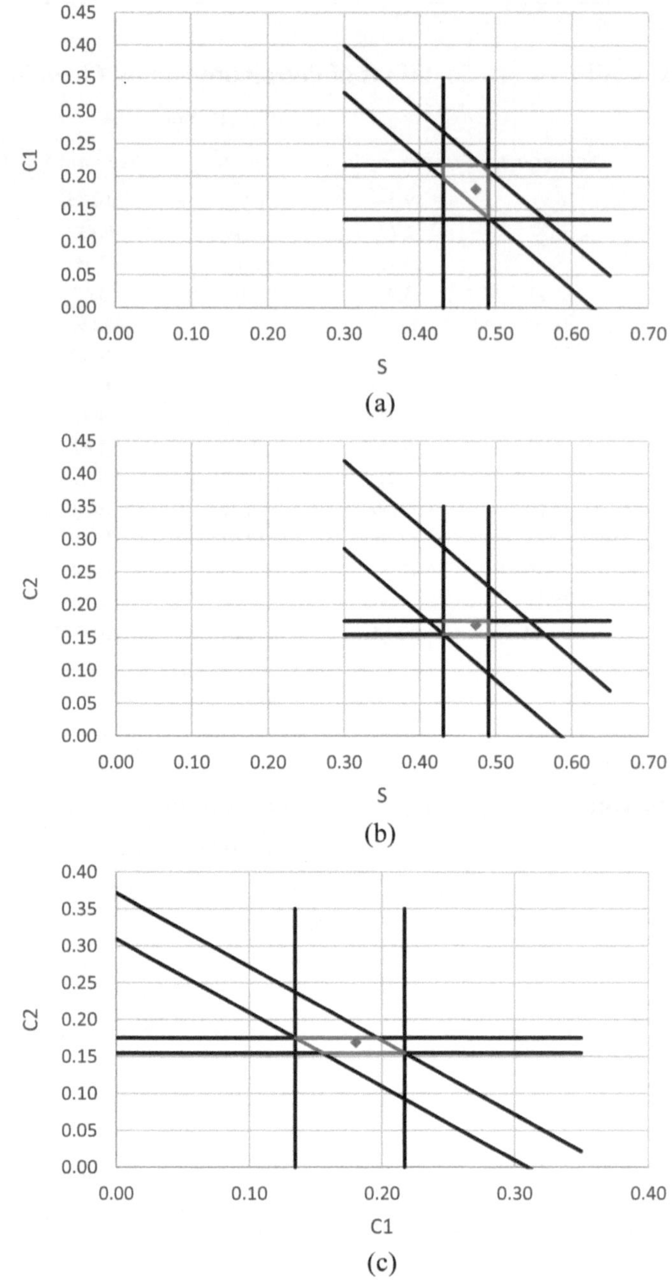

FIGURE 9.13 Core and Shapley values in scenario 1 of case 3: (a) Projection to *S*-*C*1 plane; (b) Projection to *S*-*C*2 plane; (c) Projection to *C*1-*C*2 plane.

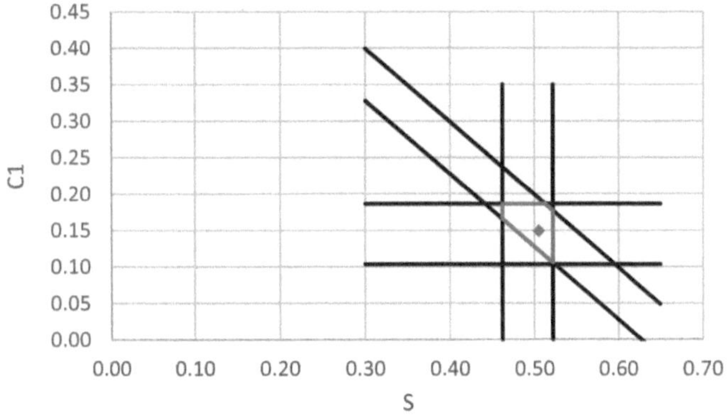

(a)

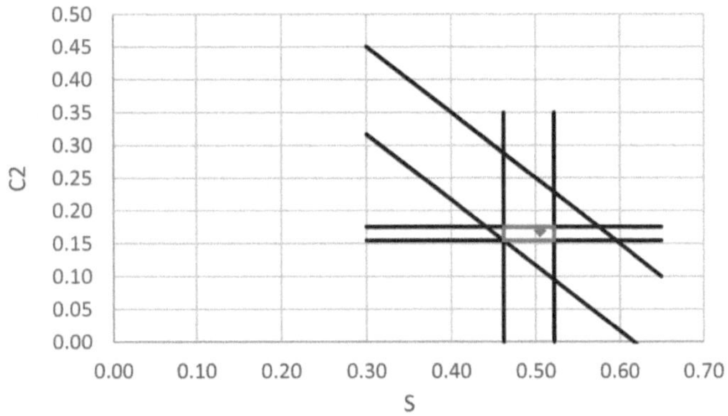

(b)

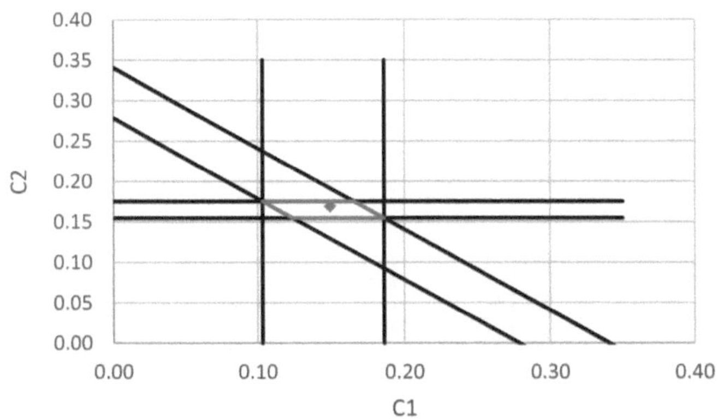

(c)

FIGURE 9.14 Core and Shapley values in scenario 2 of case 3: (a) Projection to *S*-*C*1 plane; (b) Projection to *S*-*C*2 plane; (c) Projection to *C*1-*C*2 plane.

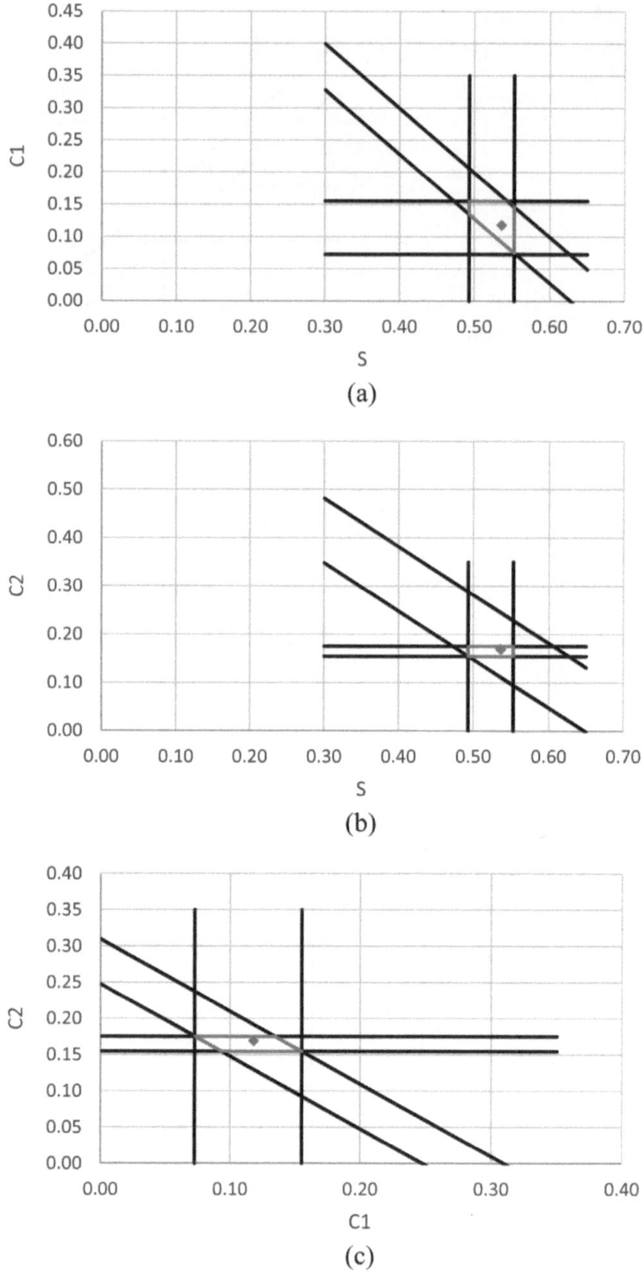

FIGURE 9.15 Core and Shapley values in scenario 3 of case 3: (a) Projection to S-$C1$ plane; (b) Projection to S-$C2$ plane; (c) Projection to $C1$-$C2$ plane.

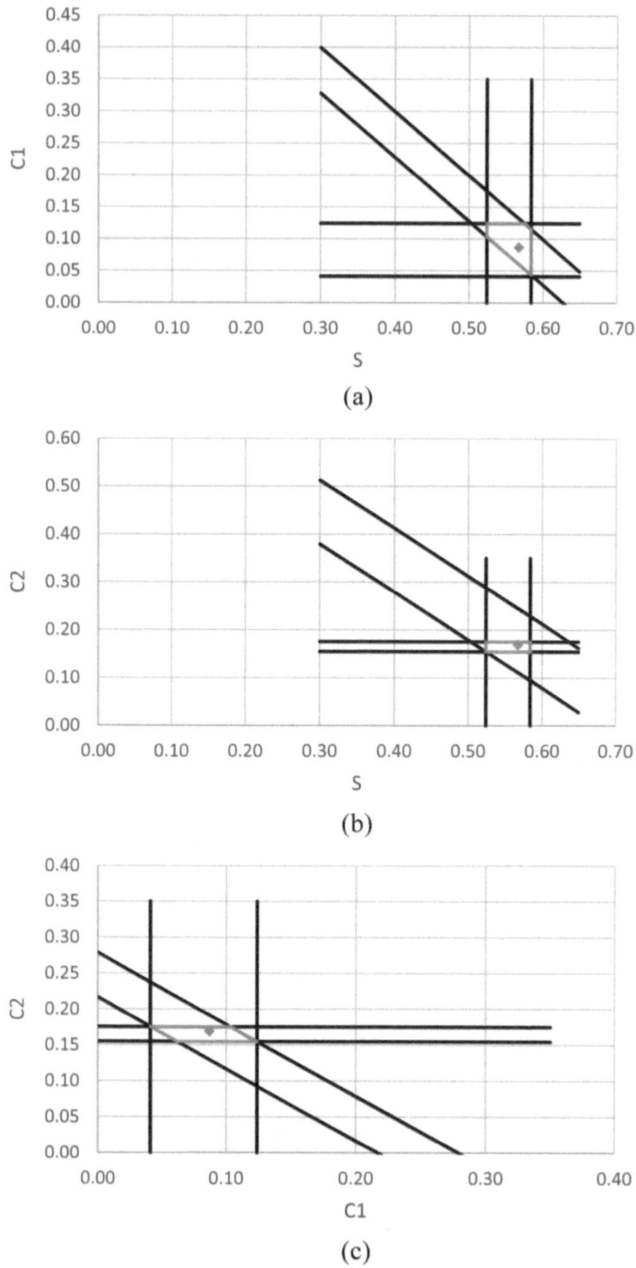

FIGURE 9.16 Core and Shapley values in scenario 4 of case 3: (a) Projection to S-$C1$ plane; (b) Projection to S-$C2$ plane; (c) Projection to $C1$-$C2$ plane.

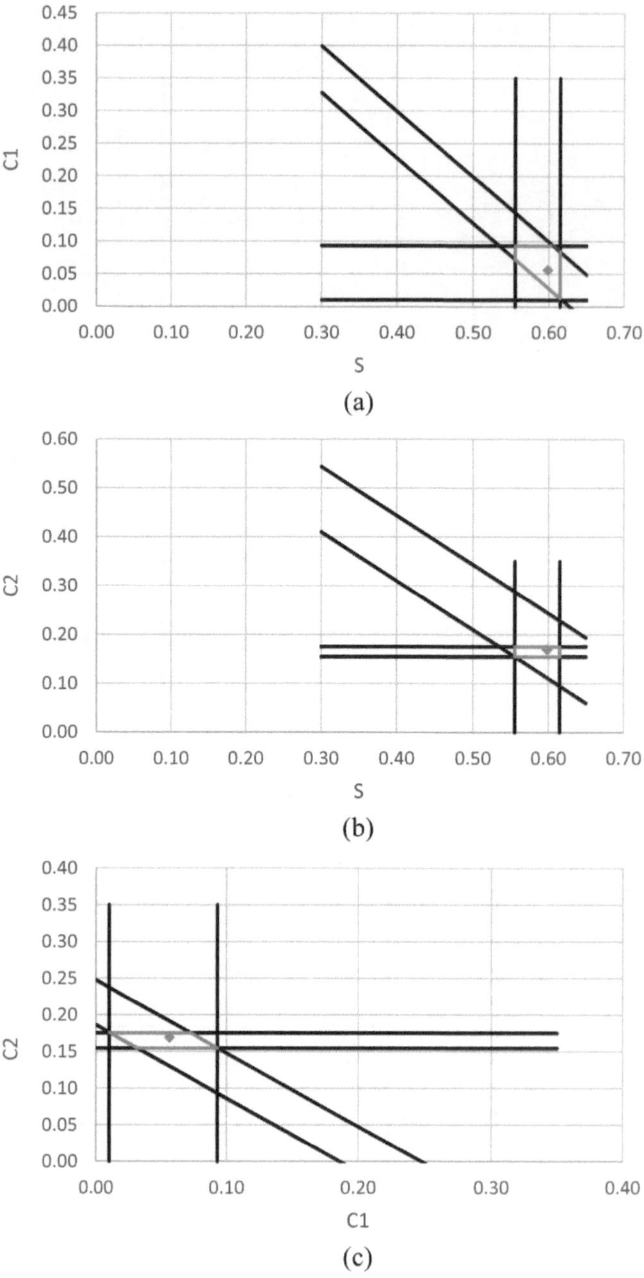

FIGURE 9.17 Core and Shapley values in scenario 5 of case 3: (a) Projection to S-$C1$ plane; (b) Projection to S-$C2$ plane; (c) Projection to $C1$-$C2$ plane.

9.3.4 Case 4

In the fourth case, the import and export price differences of naphtha, diesel, and mixed xylenes are fixed to be 3000, 4500, and 2160 NTD/m^3, respectively. The export price of naphtha varies from 7500 to 13,500 NTD/m^3 at 1500-NTD/m^3 intervals in five (5) scenarios, while those of diesel and mixed xylenes are kept unchanged at 3000 and 15,840 NTD/m^3 respectively. Notice that, in the above data, only the price difference of diesel is greater than that in case 3.

Based on the mathematical models presented in Chapter 8 and the aforementioned prices and price differences, the maximum profits of all possible coalitions in five different scenarios can be determined, and the results are presented in Table 9.13. Notice that only the bold-faced values are different from those in Table 9.11. It can also be observed that, as the naphtha price increases, the profit of actor S increases but the profit of actor $C1$ decreases. Since the prices of diesel and mixed xylenes are kept constant, the profits of actors $C2$ and $C3$ remain unchanged.

On the other hand, based upon the calculation procedure described in Section 9.1.1 and the data listed in Table 9.13, one can determine the Shapley values in every scenario (see Table 9.14). By comparing the profit of each actor working independently in every scenario (see Table 9.13) and the corresponding Shapley value in Table 9.14, one can see that the former is always smaller than the latter. Thus, the corresponding profit allocation plan may be acceptable to all actors.

As stated before in Section 9.1.2, the next item to be checked is coalition stability. Since a nonempty core is present in every scenario in the current case,

TABLE 9.13
Total Profits of All Possible Coalitions of Petroleum Supply Chain in Case 4

	Profit (10^8NTD/mon)				
Coalition	**Scenario 1**	**Scenario 2**	**Scenario 3**	**Scenario 4**	**Scenario 5**
$\{S\}$	8.91	9.54	10.17	10.83	11.46
$\{C1\}$	2.77	2.13	1.49	0.852	0.212
$\{C2\}$	3.18	3.18	3.18	3.18	3.18
$\{C3\}$	**2.50**	**2.50**	**2.50**	**2.50**	**2.50**
$\{S, C1\}$	12.96	12.96	12.96	12.96	12.96
$\{S, C2\}$	12.09	12.72	13.35	14.01	14.64
$\{S, C3\}$	12.96	13.59	14.25	14.88	15.51
$\{C1, C2\}$	6.39	5.76	5.10	4.47	3.84
$\{C1, C3\}$	**5.28**	**4.62**	**3.99**	**3.36**	**2.71**
$\{C2, C3\}$	**5.70**	**5.70**	**5.70**	**5.70**	**5.70**
$\{S, C1, C2\}$	16.56	16.56	16.56	16.56	16.56
$\{S, C1, C3\}$	17.01	17.01	17.01	17.01	17.01
$\{S, C2, C3\}$	16.14	16.80	17.43	18.06	18.72
$\{C1, C2, C3\}$	**9.78**	**9.15**	**8.52**	**7.86**	**7.23**
$\{S, C1, C2, C3\}$	20.64	20.64	20.64	20.64	20.64

TABLE 9.14
Shapley Values of All Actors of Petroleum Supply Chain in Case 4

	Shapley Values (10^8NTD/mon)			
Scenario No.	Actor *S*	Actor *C*1	Actor *C*2	Actor *C*3
1	10.08	3.69	3.48	3.36
2	10.74	3.06	3.48	3.36
3	11.37	2.42	3.48	3.36
4	12.00	1.78	3.48	3.36
5	12.66	1.14	3.48	3.36

the corresponding Shapley values are plotted against the backdrop of core in Figures 9.18–9.22, respectively. From these figures, it can be observed that the point corresponding to the Shapley values is always located within the core. This observation implies that the coalition {*S*, *C*1, *C*2, *C*3} is stable in all scenarios of case 4.

The last task in stipulating a profit allocation plan is to determine the negotiation-power indicators of all actors by substituting the data in Tables 9.13 and 9.14 into equation (9.23). It can be found that the resulting actor powers remain unchanged, i.e., $\gamma_S = 0.37$, $\gamma_{C1} = 0.28$, $\gamma_{C2} = 0.09$, and $\gamma_{C3} = 0.26$, in all scenarios. Since *C*1 cooperates with both *S* and *C*2 and, thus, gains extra savings in its feedstock and revenue in its product, the participation of *C*1 in the coalition should cause the greater improvement in the overall profit. On the other hand, since actor *S* is the provider of intermediates, i.e., naphta and diesel, to actors *C*1 and *C*3, its impact on the overall profit is equally great. However, since the price difference of diesel is greater than that in case 3 in the present case and the price difference in naphtha is relatively low, the role of actor *S* becomes more important than that of actor *C*1. Finally, the negotiation power of *C*3 should be higher than that of *C*2 since the import and export price difference of diesel is larger than that of mixed xylenes. As a result, the precedence order of negotiation-power indicators is: $\gamma_S > \gamma_{C1} > \gamma_{C3} > \gamma_{C2}$ and the corresponding profit allocation plan should be reasonable and fair.

9.3.5 Case 5

In the fifth case, the import and export price differences of naphtha, diesel, and mixed xylenes are fixed to be 4500, 3000, and 2160 NTD/m^3, respectively. The export price of naphtha varies from 7500 to 12,000 NTD/m^3 at 1500-NTD/m^3 intervals in four (4) scenarios, while those of diesel and mixed xylenes are kept unchanged at 3000 and 15,840 NTD/m^3 respectively. Notice that, in the above data, only the price difference of naphtha is greater than that in case 3.

Based on the mathematical models presented in Chapter 8 and the aforementioned prices and price differences, the maximum profits of all possible coalitions in four different scenarios can be determined and the results are presented in Table 9.15. Notice that only the bold-faced values are not the same as those in Table 9.11. It can

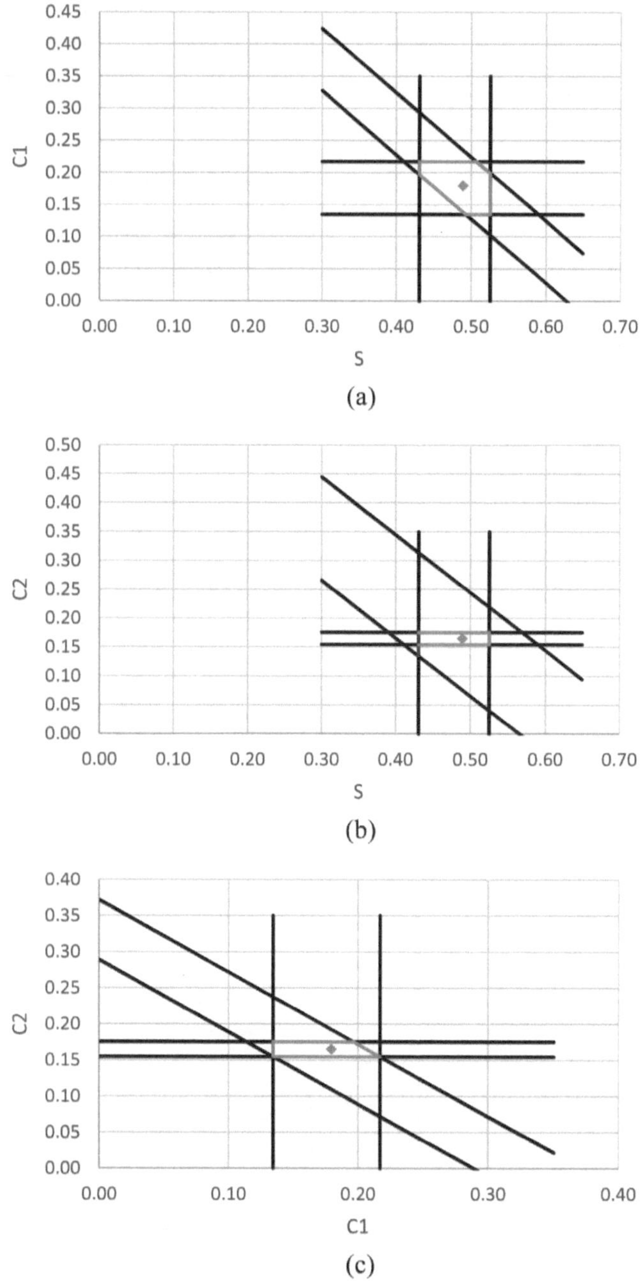

FIGURE 9.18 Core and Shapley values in scenario 1 of case 4: (a) Projection to S-$C1$ plane; (b) Projection to S-$C2$ plane; (c) Projection to $C1$-$C2$ plane.

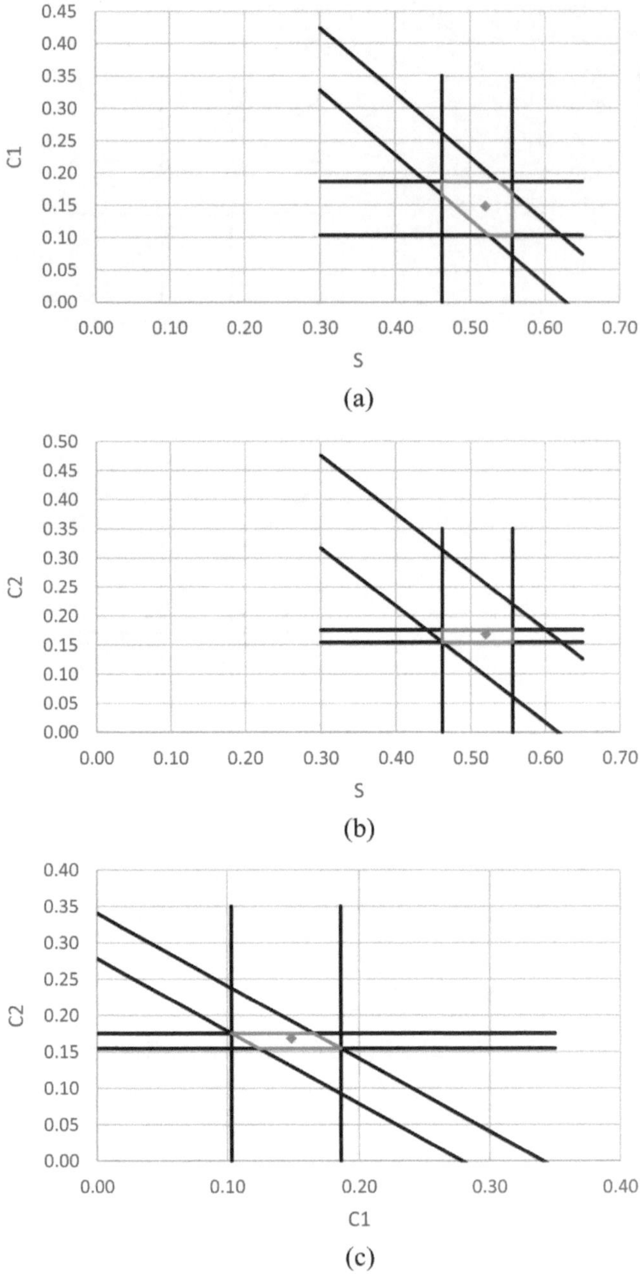

FIGURE 9.19 Core and Shapley values in scenario 2 of case 4: (a) Projection to S-$C1$ plane; (b) Projection to S-$C2$ plane; (c) Projection to $C1$-$C2$ plane.

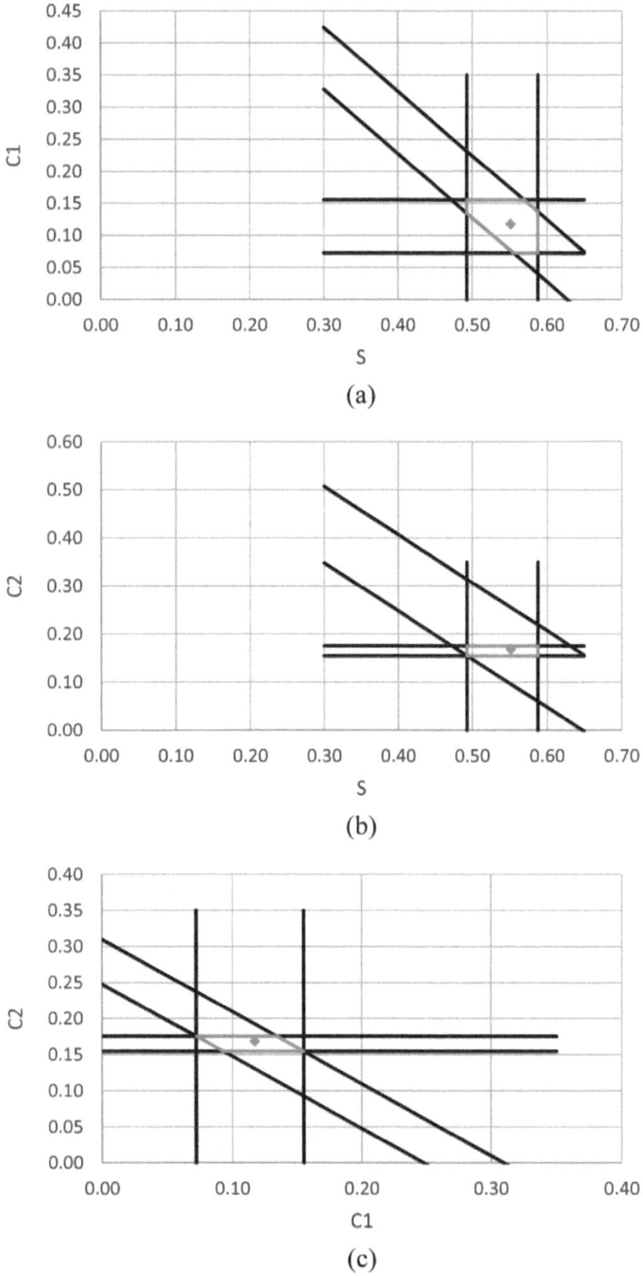

FIGURE 9.20 Core and Shapley values in scenario 3 of case 4: (a) Projection to *S*-*C*1 plane; (b) Projection to *S*-*C*2 plane; (c) Projection to *C*1-*C*2 plane.

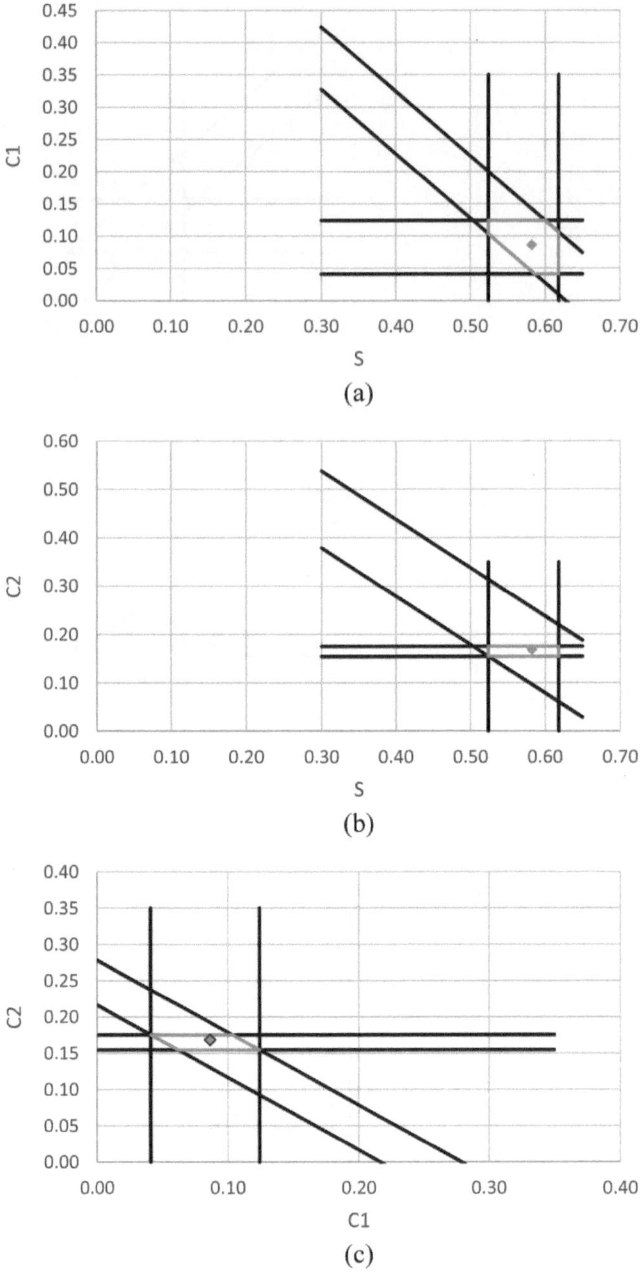

FIGURE 9.21 Core and Shapley values in scenario 4 of case 4: (a) Projection to *S*-*C*1 plane; (b) Projection to *S*-*C*2 plane; (c) Projection to *C*1-*C*2 plane.

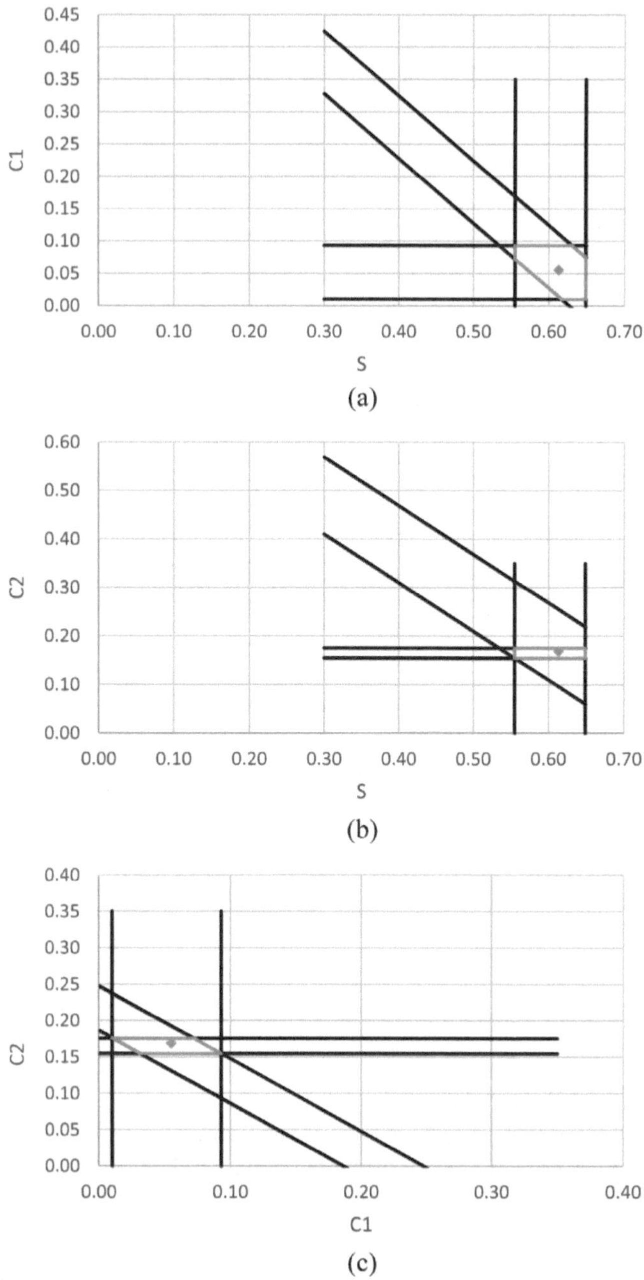

FIGURE 9.22 Core and Shapley values in scenario 5 of case 4: (a) Projection to *S*-*C*1 plane; (b) Projection to *S*-*C*2 plane; (c) Projection to *C*1-*C*2 plane.

TABLE 9.15
Total Profits of All Possible Coalitions of Petroleum Supply Chain in Case 5

	Profit (10^8NTD/mon)			
Coalition	**Scenario 1**	**Scenario 2**	**Scenario 3**	**Scenario 4**
{*S*}	8.91	9.54	10.17	10.83
{*C*1}	**2.13**	**1.49**	**0.852**	**0.212**
{*C*2}	3.18	3.18	3.18	3.18
{*C*3}	3.03	3.03	3.03	3.03
{*S*, *C*1}	12.96	12.96	12.96	12.96
{*S*, *C*2}	12.09	12.72	13.35	14.01
{*S*, *C*3}	12.96	13.59	14.25	14.88
{*C*1, *C*2}	**5.76**	**5.10**	**4.47**	**3.84**
{*C*1, *C*3}	**5.16**	**4.50**	**3.87**	**3.24**
{*C*2, *C*3}	6.21	6.21	6.21	6.21
{*S*, *C*1, *C*2}	16.56	16.56	16.56	16.56
{*S*, *C*1, *C*3}	17.01	17.01	17.01	17.01
{*S*, *C*2, *C*3}	16.14	16.80	17.43	18.06
{*C*1, *C*2, *C*3}	**9.87**	**9.21**	**8.58**	**7.95**
{*S*, *C*1, *C*2, *C*3}	20.64	20.64	20.64	20.64

TABLE 9.16
Shapley Values of All Actors of Petroleum Supply Chain in Case 5

	Shapley Values (10^8NTD/mon)			
Scenario No.	**Actor *S***	**Actor *C*1**	**Actor *C*2**	**Actor *C*3**
1	10.11	3.39	3.48	3.63
2	10.74	2.76	3.48	3.63
3	11.40	2.12	3.48	3.63
4	12.03	1.48	3.48	3.63

also be observed that, as the naphtha price increases, the profit of actor *S* increases but the profit of actor *C*1 decreases. Since the prices of diesel and mixed xylenes are kept constant, the profits of actors *C*1 and *C*3 remain unchanged.

On the other hand, based upon the calculation procedure described in Section 9.1.1 and the data listed in Table 9.15, one can determine the Shapley values in every scenario (see Table 9.16). By comparing the profit of each actor working independently in every scenario (see Table 9.15) and the corresponding Shapley value in Table 9.16, one can see that the former is always smaller than the latter. Thus, the corresponding profit allocation plan may be acceptable to all actors.

According to the description in Section 9.1.2, the next item to be checked is coalition stability. Since a nonempty core is present in every scenario in case 5, the corresponding Shapley values are plotted against the backdrop of core in Figures 9.23–9.26, respectively. From these figures, it can be observed that the point corresponding to the Shapley values is always located within the core. This again implies that the coalition {*S*, *C*1, *C*2, *C*3} is stable in all scenarios of case 5.

The step to finalize a profit allocation plan is to determine the negotiation-power indicators of all actors by substituting the data in Tables 9.15 and 9.16 into equation (9.23). It can be found again that the resulting actor powers, i.e., $\gamma_S = 0.36$, $\gamma_{C1} = 0.37$, $\gamma_{C2} = 0.09$, and $\gamma_{C3} = 0.18$, remain unchanged in all scenarios. The interpretations of these indicators are similar to those for case 3. Since *C*1 cooperates with both *S* and *C*2 and, thus, gains extra savings in its feedstock and revenue in its product, the participation of *C*1 in the coalition should cause the greater improvement in the overall profit. On the other hand, since actor *S* is the provider of two intermediates, i.e., naphta and diesel, to be consumed respectively by actors *C*1 and *C*3, its impact on the overall profit should be equally great. However, since the price difference of naphtha is larger than that of diesel in the present case, the role of actor *C*1 becomes slightly more important than that of actor *S*. Finally, the negotiation power of *C*3 should be higher than that of *C*2 since the import and export price difference of diesel is larger than that of mixed xylenes. As a result, the precedence order of negotiation-power indicators is: $\gamma_{C1} > \gamma_S > \gamma_{C3} > \gamma_{C2}$ and the corresponding profit allocation plan should be reasonable and fair.

9.3.6 Case 6

In this last case, the import and export price differences of naphtha, diesel, and mixed xylenes are fixed to be 4500, 4500, and 2160 NTD/m^3, respectively. The export price of naphtha varies from 7500 to 12,000 NTD/m^3 at 1500-NTD/m^3 intervals in four (4) scenarios, while those of diesel and mixed xylenes are kept unchanged at 3000 and 15,840 NTD/m^3 respectively. Notice that, in the above data, only the price difference of diesel is greater than that in case 5.

Based on the mathematical models presented in Chapter 8 and the aforementioned prices and price differences, the maximum profits of all possible coalitions in four different scenarios can be determined, and the results are presented in Table 9.17. Notice that only the bold-faced values are not the same as those in Table 9.15. It can also be observed that, as the naphtha price increases, the profit of actor *S* increases but the profit of actor *C*1 decreases. Since the prices of diesel and mixed xylenes are kept constant, the profits of actors *C*2 and *C*3 remain unchanged.

On the other hand, based upon the calculation procedure described in Section 9.1.1 and the data listed in Table 9.17, one can determine the Shapley values in every scenario (see Table 9.18). By comparing the profit of each actor working independently in every scenario (see Table 9.17) and the corresponding Shapley value in Table 9.18, one can see that the former is always smaller than the latter. Thus, the corresponding profit allocation plan may be accepted by all actors.

Next, let us check coalition stability according to the description in Section 9.1.2. Since a nonempty core is present in every scenario in the current case, the corresponding Shapley values are plotted against the backdrops of cores in

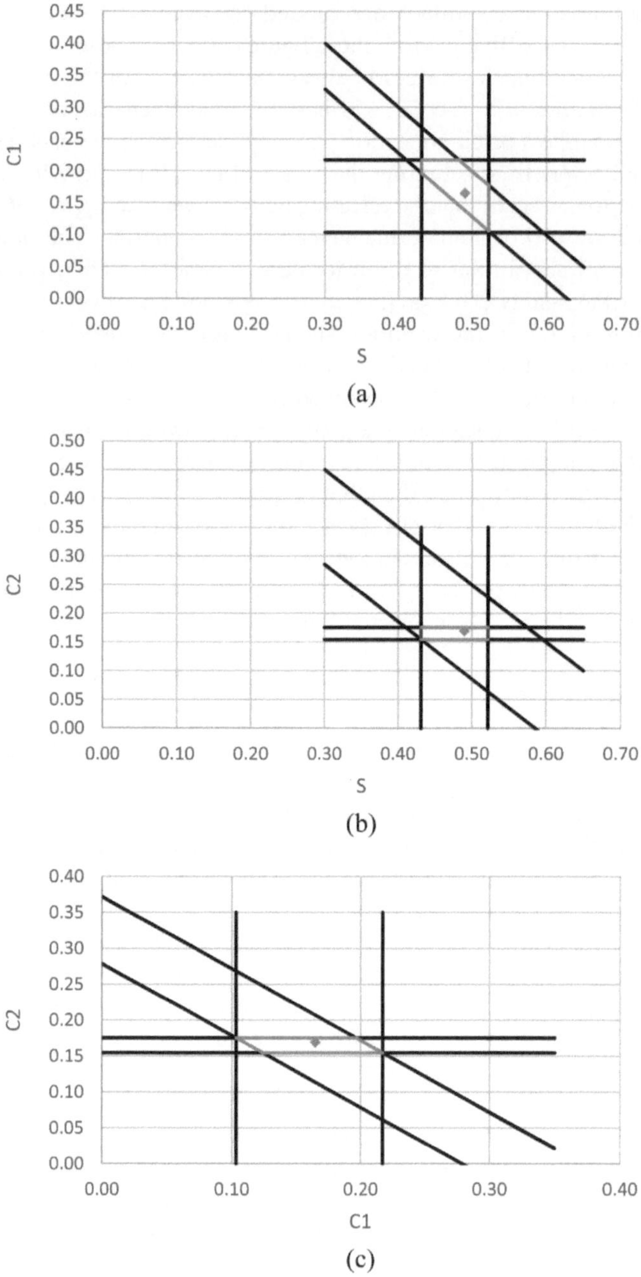

FIGURE 9.23 Core and Shapley values in scenario 1 of case 5: (a) Projection to S-$C1$ plane; (b) Projection to S-$C2$ plane; (c) Projection to $C1$-$C2$ plane.

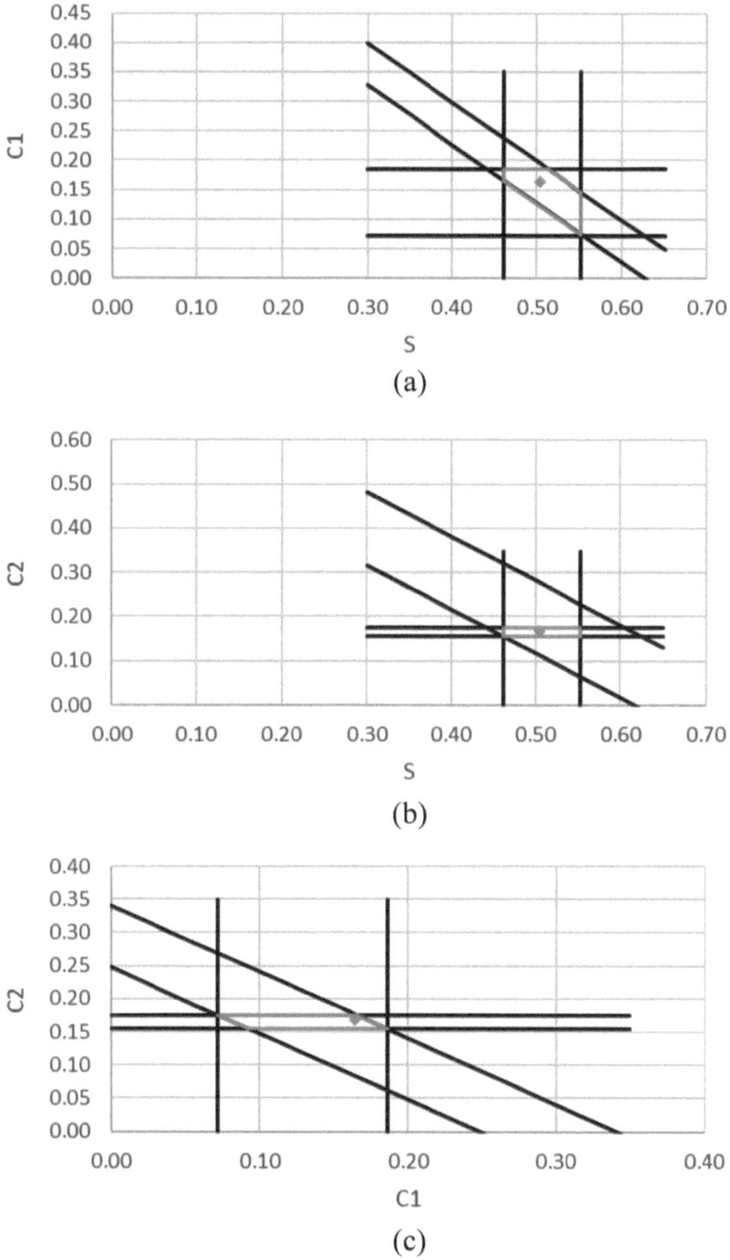

FIGURE 9.24 Core and Shapley values in scenario 2 of case 5: (a) Projection to S-$C1$ plane; (b) Projection to S-$C2$ plane; (c) Projection to $C1$-$C2$ plane.

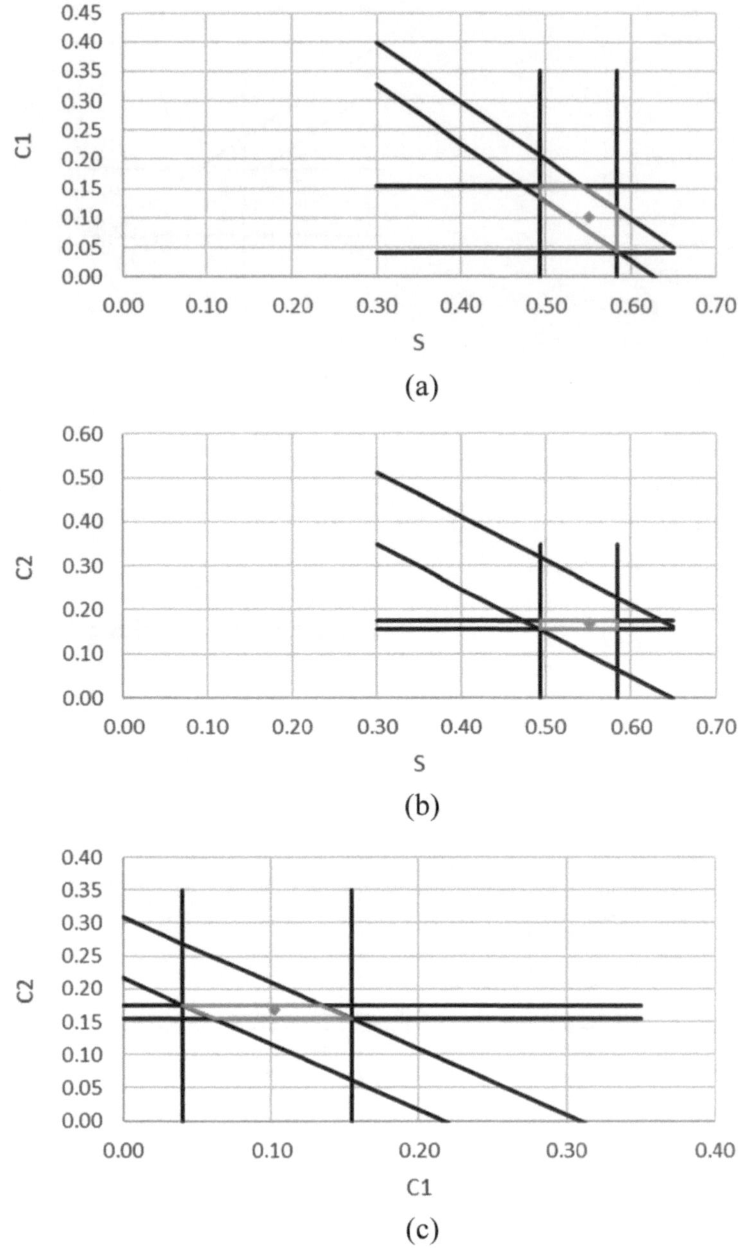

FIGURE 9.25 Core and Shapley values in scenario 3 of case 5: (a) Projection to *S*-*C*1 plane; (b) Projection to *S*-*C*2 plane; (c) Projection to *C*1-*C*2 plane.

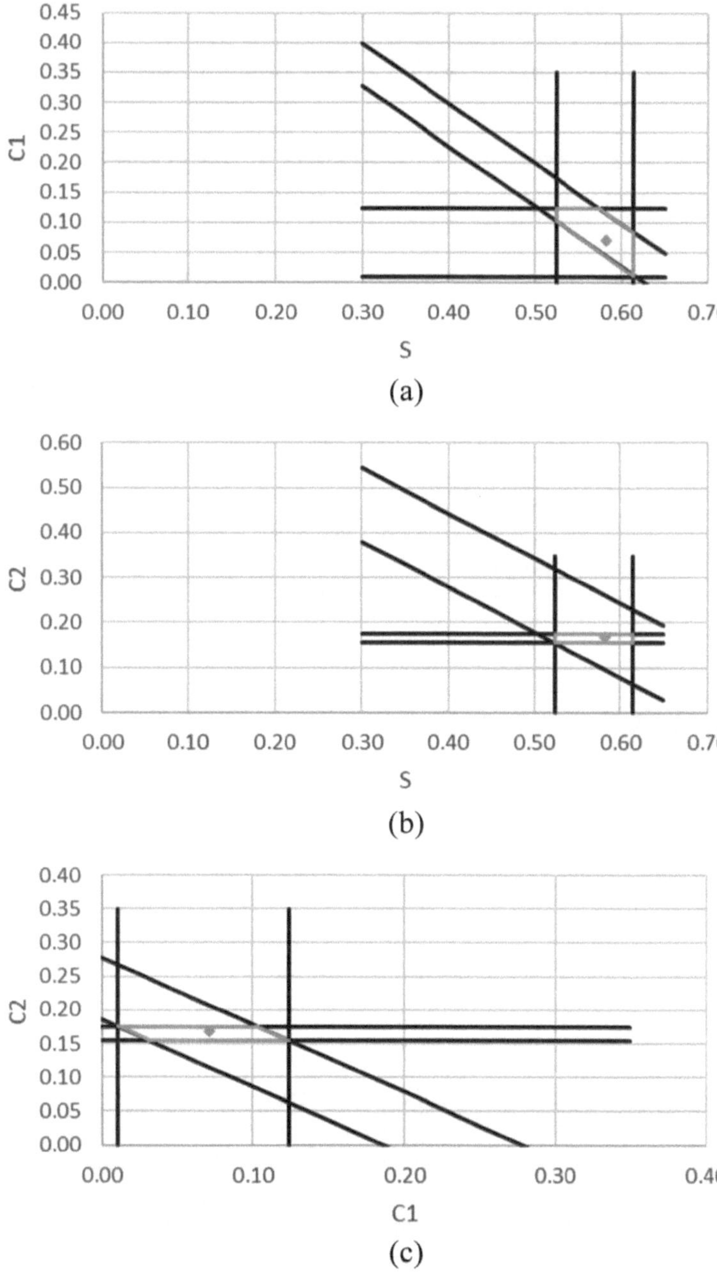

FIGURE 9.26 Core and Shapley values in scenario 4 of case 5: (a) Projection to S-$C1$ plane; (b) Projection to S-$C2$ plane; (c) Projection to $C1$-$C2$ plane.

TABLE 9.17
Total Profits of All Possible Coalitions of Petroleum Supply Chain in Case 6

	Profit (10^8NTD/mon)			
Coalition	**Scenario 1**	**Scenario 2**	**Scenario 3**	**Scenario 4**
{*S*}	8.91	9.54	10.17	10.83
{*C*1}	2.13	1.49	0.852	0.212
{*C*2}	3.18	3.18	3.18	3.18
{*C*3}	**2.50**	**2.50**	**2.50**	**2.50**
{*S*, *C*1}	12.96	12.96	12.96	12.96
{*S*, *C*2}	12.09	12.72	13.35	14.01
{*S*, *C*3}	12.96	13.59	14.25	14.88
{*C*1, *C*2}	5.76	5.10	4.47	3.84
{*C*1, *C*3}	**4.62**	**3.99**	**3.36**	**2.71**
{*C*2, *C*3}	**5.70**	**5.70**	**5.70**	**5.70**
{*S*, *C*1, *C*2}	16.56	16.56	16.56	16.56
{*S*, *C*1, *C*3}	17.01	17.01	17.01	17.01
{*S*, *C*2, *C*3}	16.14	16.80	17.43	18.06
{*C*1, *C*2, *C*3}	**9.15**	**8.52**	**7.86**	**7.23**
{*S*, *C*1, *C*2, *C*3}	20.64	20.64	20.64	20.64

TABLE 9.18
Shapley Values of All Actors of Petroleum Supply Chain in Case 6

	Shapley Values (10^8NTD/mon)			
Scenario No.	**Actor *S***	**Actor *C*1**	**Actor *C*2**	**Actor *C*3**
1	10.41	3.39	3.48	3.36
2	11.04	2.74	3.48	3.36
3	11.70	2.10	3.48	3.36
4	12.33	1.46	3.48	3.36

Figures 9.27–9.30, respectively. From these figures, it can be observed that the point corresponding to the Shapley values is located within the core in each scenario. These phenomena again suggest that the coalition {*S*, *C*1, *C*2, *C*3} is always stable in case 6.

The final step to validate a profit allocation plan is to determine the negotiation-power indicators of all actors by substituting the data in Tables 9.17 and 9.18 into equation (9.23). It can be found again that the resulting actor powers, i.e., $\gamma_S = 0.39$, $\gamma_{C1} = 0.32$, $\gamma_{C2} = 0.07$, and $\gamma_{C3} = 0.22$, remain unchanged in all scenarios. The interpretations of these indicators can be derived from those for case 5. Since *C*1 cooperates with both *S* and *C*2 and, thus, gains extra savings in its feedstock and revenue in its product, the participation of *C*1 in the coalition should cause a significant

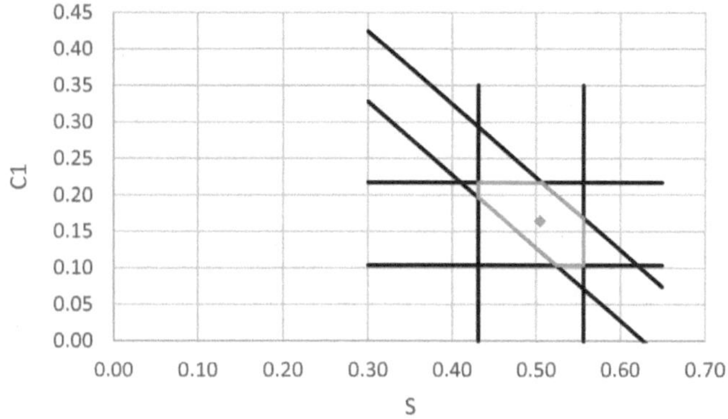

(a)

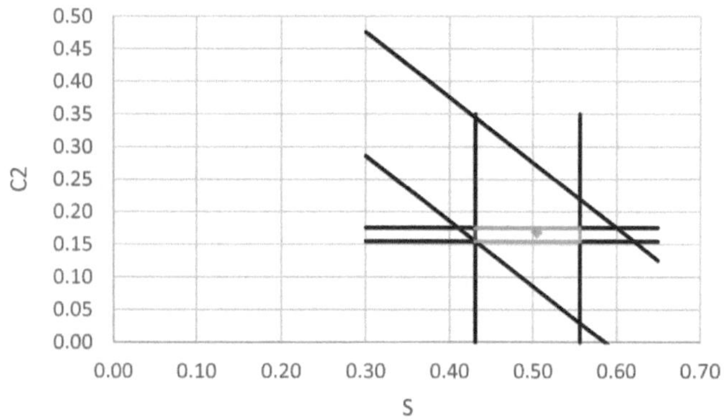

(b)

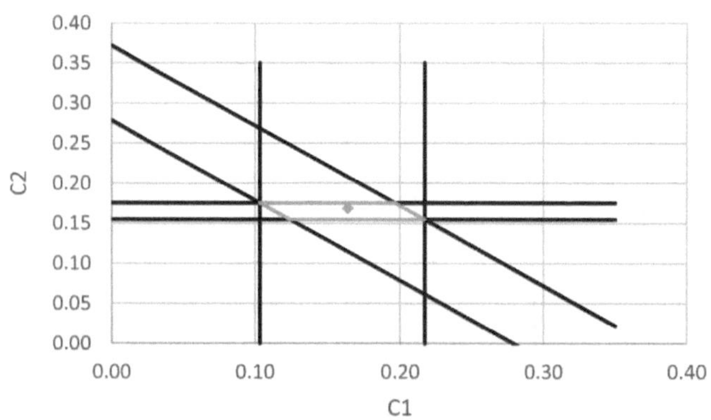

(c)

FIGURE 9.27 Core and Shapley values in scenario 1 of case 6: (a) Projection to S-$C1$ plane; (b) Projection to S-$C2$ plane; (c) Projection to $C1$-$C2$ plane.

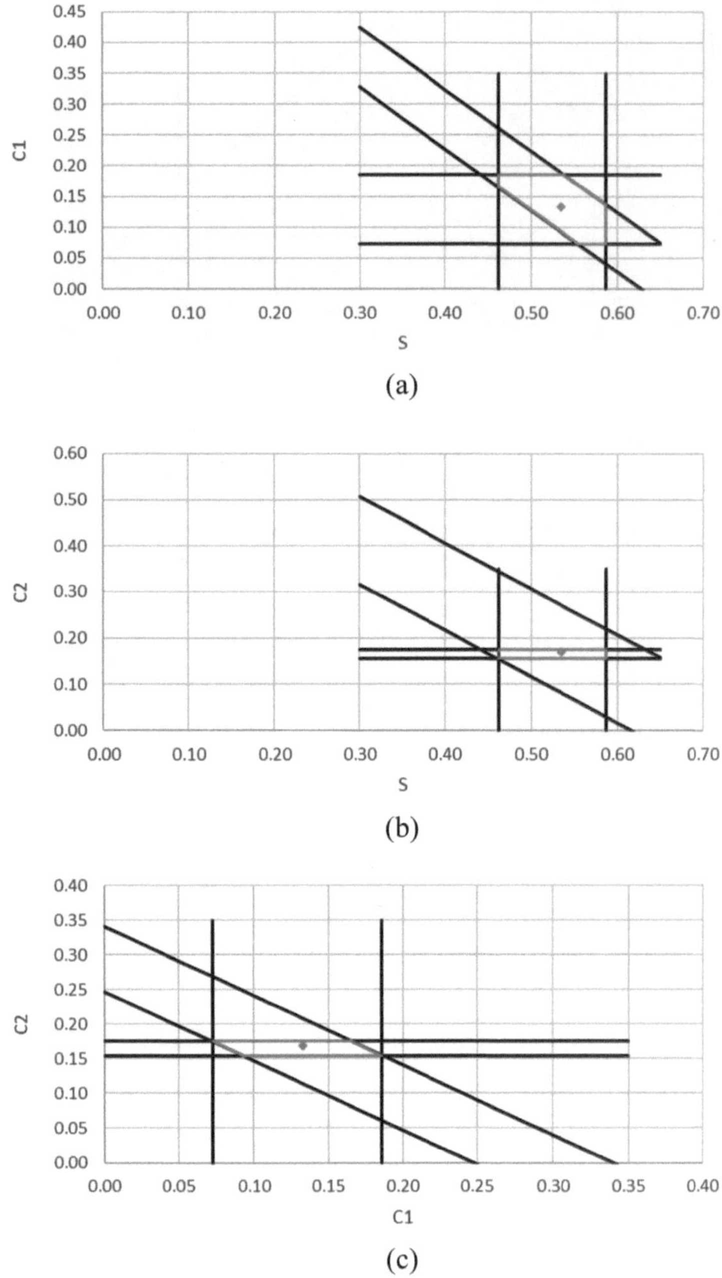

FIGURE 9.28 Core and Shapley values in scenario 2 of case 6: (a) Projection to *S*-*C*1 plane; (b) Projection to *S*-*C*2 plane; (c) Projection to *C*1-*C*2 plane.

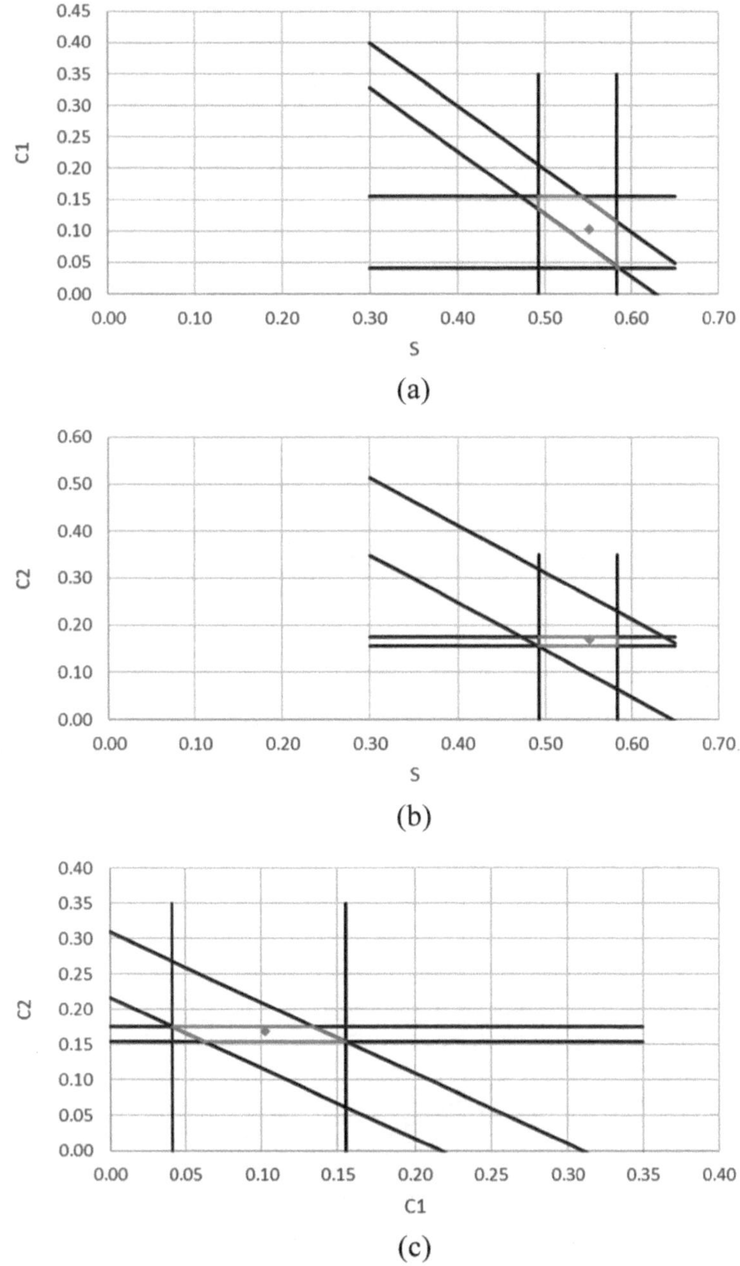

FIGURE 9.29 Core and Shapley values in scenario 3 of case 6: (a) Projection to *S*-*C*1 plane; (b) Projection to *S*-*C*2 plane; (c) Projection to *C*1-*C*2 plane.

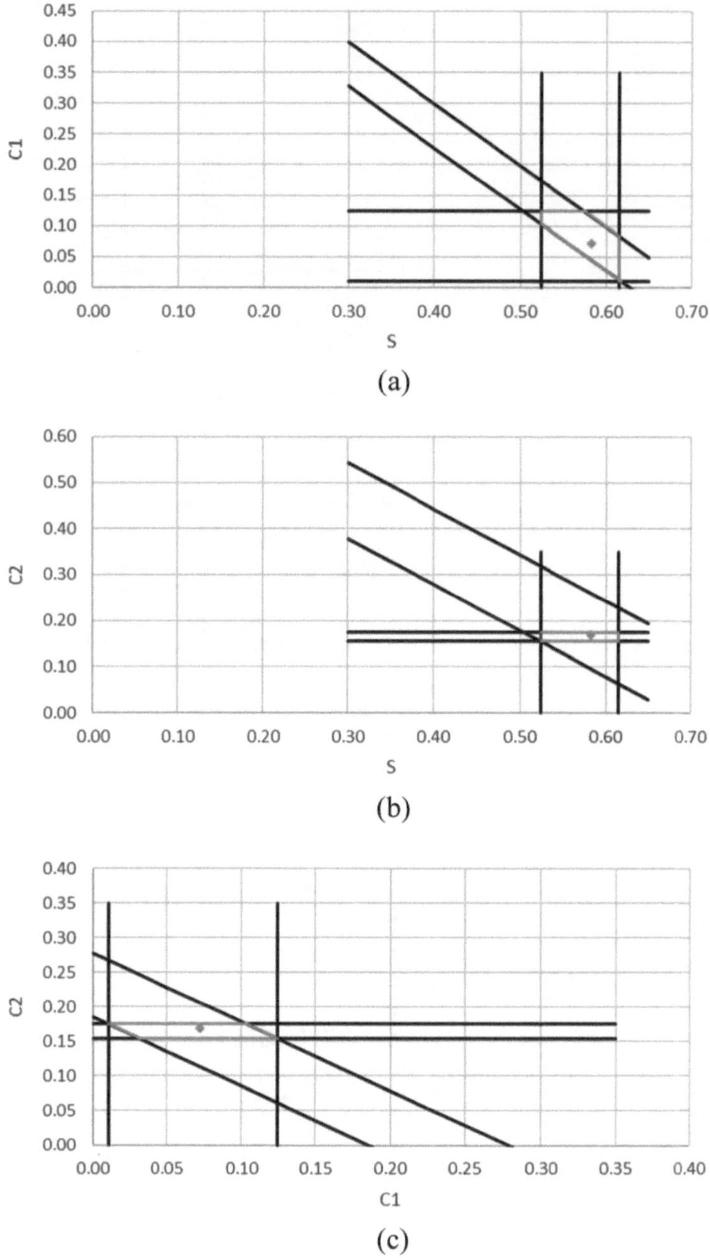

FIGURE 9.30 Core and Shapley values in scenario 4 of case 6: (a) Projection to *S*-*C*1 plane; (b) Projection to *S*-*C*2 plane; (c) Projection to *C*1-*C*2 plane.

improvement in the overall profit. On the other hand, since actor S is the provider of two intermediates, i.e., naphta and diesel, to be consumed respectively by actors $C1$ and $C3$, its impact on the overall profit should be equally great. However, since the price difference of diesel is raised to a level that is higher than that in case 5, the role of actor S becomes slightly more prominent than that of actor $C1$. Finally, the negotiation power of $C3$ should be higher than that of $C2$ since the import and export price difference of diesel is larger than that of mixed xylenes. As a result, the precedence order of negotiation-power indicators is: $\gamma_S > \gamma_{C1} > \gamma_{C3} > \gamma_{C2}$ and the corresponding profit allocation plan should be reasonable and fair.

9.4 CONCLUDING REMARKS

On the basis of case studies presented in Section 9.4, the following conclusions can be drawn:

- If the import-export price difference of an intermediate, e.g., naphtha, equals zero before cooperation, the Shapley value-based profit allocation plan of the corresponding coalition may not facilitate stability. This feature is demonstrated in case 1 and case 2.
- As shown in cases 2–6, the volume of core is in general directly proportional to the import-export price differences of intermediates before forming coalition. As also demonstrated in case 1, this volume may become zero if a price difference is zero.
- Since the negotiation power of an actor is determined according to equation (9.23), it does not depend upon the corresponding Shapley value.
- An increase in the import-export price difference of an intermediate product before forming a coalition results in an increase of the negotiation space between the actors providing and consuming this product and, consequently, an improvement of the overall profit of the supply chain. For this reason, the negotiation-power indicators of these two actors can also be raised to higher values. On the other hand, the negotiation powers of other actors not directly related to the aforementioned intermediate should be brought down to lower levels.
- For all scenarios in any case presented in the previous section, the negotiation-power indicator of each actor remains unchanged. This result is again due to equation (9.23). Notice that the denominator of the right side of this equation is constant in any given case because the overall profit of coalition (TP) and the sum of profits of all actors working independently, i.e., $\sum_i Profit_i$, remain the same in all scenarios. On the other hand, it should also be noted that the corresponding numerator, i.e., $AP_i - Profit_i$, represents the extra profit of actor i gained by forming the coalition and it is essentially caused by eliminating the price differences of intermediates before cooperation. Finally, note that these price differences are not altered in each case.

REFERENCES

Nash, J. 1951. Non-cooperative games. *Annals of Mathematics* 54(2): 286–295.

Nash, J. 1953. Two-person cooperative games. *Econometrica: Journal of the Econometric Society* 21(1): 128–140.

Roth, A.E. 1979. *Axiomatic Models of Bargaining*. Vol. 170. New York: Springer Science & Business Media.

10 Coordinated Supply Chain Management

Biomass

10.1 INTRODUCTION

Global concern about the climate change and environmental pollution that result from the exploitation of fossil fuels has urged the world to seek sustainable energy alternatives. Biomass is an important source of renewable energy, accounting for 10% of the world's total primary energy supply today (International Energy Agency, 2020). Examples of biomass include wood and wood processing waste, agricultural crops and waste materials, biogenic materials in municipal solid waste, and animal manure and human sewage (U.S. Energy Information Administration, 2020). Biomass is usually locally available and requires an infrastructure for its harvesting, transport, storage, and processing. The design and management of biomass supply chains involve the consideration of local conditions and constraints. Although biomass itself is considered virtually carbon-neutral with the offsetting of direct emissions from combustion, there are indirect carbon emissions along the supply chain. In particular, the transport of biomass has been identified as the main contributor of carbon emissions in the regional bioenergy supply network (Forsberg, 2000). Therefore, a common objective of biomass supply chain synthesis is to minimize carbon emissions from biomass transport.

In this chapter, a simple yet generic mathematical model is presented for the synthesis of regional bioenergy supply chains. The model considers all possible connections between biomass sources (farms, forest, etc.) and sinks (e.g., zones or sites with demand), and can be modified to address topological constraints and other case-specific constraints, as well as different scenarios. Three literature case studies are used to demonstrate the application of the model.

10.2 PROBLEM STATEMENT

The general problem addressed in this chapter can be stated as follows. Given:

- A set of biomass sources $i \in \mathbf{I}$. Each source has an available amount of biomass to be utilized.
- A set of biomass sinks $j \in \mathbf{J}$. Each sink requires a certain amount of biomass.

- The distances between biomass sources and sinks, together with the carbon emission factor of biomass transport.

The objective is to determine the optimal biomass supply chain network that meets the demand while minimizing the associated carbon emissions.

10.3 MODEL FORMULATION

The basic model for the synthesis of biomass supply chain networks is based on a source-sink superstructure as shown in Figure 10.1, considering all possible connections between biomass sources and sinks. The formulation consists of mass balances or equivalent energy balances, as presented below.

Equation (10.1) ensures that the amount of biomass/bioenergy allocated from source i to all sinks ($\sum_{j \in \mathrm{J}} f_{ij}$) does not exceed the available amount (F_i).

$$\sum_{j \in \mathbf{J}} f_{ij} \leq F_i \quad \forall i \in \mathbf{I} \tag{10.1}$$

Equation (10.2) ensures that the amount of biomass/bioenergy allocated from all sources to sink j ($\sum_{i \in \mathbf{I}} f_{ij}$) meets its demand ($F_j$).

$$\sum_{i \in \mathbf{I}} f_{ij} \geq F_j \quad \forall j \in \mathbf{J} \tag{10.2}$$

A common objective for environmental protection is to minimize carbon emissions (g_{CE}) from the allocation of biomass/bioenergy, as given in equation (10.3).

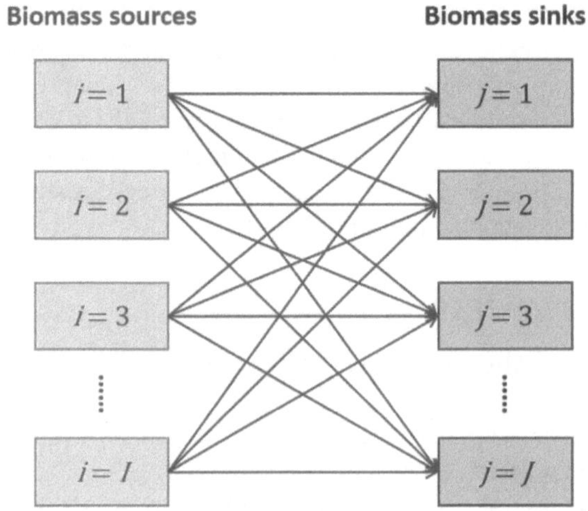

FIGURE 10.1 Source-sink superstructure for a biomass supply chain.

$$\min g_{\text{CE}} = \sum_{i \in \mathbf{I}} \sum_{j \in \mathbf{J}} C_{ij} f_{ij} D_{ij} \tag{10.3}$$

where C_{ij} is the emission factor of biomass/bioenergy supply from source i to sink j; D_{ij} is the distance between source i and sink j.

Equations (10.1)–(10.3) constitute a linear programming (LP) model, for which finding the global optimum can be guaranteed.

10.4 CASE STUDIES

In this section, the use of the basic model and its variants is demonstrated through three case studies. The linear models are solved in GAMS.

10.4.1 Case Study 1

The first case study is taken from Lam et al. (2010), focusing on the biomass transport network involving ten zones. Table 10.1 shows the location and energy balance for each zone. These data are collected from a generalized central European region. For simplicity, the distances for biomass transport can be approximated by the straight-line distances between the source and sink zones. The emission factor of bioenergy delivered from source i to sink j is then calculated using equation (10.4). Tables 10.2 and 10.3 present the distances and emission factors used in this case study.

$$C_{ij} = \frac{FC}{TC \times HV_i (1 - LF)} CEF \quad \forall i \in \mathbf{I}, j \in \mathbf{J} \tag{10.4}$$

where FC is the fuel consumption by a 20-t truck (= 0.3 L/km); TC is the truck capacity (= 20 t); HV_i is the heating value for the biomass from source i; LF is the fraction of energy lost during the conversion of biomass to bioenergy (= 0.5); CEF is the carbon emission factor of fuel (= 2.69 kg CO_2/L).

Solving the LP model (equations (10.1)–(10.3)) gives the minimum carbon emissions of 779.5 kg CO_2/y. The optimal bioenergy flows are illustrated in Figure 10.2. It can be seen that the region may be partitioned into two clusters, with zones SR1-SR3 and SK1 in cluster 1 and the others in cluster 2. Because the cluster is a group of zones with stronger links, further energy planning can be carried out within each cluster, without having to consider the whole region. The resulting bioenergy supply network is different from that of Lam et al. (2010), which involves infeasible bioenergy exchange flows.

10.4.2 Case Study 2

The second case study is taken from Li et al. (2016), considering energy planning in Laixi, a county of Qingdao. Laixi is divided into 15 zones, of which the locations and potential biomass are given in Table 10.4. The heating values vary from zone

TABLE 10.1
Regional Data for Case Study 1

Source Zone	Location (km,km)	Heating Value (GJ/t)	Bioenergy Supply (TJ/y)
SR1	(0,0)	18	22.3
SR2	(4.4,2.5)	18.5	5.7
SR3	(5.3,2.4)	17.3	3.5
SR4	(7.9,5.1)	15.5	8.6
SR5	(2.4,6.8)	18.2	19.7
SR6	(9.4,5.5)	16.8	6.7
SR7	(3.2,6.6)	17.3	10.5
Sink Zone	**Location (km,km)**	**Bioenergy Demand (TJ/y)**	
SK1	(4.1,0.2)	30	
SK2	(6.4,5.5)	19.8	
SK3	(2.3,7.3)	22.8	

TABLE 10.2
Distances (km) between Source and Sink Zones

	Sink Zone		
Source Zone	**SK1**	**SK2**	**SK3**
SR1	4.105	8.439	7.654
SR2	2.319	3.606	5.239
SR3	2.506	3.289	5.745
SR4	6.201	1.552	6.017
SR5	6.815	4.206	0.51
SR6	7.495	3	7.325
SR7	6.463	3.384	1.14

TABLE 10.3
Emission Factors (kg CO_2/TJ/km) of Bioenergy Delivery

	Sink Zone		
Source Zone	**SK1**	**SK2**	**SK3**
SR1	4.483	4.483	4.483
SR2	4.362	4.362	4.362
SR3	4.665	4.665	4.665
SR4	5.206	5.206	5.206
SR5	4.434	4.434	4.434
SR6	4.804	4.804	4.804
SR7	4.665	4.665	4.665

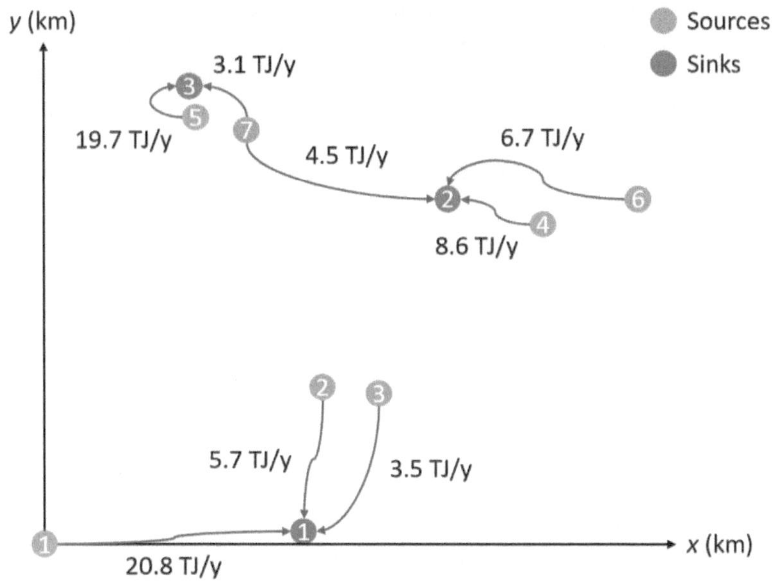

FIGURE 10.2 Optimal bioenergy flows for case study 1.

TABLE 10.4
Data for the Zones in Laixi

Zone	Location (km, km)	Biomass (t/y)	Low Heating Value (GJ/t)	Potential Electricity (GWh/y)		
				λ = 100%	λ = 80%	λ = 40%
1	(16,16)	39,136	15.23	49.68	39.74	19.87
2	(16,22)	29,007	15.28	36.94	29.55	14.78
3	(18,28)	31,968	15.22	40.55	32.44	16.22
4	(3,21)	43,625	15.17	55.14	44.12	22.06
5	(8,10)	28,540	15.29	36.37	29.1	14.55
6	(16,2)	76,320	15.19	96.59	77.27	38.64
7	(10,1)	65,147	15.14	82.19	65.75	32.88
8	(0,10)	33,885	15.16	42.82	34.25	17.13
9	(1,30)	65,349	15.17	82.6	66.08	33.04
10	(1,40)	48,717	15.22	61.8	49.44	24.72
11	(15,38)	57,614	15.26	73.25	58.6	29.3
12	(0,0)	33,357	15.28	42.47	33.97	16.99
13	(20,0)	63,669	15.17	80.47	64.38	32.19
14	(0,20)	58,026	15.22	73.58	58.86	29.43
15	(5,42)	73,036	15.28	92.99	74.39	37.2

*λ is the collection rate.

to zone because of the difference in biomass resources. With different willingness of the rural residents to collect biomass, Li et al. (2016) considered three simplified scenarios with different mean collection rates: 100%, 80%, and 40%, assuming fixed thermal efficiency of 30% for power generation. Given an electricity target of 750 GWh/y for biomass, the optimal biomass supply chain network with minimum carbon emissions from transport can be synthesized. For simplicity, a common road factor of 1.2 is assumed in this case study to estimate the distance between two zones. It is further assumed that the biomass power plant is to be built in either zone 2 or 13, based on the government's intention. The emission factor of biomass transport using diesel-fueled vehicles is set to 0.04 $kgCO_2/t/km$. The emission factor of electricity from biomass transported from source i to sink j is then calculated using equation (10.5).

$$C_{ij} = 0.04 \times \frac{3600}{LHV_i \times \eta} \quad \forall i \in \mathbf{I}, j \in \mathbf{J} \tag{10.5}$$

where LHV_i is the low heating value of biomass from source i; η is the thermal efficiency (= 30%); 3600 (s/h) is a conversion factor converting GJ to GWh.

In this case study, each of the 15 zones is taken as both a biomass source and a biomass sink. However, with the assumption that the biomass power plant is to be built only in either zone 2 or 13, biomass will be collected and sent to only one of these two zones. To account for such topological constraints, equations (10.6)–(10.8) are included.

$$\sum_{j \in \mathbf{J}} y_j \le N^{\mathrm{P}} \tag{10.6}$$

$$y_j \le M_j \quad \forall j \in \mathbf{J} \tag{10.7}$$

$$f_{ij} \le F_i y_j \quad \forall i \in \mathbf{I}, j \in \mathbf{J} \tag{10.8}$$

where y_j is a binary variable denoting if a biomass power plant is built at sink j; N^{P} is the number of biomass power plants to be built; M_j is a binary parameter denoting whether sink j is suitable for a biomass power plant ($M_j = 1$) or not ($M_j = 0$). In addition, equation (10.9) is used instead of equation (10.2) to account for the target of electricity from biomass (F^{BE}).

$$\sum_{i \in \mathbf{I}} \sum_{j \in \mathbf{J}} f_{ij} \ge F^{\mathrm{BE}} \tag{10.9}$$

With the introduction of binary variables (y_j), the modified model (equation (10.3) subject to equations (10.1) and (10.6)–(10.9)) is a mixed-integer linear program (MILP).

TABLE 10.5
Optimal Bioenergy Supply for Case Study 2

Biomass collection rate	**λ = 100%**	**λ = 80%**	**λ = 400%**
Biomass power plant location	**Zone 2**	**Zone 2**	**Zone 2**
Bioenergy delivered* (GJ/y):			
Zone 1 to zone 2	49.68	39.74	19.87
Zone 2 to zone 2	36.94	29.55	14.78
Zone 3 to zone 2	40.55	32.44	16.22
Zone 4 to zone 2	55.14	44.12	22.06
Zone 5 to zone 2	36.37	29.1	14.55
Zone 6 to zone 2	96.59	77.27	38.64
Zone 7 to zone 2	82.19	65.75	32.88
Zone 8 to zone 2	42.82	34.25	17.13
Zone 9 to zone 2	82.6	66.08	33.04
Zone 10 to zone 2	0	49.44	24.72
Zone 11 to zone 2	73.25	58.6	29.3
Zone 12 to zone 2	0	26.03	16.99
Zone 13 to zone 2	80.29	64.38	32.19
Zone 14 to zone 2	73.58	58.86	29.43
Zone 15 to zone 2	0	74.39	37.2
Emissions from biomass transport (t/y)	452.83	496.93	252.56

*The actual amount of biomass transported (B_{ij}) can be calculated using Equation (10.10).

In the scenarios of 100% and 80% biomass collection rates, the biomass collected has sufficient potential electricity to fulfill the target of 750 GWh/y. Solving the MILP model (setting $N^P = 1$, $M_j = 1$ for zones 2 and 13, and $M_j = 0$ for all the other zones) for both scenarios gives the results in Table 10.5. The biomass power plant is to be built in zone 2 in both scenarios, resulting in carbon emissions of 452.83 and 496.93 t/y from biomass transport, respectively. The resulting emission level for λ = 80% is higher than that for λ = 100%, because in the scenario of a lower biomass collection rate, biomass has to be transported from more zones to the biomass power plant to meet the electricity target. However, in the scenario of a 40% biomass collection rate, the biomass collected has only 379 GWh/y of potential electricity. Therefore, the electricity target is revised accordingly to fully utilize the available bioenergy. Solving the same model but setting $D^E = 379$ GWh/y (instead of 750 GWh/y) yields the results in Table 10.5. Still, the biomass power plant is to be built in zone 2, incurring 252.56 t/y of carbon emissions from biomass transport. This scenario involves the lowest emission level because of the smallest total amount of biomass transported. The results obtained for all three scenarios are consistent with those reported by Li et al. (2016), confirming the validity of the MILP model.

$$B_{ij} = f_{ij} \times \frac{3600}{LHV_i \times \eta} \quad \forall i \in \mathbf{I}, j \in \mathbf{J} \tag{10.10}$$

So far the results obtained are based on the assumption that the biomass power plant is to be built only in either zone 2 or 13, and may not be truly optimal in terms of carbon emissions from biomass transport. Suppose that this assumption can be relaxed and it is now assumed that the biomass power plant can be built in any of the 15 zones. Assuming also a biomass collection rate of 80% with the original electricity target of 750 GWh/y, solving the MILP model (setting $N^P = 1$ and $M_j = 1$ for all zones) gives the minimum emission level of 460.49 t/y, which is 7.3% lower than that previously determined for $\lambda = 80\%$ (496.93 t/y). Figure 10.3 shows the resulting biomass supply chain network, where the biomass power plant is built in zone 4.

The impact of decentralizing biomass energy conversion on carbon emissions from biomass transport is further analyzed. Assuming $\lambda = 80\%$ and that up to two biomass power plants may be built, solving the MILP model (setting $N^P = 2$ and $M_j = 1$ for all zones) gives the optimal sites in zones 6 and 9, with the minimum emission level of 265.75 t/y. This corresponds to a significant further reduction of 39.2% (as compared to 496.93 t/y). Figure 10.4 shows the resulting biomass supply chain

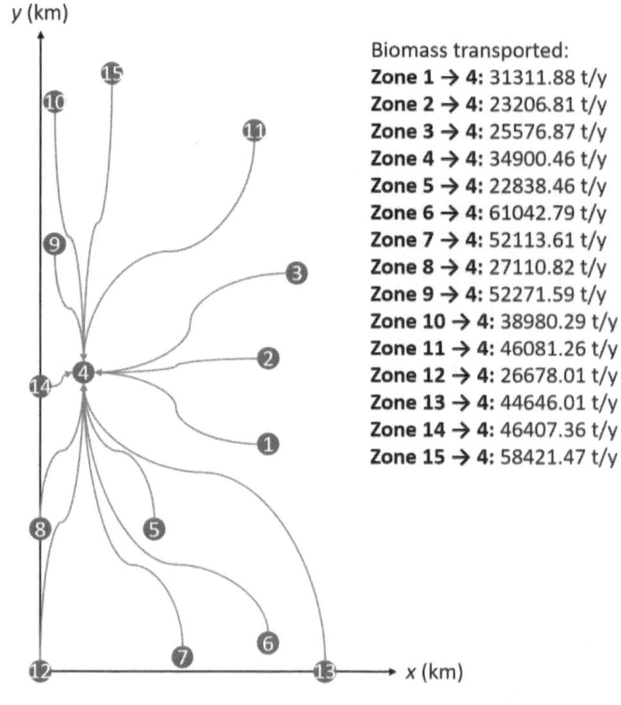

FIGURE 10.3 Optimal biomass supply network with one power plant. (© 2017 IEEE. Adapted, with permission, from 2017 6th AdCONIP, Taipei, 2017, 395–399.)

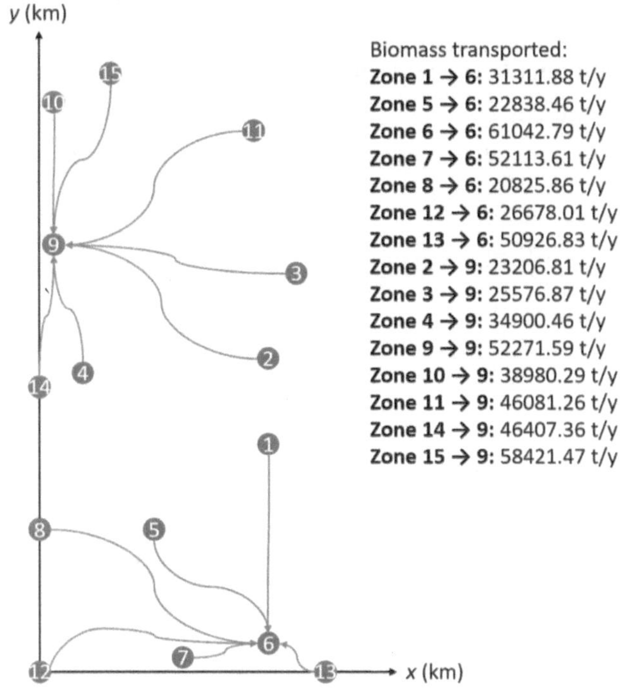

FIGURE 10.4 Optimal biomass supply network with two power plants. (© 2017 IEEE. Adapted, with permission, from 2017 6th AdCONIP, Taipei, 2017, 395–399.)

network. It can be seen that decentralized energy conversion with two power plants shortens the distance of biomass transport, thus reducing the associated carbon emissions. However, building more power plants of smaller capacity could increase the cost, suggesting a trade-off between the number of biomass power plants and carbon emissions from biomass transport.

10.4.3 Case Study 3

The third case study is taken from Foo et al. (2013). This case study considers the optimal allocation of empty fruit bunches (EFBs) between the suppliers (sources) and consumers (sinks) in Sabah, Malaysia. The EFB suppliers are palm oil mills, where EFBs are produced as a byproduct, while the EFB consumers are process plants where combined heat and power (CHP) systems are used to produce energy from EFBs. There are two equally probable anticipated scenarios with different EFB yields in the palm oil mills. Both scenarios account for factors such as seasonal variations. Table 10.6 shows the EFB availability and requirements of the suppliers and consumers. The two values given in the second column represent the EFB availability in the two scenarios. For both scenarios, it is desired to allocate 640 kt/y of EFBs to all consumers (for cogeneration). Table 10.7 shows the distances between

TABLE 10.6
Data for EFB Suppliers and Consumers

Supplier	EFB Availability (kt/y)	Consumer	EFB Requirement (kt/y)
S1	90; 80	C1 (oleochemical plant)	200–300
S2	75; 86	C2 (palm oil mill, S8)	200–250
S3	80; 82	C3 (palm oil refinery)	180–240
S4	85; 88		
S5	82; 70		
S6	86; 90		
S7	92; 80		
S8	78; 70		
S9	80; 94		
S10	88; 84		
S11	84; 80		

TABLE 10.7
Distances (in km) between EFB Suppliers and Consumers

	Sink		
Source	C1	C2	C3
S1	11.9	65.7	164
S2	9.9	69.9	168
S3	12.3	66.5	165
S4	12.3	66.5	165
S5	11.9	65.7	164
S6	8	77.8	176
S7	8	77.8	176
S8	75.6	0	98
S9	172	96.2	1.9
S10	170	94.5	5.5
S11	168	92.5	7.1

the EFB sources and sinks. Lorries are used to transport EFBs. The emission factor of transport is assumed to be 0.092 kg CO_2/km/t.

To allow for multiple scenarios, equations (10.1)–(10.3) are modified by introducing index k:

$$\sum_{j \in \mathbf{J}} f_{ijk} \leq F_{ik} \quad \forall i \in \mathbf{I}, k \in \mathbf{K} \tag{10.11}$$

$$\sum_{i \in \mathbf{I}} f_{ijk} \geq f_{jk} \quad \forall j \in \mathbf{J}, k \in \mathbf{K} \tag{10.12}$$

$$\min g_{\mathrm{CE}} = TEF \sum_{i \in \mathbf{I}} \sum_{j \in \mathbf{J}} \sum_{k \in \mathbf{K}} P_k f_{ijk} D_{ij} \tag{10.13}$$

where *TEF* is the emission factor of transport; P_k is the probability of scenario k. If the scenarios are equally probable, the probability of each scenario is equal to one divided by the number of anticipated scenarios.

With different scenarios, the amount of EFBs consumed by sink j (f_{jk}) is treated as a variable bounded by its lower (F_j^{L}) and upper limits (F_j^{U}):

$$F_j^{\mathrm{L}} \le f_{jk} \le F_j^{\mathrm{U}} \quad \forall j \in \mathbf{J}, k \in \mathbf{K} \tag{10.14}$$

Equation (10.15) is then used to impose the desired amount of EFBs allocated to consumers.

$$\sum_{j \in \mathbf{J}} f_{jk} = F_k^{\mathrm{CHP}} \quad \forall k \in \mathbf{K} \tag{10.15}$$

The modified model for this case study (equations (10.11)–(10.15)) is still an LP. Solving the modified model gives the optimal allocation of EFBs in Table 10.8, where the two numbers given for each source-sink pair represent the EFB allocations in the two scenarios. The minimum level of carbon emissions from EFB transport is determined to be 1015.83 t/y, with 200, 200, and 240 kt/y of EFBs consumed by C1–C3, respectively, in both scenarios.

If the regional biomass supply chain has to meet an increased energy demand, building a new CHP plant (C4) will be considered, without shutting down any existing

TABLE 10.8
Allocation of EFBs to Three Consumers (Numbers in kt/y)

	Sink			
Source	**C1**	**C2**	**C3**	**Unutilized***
S1		90; 80		
S2	22; 30			53; 56
S3				80; 82
S4				85; 88
S5		32; 50		50; 20
S6	86; 90			
S7	92; 80			
S8		78; 70		
S9			80; 94	
S10			88; 84	
S11			72; 62	12; 18

*Calculated from optimization results.

TABLE 10.9
Distances (km) between the New EFB Sink and Nearby EFB Sources

Sources	Sink (C4)
S1	8
S2	7.6
S3	8.8
S4	6.8

CHP systems (in C1–C3). If built, the new CHP plant should generate 5–10 MW of power using 100–200 kt/y of EFBs. It is assumed that the EFBs for C4 are sourced from S1–S4, which are nearer to C4. The distances between S1–S4 and C4 are given in Table 10.9. Because significant improvements in the roads are required for EFBs to be transported to the new CHP plant, any source-sink match between S1–S4 and C4 has a lower limit of 50 kt/y to justify the investment in the infrastructure. With the increased energy demand, it is desired to allocate 840 kt/y of EFBs to all consumers in both scenarios.

To incorporate the lower limit to the amount of EFBs transported (F_{ij}^{L}) and the decision on whether or not to build a new CHP plant, the following constraints are needed.

$$F_{ij}^{\mathrm{L}} y_{ij} \leq f_{ijk} \leq F_{ik} y_{ij} \quad \forall i \in \mathbf{I}, j \in \mathbf{J}, k \in \mathbf{K} \tag{10.16}$$

$$F_j^{\mathrm{L}} y_j \leq f_{jk} \leq F_j^{\mathrm{U}} y_j \quad \forall j \in \mathbf{J}, k \in \mathbf{K} \tag{10.17}$$

where y_{ij} indicates the existence of the match between source i and sink j, and y_j denotes the existence of a CHP plant in sink j. Equation (10.17) is derived from equation (10.14). The introduction of binary variables results in an MILP model.

Solving the MILP model (equations (10.11)–(10.13) and (10.15)–(10.17)) gives the optimal allocation of EFBs in Table 10.10. Again, the two numbers given for each source-sink pair represent the EFB allocations in the two anticipated scenarios. The minimum level of carbon emissions from EFB transport is determined to be 1153.24 t/y, with 200, 200, 240, and 200 kt/y of EFBs consumed by C1–C4, respectively, in both scenarios.

10.5 SUMMARY

Superstructure-based LP and MILP models for the synthesis of regional bioenergy supply chain networks have been developed in this chapter. With the objective of minimizing carbon emissions from biomass transport, the LP models determine the optimal allocation of biomass from sources to sinks for single and multiple planning

TABLE 10.10
Allocation of EFBs to Four Consumers (Numbers in kt/y)

	Sink				
Source	**C1**	**C2**	**C3**	**C4**	**Unutilized***
S1		28; 24		62; 56	
S2	22; 30			53; 56	
S3		12; 36			68; 46
S4				85; 88	
S5		82; 70			
S6	86; 90				
S7	92; 80				
S8		78; 70			
S9			80; 94		
S10			88; 84		
S11			72; 62		12; 18

*Calculated from optimization results.

scenarios, while the MILP models further allow for topological constraints, biomass allocation rate limits, and the location of thermal plants. Using these models provides decision support for the planning and implementation of bioenergy systems, even under some degree of uncertainty. Three case studies based on data from central Europe, China, and Malaysia were solved and results discussed to illustrate the application of the developed models.

NOMENCLATURE

Indices and Sets

$i \in \mathbf{I}$ biomass sources
$j \in \mathbf{J}$ biomass sinks
$k \in \mathbf{K}$ anticipated scenarios

Parameters

C_{ij} emission factor of biomass/bioenergy allocated from source i to sink j
D_{ij} distance between source i and sink j
F_i amount of biomass/bioenergy available from source i
F_{ij}^{L} lower limit to the amount of biomass allocated from source i to sink j
F_{ij}^{U} upper limit to the amount of biomass allocated from source i to sink j
F_{ik} amount of biomass available from source i in scenario k
F_j amount of biomass/bioenergy required for sink j

F_j^L lower limit to the amount of biomass consumed by sink j
F_j^U upper limit to the amount of biomass consumed by sink j
F^{BE} target of electricity from biomass
F^{CHP} required amount of biomass for cogeneration
M_j binary parameter indicating if sink j is suitable for a biomass power plant.
N^P maximum number of biomass power plants

Variables

f_{ij} amount of biomass/bioenergy allocated from source i to sink j
f_{ijk} amount of biomass allocated from source i to sink j in scenario k
f_{jk} amount of biomass required for sink j in scenario k
y_j binary variable indicating if there is a thermal plant in sink j
y_{ij} binary variable indicating if there is a match between source i and sink j

REFERENCES

Foo, D.C.Y., Tan, R.R., Lam, H.L., Abdul Aziz, M.K., Klemeš, J.J. 2013. Robust models for the synthesis of flexible palm oil-based regional bioenergy supply chain. *Energy* 55: 68–73.

Forsberg, G. 2000. Biomass energy transport. Analysis of bioenergy transport chains using life cycle inventory method. *Biomass and Bioenergy* 19(1): 17–30.

International Energy Agency 2020. Bioenergy. https://www.iea.org/fuels-and-technologies/bioenergy.

Lam, H.L., Varbanov, P.S., Klemeš, J.J. 2010. Optimisation of regional energy supply chains utilising renewables: P-graph approach. *Computers & Chemical Engineering* 34(5): 782–792.

Li, Z., Jia, X., Foo, D.C.Y., Tan, R.R. 2016. Minimizing carbon footprint using pinch analysis: The case of regional renewable electricity planning in China. *Applied Energy* 184: 1051–1062.

U.S. Energy Information Administration 2020. Biomass explained. https://www.eia.gov/energyexplained/biomass/.

Index

H

I

R

S

T

U

V

W

X